machine trades blueprint reading

RUSSEL W. IHNE

WALTER E. STREETER

AMERICAN TECHNICAL SOCIETY

Chicago, Illinois • 60637

Library of Congress Card Catalog No.: 74-180917
ISBN: 0-8269-1862-X

FIRST EDITION

1st Printing 1941
2nd Printing August, 1941
3rd Printing October, 1941
4th Printing January, 1942
5th Printing April, 1942
6th Printing October, 1942
7th Printing 1943
8th Printing 1945
9th Printing 1946
10th Printing 1947

SECOND EDITION

11th Printing 1948
12th Printing 1951
13th Printing 1952
14th Printing 1954

THIRD EDITION

15th Printing 1956
16th Printing 1957
17th Printing 1960
18th Printing 1961

FOURTH EDITION

19th Printing 1962
20th Printing 1963
21st Printing 1964
22nd Printing 1966

FIFTH EDITION

23rd Printing 1966
24th Printing 1967
25th Printing 1968
26th Printing 1970
27th Printing 1970

SIXTH EDITION

28th Printing 1972
29th Printing 1974
30th Printing 1975

Printed in the United States of America

7

WALTER E. STREETER owns and operates the Streeter Engineering and Sales Company in Erie, Pennsylvania. He holds three patents basic to the tool industry that are internationally famous. He was formerly: Plant Manager, Kaiser Aluminum Chemical Corp., Erie, Pennsylvania; and President, Paragon Aluminum Corp., Monroe, Michigan.

Throughout his industrial work he has been intensely interested in the training of personnel for the machine trades and has given liberally of his time both to teaching and advising on apprenticeship, trade, vocational, and adult education.

RUSSEL W. IHNE is presently in Vocational and Industrial Education work in Orange County, California. He has wide experience in manufacturing and industrial management and has personally established and put into operation new manufacturing facilities. He was formerly: Director of Manufacturing for the Sta-Hi Corporation in Newport Beach, California; West Coast Manager of Manufacturing, Taylor Forge & Pipe Works, Fontana, California; Assistant to the President, Cornell Forge Co., Chicago, Illinois; Consultant, Drop Forge Industry; and Assistant Manager, Industrial Engineering, P. R. Mallory & Co., Indianapolis, Indiana.

He has extensive experience in vocational-trade and adult education and was formerly Director of Trade and Industrial Education at New Castle, Indiana, and State Supervisor of War Production Training for the State of Indiana.

Publisher's Acknowledgment

MACHINE TRADES BLUEPRINT READING, first published over thirty years ago was an instant success. Over the years it has gained wide recognition and acceptance.

A great many changes have been incorporated in the text and illustrations through many revisions in order to keep abreast with technical advances and teaching methods. The 6th edition, in addition to updating existing material and adding significant new material, has been completely reset in larger more legible type. This improvement is appropriate to the thoroughly updated content.

Throughout the years Mr. Streeter, as an industrialist and teacher, has actively followed all trends in drawing practices as related to industry, and has helped keep the book in the forefront through successive editions. We feel he deserves special recognition for his individual efforts in this respect.

Mr. Ihne, a teacher, a Director of Trade and Industrial Education in a large school system, and an industrialist has likewise kept in constant touch with modern trends, and his efforts and contributions to the tremendous success of this publication also deserve special recognition.

THE PUBLISHERS

Preface

There are many reasons for the accelerated rate of progress which we find in the Machine Trades and certainly none more important than the increased knowledge and skill of the craftsmen. A large measure of credit for this important development goes to the high quality of instruction and training which craftsmen and potential craftsmen have been receiving.

To help maintain this high standard of training, the authors have prepared this New Edition. The authors, who work in the industrial field, are fully aware of the changing requirements and new developments in industry, and of the specific problems involved in training skilled craftsmen. Through reports on the tens of thousands of students who have benefited from this book, they have devised means of simplifying the learning process, and overcoming certain obstacles which may delay the development of blueprint reading skill.

Typical of their work is the new method introduced to teach the student to visualize an object from a blueprint. It has long been recognized that the ability to visualize is basic to the development of skill in reading blueprints, but in most cases such training has been based on teaching methods more appropriate to the training of draftsmen. The function of the draftsman is, of course, contrary to that of the Machine Tradesman. The draftsman *represents an object* by means of a blueprint on the basis of the actual object, a model, a sketch, or an idea. The Machine Tradesman *produces the object* on the basis of the information contained in the blueprint. To do this effectively, he must first *visualize the object*. In this New Edition the authors have devised a step-by-step method of guiding the student in visualizing from a blueprint.

To prepare the student for the actual conditions found in the shop and to give him practical experience in print reading, new industrial blueprints are added in this edition. These are carefully analyzed and difficulties that the student might encounter are explained in detail. The new industrial prints also serve to introduce the student to the latest technological innovations. Both new materials (such as plastics) and new processes (such as numerical control) are discussed at length. Explanations are also given of the shop practices involved in the actual production of machine parts. This enables the student to see the relationship between the working drawings and the manufacturing processes involved in making or finishing the part.

This new edition also treats two relatively new trends in drawing practice: 1) dimensioning of drawings in both decimal inches and millimeters (in Chapter 9), and 2) the newly introduced system of geometric dimensioning and tolerancing described and illustrated in Appendix A.

The practice of metric dimensioning is one certain to increase because of the growing importance of international business. Presently the practice of geometric dimensioning and tolerancing is chiefly of interest to companies with government contracts.

Russel W. Ihne
Walter E. Streeter

Contents

The production of the finished product from the original design requires the co-operation and knowledge of many skilled craftsmen.

CHAPTER 1

Blueprints: From Design to Product

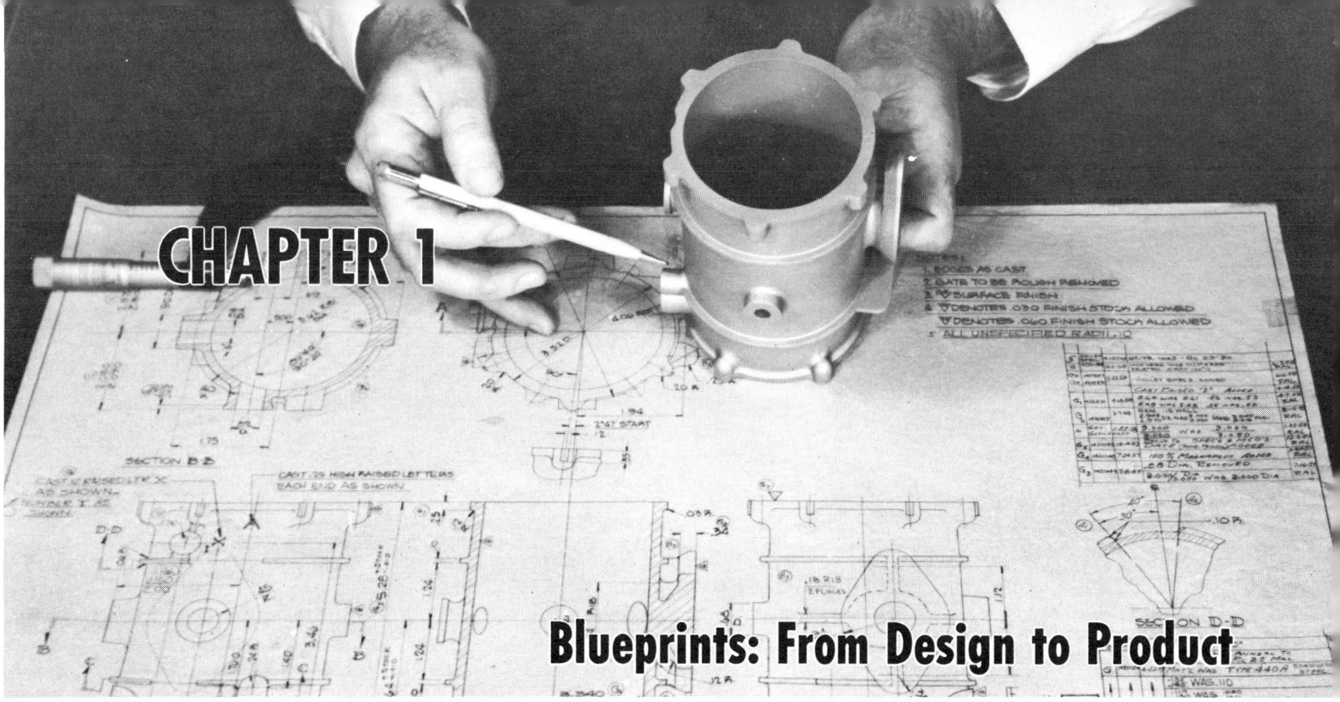

Tooling and Production Magazine

The blueprint is a photoprint made by exposing sensitized paper to a brilliant light through a drawing made on transparent material. The result is a negative print on which the lines of the drawing appear as white lines on a dark blue background. In recent years, since the introduction of new sensitized papers and new chemical treatments, most prints are now produced with black, dark brown, blue or sepia lines on a white background. The name "blueprint," though, is still generally applied to most prints made of engineering drawings.

The blueprint becomes a graphic language to those involved in the manufacture of machines and mechanical structures. It contains exact and positive information regarding every detail of a machine or structure. The blueprint is not a pictorial representation of an object as the artist would sketch it or the eye would see it; however, those who understand this graphic language are able to visualize the exact shape and nature of an object from the mass of lines.

Did you ever wonder how an automobile, an airplane, an electric toaster, or any number of industrial products were made and put together? Consider the immense amount of accurate and detailed information the designers of any one of these items must convey to the people who actually participate in the manufacturing, assembling and selling of such products. Consider, for instance, that every automobile consists of thousands of parts, each of which must fit perfectly with other parts to form a unit such as an engine, a braking system, a door, or an instrument panel.

Imagine what sort of product would result if the designing, manufacturing, and assembling of these parts were left to guesswork, word of mouth, or even written information!

For each part that goes into the mass-production of a product, whether the product be a watch, a kitchen cabinet, or a power tool, there is at least one drawing or blueprint which conveys information necessary for its production. The number of blueprints required for the production of an airplane runs easily into the thousands.

From Drawing Board to Consumer. Let us continue with our example of the automobile and follow the steps in producing one of its relatively simple parts, a hub cap.

The first stage in the production of a hub cap is its origination as an idea in the mind of an automotive designer. Often the idea is expressed by the designer in the form of freehand sketches in color.

The sketches for the hub cap are then submitted to and studied by automotive engineers in charge of the development of a new model. The engineers arrive at a decision as to which design will best serve its purpose, how well it fits into the general design of the car, and how it can be economically and efficiently produced. If the design is otherwise acceptable to them, they may make final changes before submitting it to a draftsman who proceeds to make a finished drawing, or blueprint, of the hub cap like the one shown in Fig. 1.

The distribution and use of blueprints through the departments of manufacturing establishments varies with the organizational structure of the plants. We cite the following, however, to let you know the vital part a blueprint plays in the creation of one detail of an automobile.

1. *Materials Control Department.* Copies of the blueprint are first sent to the materials control department. The hub cap is classified as a small steel metal stamping. It is possible that this part might be purchased from a vendor that specializes in this type of stamping at less cost than it could be produced in the car maker's plant.

2. *Purchasing Department.* It is the responsibility of the purchasing department to get the information on the cost of procuring this part from an outside source. The blueprint, showing complete information on the part, makes it possible for the purchasing department to query several companies to determine the most economical place to obtain the hub cap.

3. *Tooling Department.* When the source has been selected, the blueprint will be sent to the tool and die department. More blueprints will be made describing in minute de-

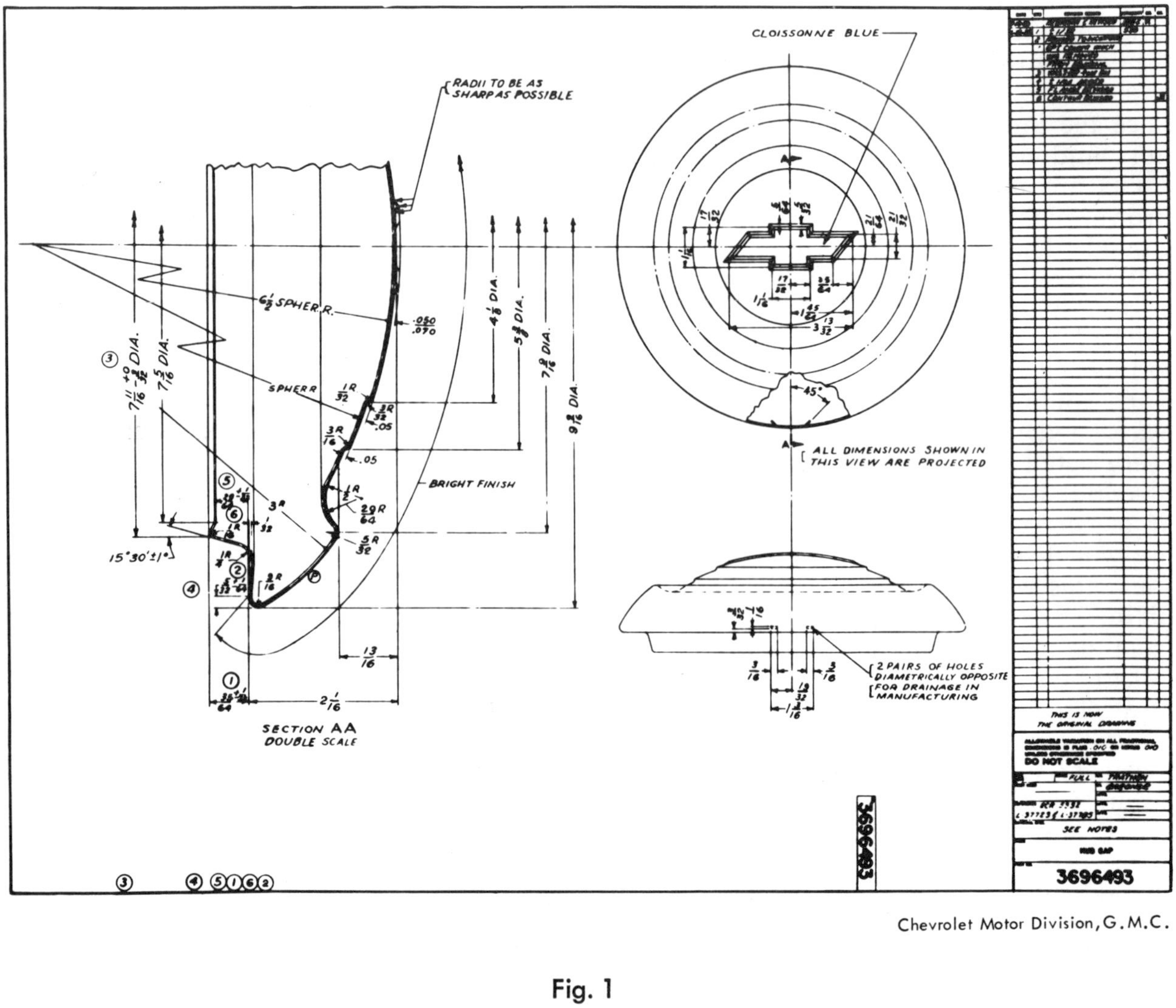

Chevrolet Motor Division, G.M.C.

Fig. 1

tail the equipment needed to manufacture and to maintain uniform quality of the product.

4. *Diemaking Department*. The diemaking department will use the die design blueprints as a basis for making dies that will shape and form the finished hub caps out of thin metal discs. In the production of parts which are machined from castings, a pattern maker, rather than a die maker, will be concerned with this phase.

5. *Production Department*. The blueprint of the part as well as the tool and die prints will be needed by the die setter in the production department. He will use them to check the assembly arrangement of the die and also, in some cases, to regulate and adjust the stroke of the machine in which the dies operate.

6. *Plating Department*. The necessary information for plating and control of this operation is listed in the specifications on the blueprint.

7. *Quality Control Department*. In mass production a careful inspection of units as they are produced is necessary to insure that they meet the specified standards. This is the responsibility of the personnel in the quality control department. A piece that is inspected is compared with the blueprint to ascertain whether or not it conforms to the precise requirements established by the designers and engineers.

Basically, what a production blueprint does, then, is to provide complete and precise information about an object for people who will be concerned with one or more steps in the production of that object.

We may summarize by saying that the blueprint serves the purpose of conveying information about an object in the following ways:

1. By indicating the true shape of the object.
2. By indicating the exact size of the object.
3. By providing the instructions necessary for producing the object.

To learn to read a blueprint quickly and accurately, the ability to *visualize* an object on the basis of the drawings and views of it should be developed. This makes it easier to *interpret* certain facts about the object on the basis of the dimensions, notes, and symbols provided on the blueprint. The procedure of the rest of this book will be to help the student develop the two basic skills of *visualization* and *interpretation*. The emphasis will be on a step-by-step treatment of the elements of blueprint reading, and practice in blueprint reading. For this purpose, many blueprints actually used in industry have been included as typical examples of blueprints that the technician, craftsman, or machinist will encounter in most phases of industry.

SELF STUDY SUGGESTION NO. 1.

In the Machine Trades, as in any other field of specialization, there are numerous technical terms which the craftsman is likely to use, hear, and read during the course of the day. In order to begin and finish a job, to carry out or give instructions, the Machine Tradesman should have a clear understanding of at least the most basic and common terms.

You are advised to review the Glossary of Common Machine Trade Terms which is contained in Appendix E at the end of this book. If you have already encountered some or all of these terms, the review will nevertheless be helpful in refreshing your memory. If the terms are new to you, your reading them will serve as part of this introduction to blueprint reading.

Also useful to review is the List of Common Abbreviations for the Machine Trades (Appendix F), which contains many abbreviations commonly used on blueprints.

The Problems in Shop Arithmetic (Appendix D) and the Decimal Equivalent Chart (Appendix C) are included for the benefit of those who would like to brush up on some basic arithmetic.

Northrop Corp.

In Chapter 1 it was pointed out that on a blueprint, the different views of an object indicate the true shape of the object. It was also pointed out that one of the basic skills of blueprint reading is the ability to *visualize* an object from the views of it given on the blueprint.

What does it mean to "visualize an object?" It means to *form a mental picture of the object*.

However, before the machine tradesman can develop the ability to visualize a blueprint, he should understand how the views of an object are obtained.

THREE-VIEW DRAWINGS

Let us assume that it is desired to make a blueprint of the object shown in the following picture-like drawing (Fig. 1).

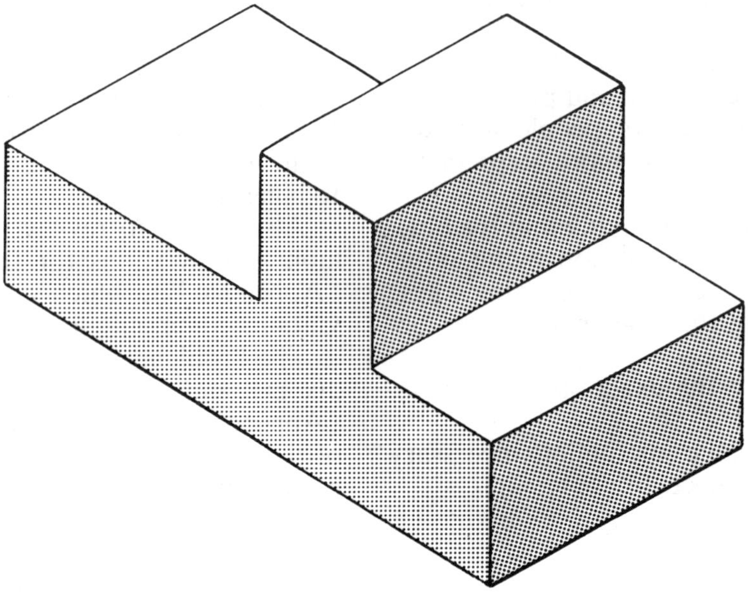

Fig. 1

Front View. Let us assume that a draftsman makes a drawing by placing a transparent sheet of paper directly between himself and the object, as in Fig. 2, left, and traces the outline of the object as it appears to him through the transparent paper. Finally, assume that he labels the drawing the "Front View." This is, of course, not the actual method of making a three-view drawing, but the idea will be helpful in explaining the principle of three-view drawings.

Top View. To get a more complete representation of the object, the draftsman shows the object as it appears from directly above. Assume that he does this by attaching a second sheet of transparent paper to the first sheet in the manner shown in Fig. 2, center. He repeats the procedure for the front view and obtains what he labels the "Top View."

Side View. Deciding that the object would be even more fully represented if it were seen from the end or side, the draftsman repeats the procedure for the front and top views and draws a "Side View" (Fig. 2, right). The result is the box-like figure shown in Fig. 3 with three views of the object, each drawn as the object appears to an observer looking directly at it from the front, the top, and the side.

If the attached sheets on which the views are drawn were laid out flat so that they appear as they do in Fig. 4, they would be arranged like the views of a typical three-view drawing.

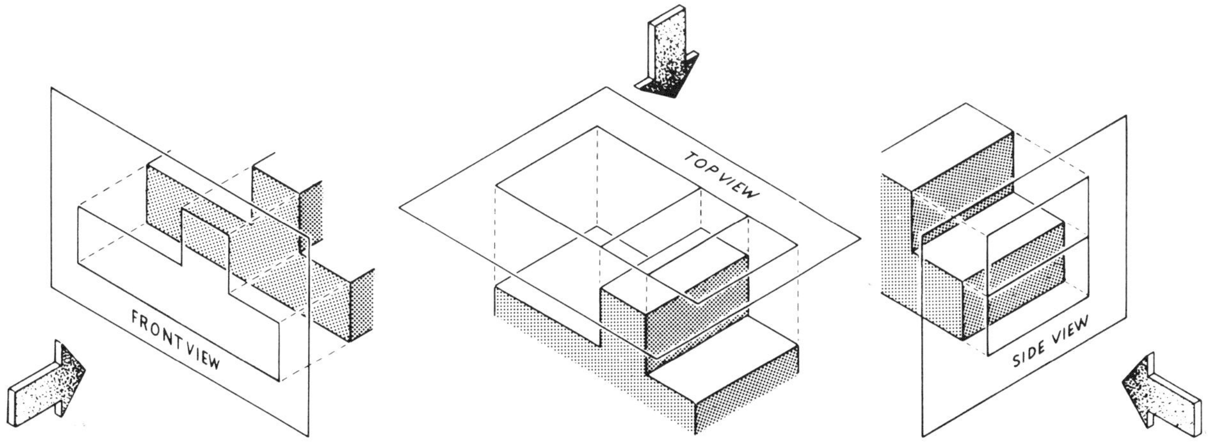

Fig. 2

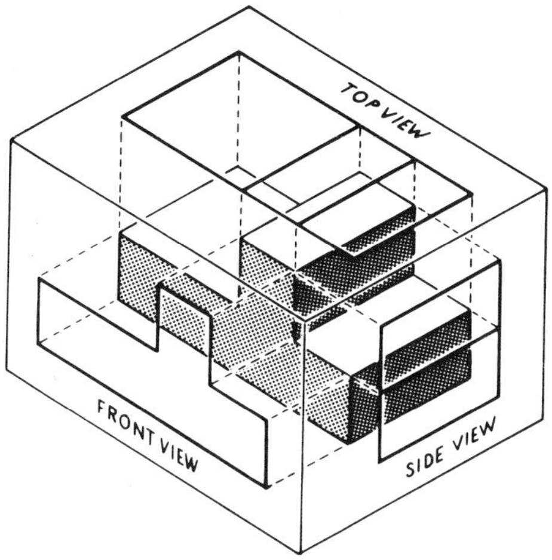

Fig. 3

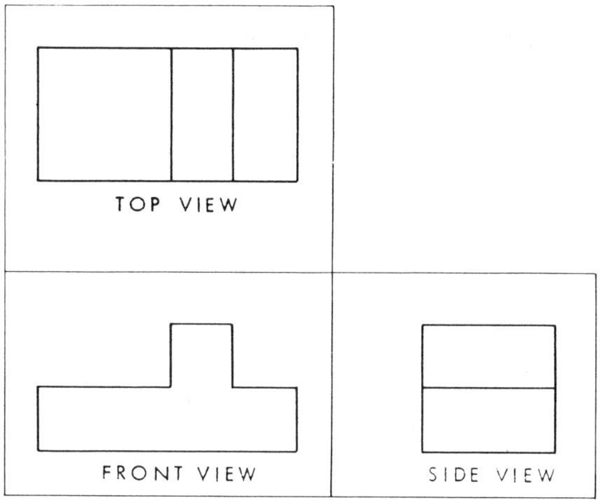

Fig. 4

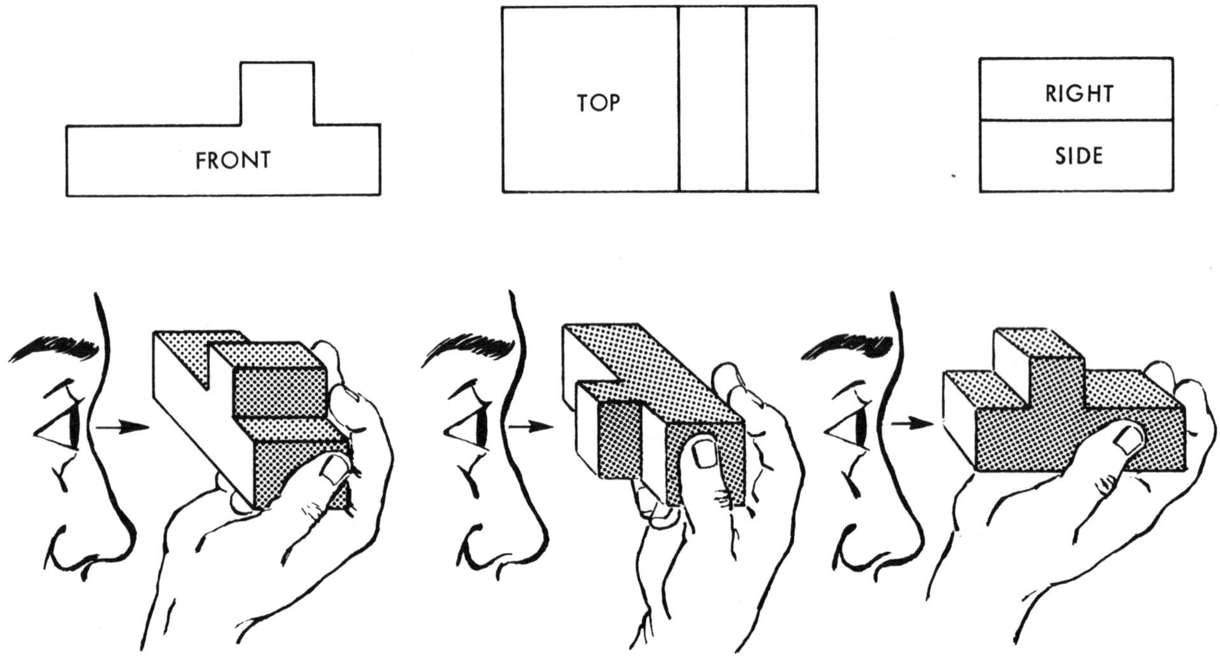

Fig. 5

Another way to visualize a three view drawing is to imagine that the draftsman held the actual object in his hand as shown in Fig. 5. To see the top view he could hold the object so as to look directly on the top (Fig. 5, top); to see the front and right side, he could hold the object so as to look directly at these sides (Fig. 5). Looking at the object directly from these three different sides would again result in the basic three-view drawing layout, as shown in Fig. 6. With this simple object the three-view drawing of Fig. 6 shows exactly what the draftsman would see if he held the object as shown in Fig. 5.

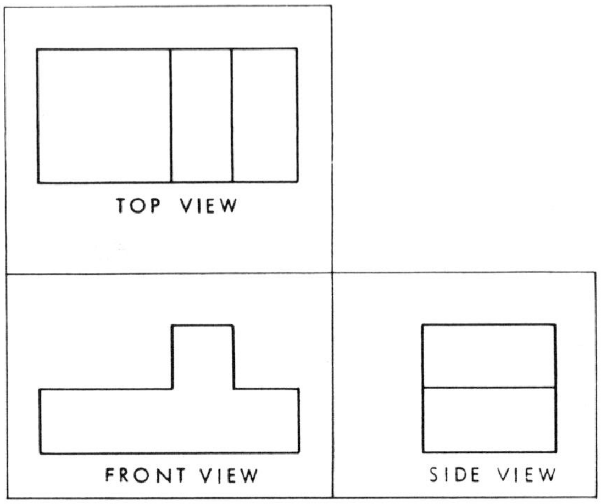

Fig. 6

PROJECTION

A draftsman does not actually fold a piece of paper around the object, draw the various views, and then unfold the paper to obtain the three views shown in Fig. 4. First, he draws the outline of that view of the object that best describes its shape. This view is usually called the front view. The draftsman then projects the necessary lines from the front view to the other views. (With long or cylindrical shaped objects, however, the side view is often drawn first.)

Consider again the pictorial shown in Fig. 1. In making a three-view drawing of this object the draftsman first finds the length of the object, the height of the right and left sides, and the height and location of the part that is slightly off-center. Now he can draw the front view.

Using the front view as shown in Fig. 7 the length and location where two surfaces join (as indicated by points A, B, C, and D) are projected upwards by means of projection lines. The projection lines transfer the object relationship from the front view to the top view. In Fig. 7 the projection lines have been left in to demonstrate how "information" is projected from one view to another. Projection lines do not appear on a finished drawing.

Points A, B, C, and D in the front view now become a part of the object lines in the top view, represented by lines I, J, K, and L. The width of the object in the top view is determined by measurements taken from the actual object.

Since there are no additional details shown on the side of the object that are not already shown in the front and top views, the right side view can be easily drawn by using projection lines. The height of the right side view is projected horizontally from the front view. You will note that the width of the side view is the same as the width of the top view. (Note that the views are located a convenient distance away from each other, suitable to the

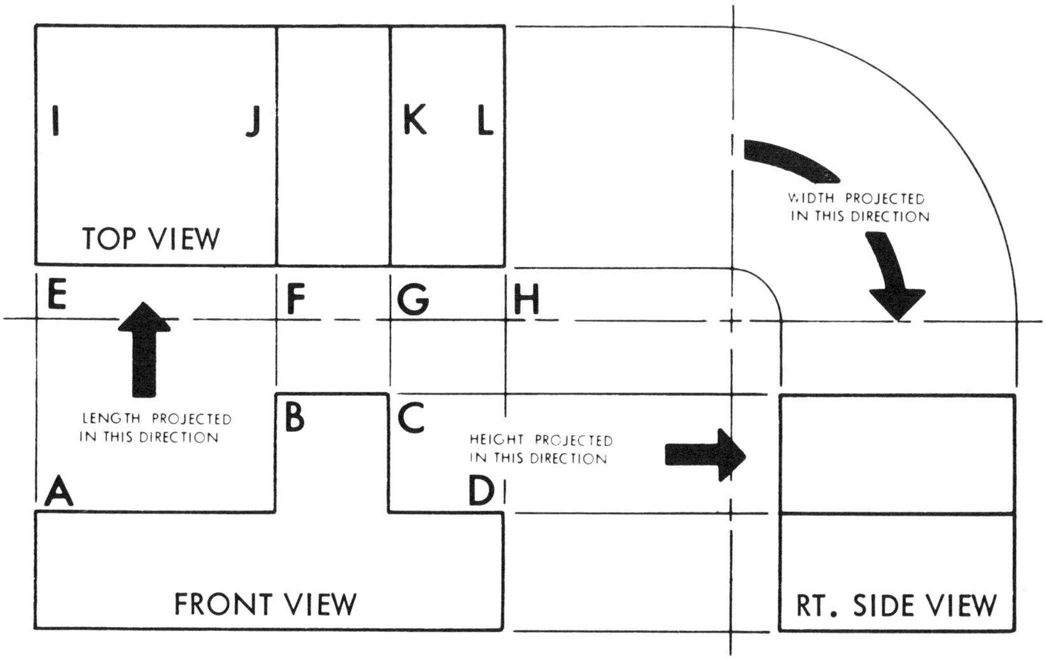

Fig. 7

size of the views and the size of the drawing paper used.)

Fig. 8 shows the three views of the object in proper relationship to each other. Let us study this relationship by means of some identifying points, lines, and surfaces.

Point A (Front View) becomes Line I (Top View).
Point B (Front View) becomes Line J (Top View).
Point D (Front View) becomes Line L (Top View).
Surface P (Front View) is represented by Line M (Side View).
Surface R (Top View) is represented by Line N (Side View).
Surface T (Side View) is represented by Line K (Top View).
Surface U (Side View) is represented by Line L (Top View).

PROJECTION AND VISUALIZATION

We have demonstrated the principle by which a three-view drawing is derived from an object. In effect, what the draftsman does is to "take apart" an object by representing it as three different views.

It is equally important, however, to clearly understand the relationship between views and to see how changes in the location and contour of surfaces in one view will affect the lines of the other views. First we must understand that lines always indicate either an abrupt change in a surface direction or the intersection of two surfaces. Then by properly associating the lines in one view with those in the other views, the true shape of the object can be visualized. It does not necessarily follow that if a change is made in an object, it will affect all views. An example of this can be seen in the object shown in Fig. 7. If the width of this object were increased or decreased, it would not affect the front view. This may be demonstrated by taking a simple object and introducing changes in the views. In visualizing the changes in the shape of object always begin with and work from the view that seems to best describe its

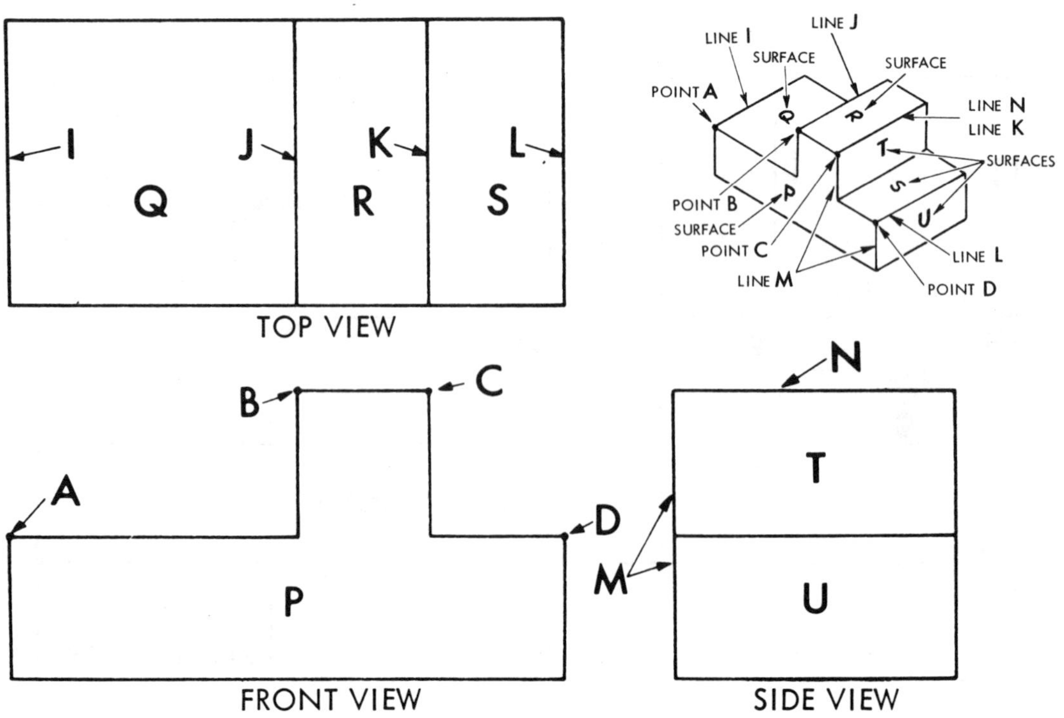

Fig. 8

general shape. This is *usually* the front view.

Let us carry out the procedure on the basis of the simple object shown in Fig. 8.

Using the object, progressive changes will be made in its shape to further demonstrate the relationship of views and emphasize the primary importance of the front view in visualizing an object.

First, examine the view that best describes the shape of the object. In this case, it is the front view. Form a mental picture of this view. In other words, be able to visualize the shape of the object from the position of this view without actually looking at the drawing. Now examine the other views and visualize the width of the object and how it relates to the length and height. With practice you will be able to visualize this object as it is shown in Fig. 1 just as easily as you can now visualize the shape and the size of your house or your favorite chair.

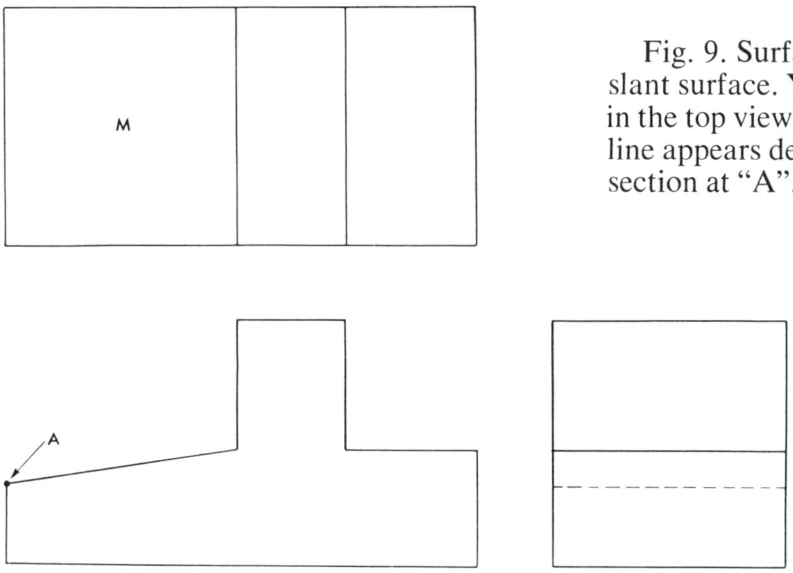

Fig. 9

Fig. 9. Surface "M" has been changed to a slant surface. You will note there is no change in the top view. In the right side view a dashed line appears denoting the hidden surface intersection at "A".

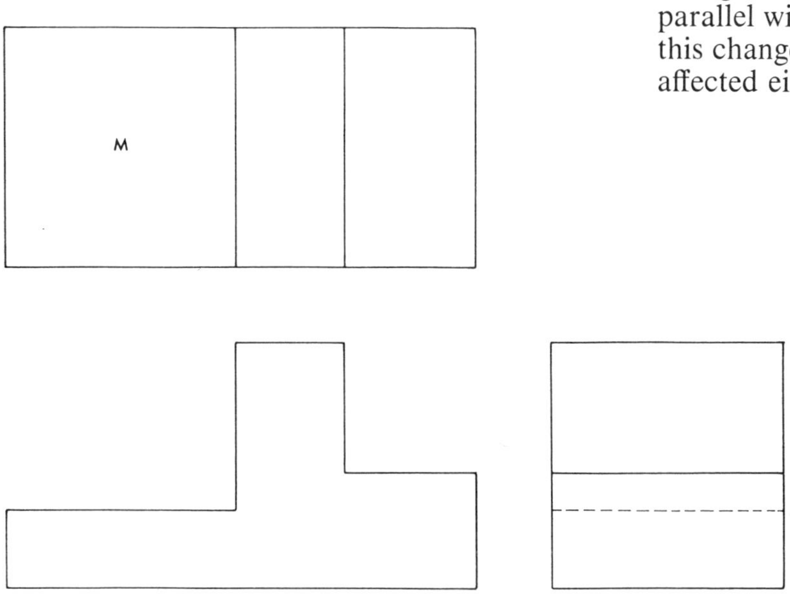

Fig. 10

Fig. 10. Surface "M" has now been made parallel with the base of the object. Note that this change in the shape of the object has not affected either the top or the right side view.

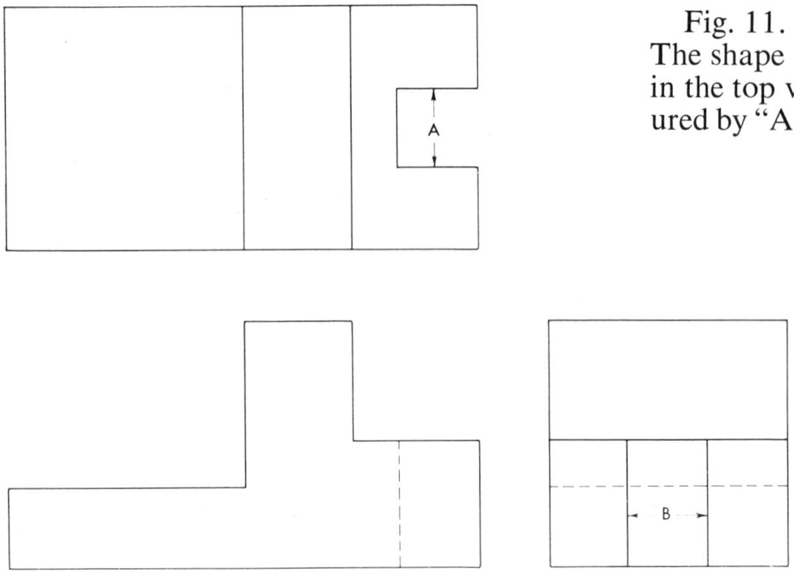

Fig. 11

Fig. 11. A slot has been cut into the object. The shape and location of this slot are shown in the top view. Note that the distances measured by "A" and "B" are the same.

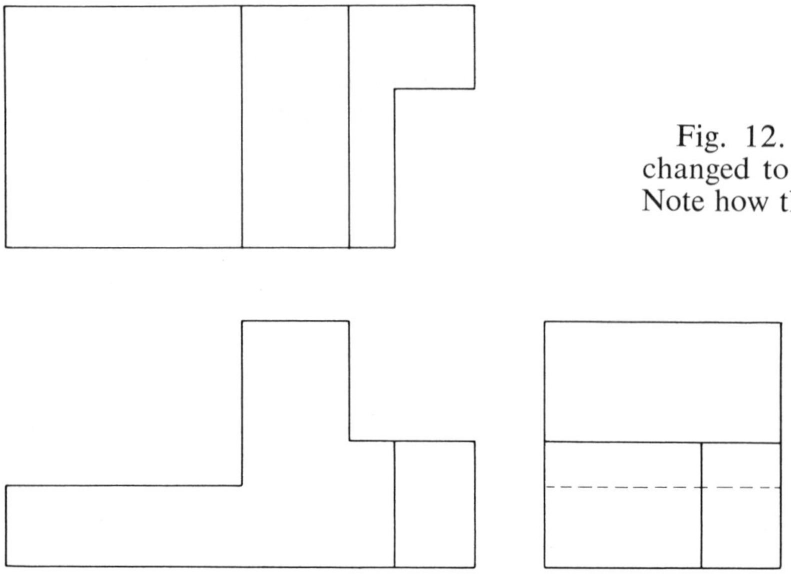

Fig. 12

Fig. 12. The shape of the slot has been changed to extend to the edge of the object. Note how this change affects the other views.

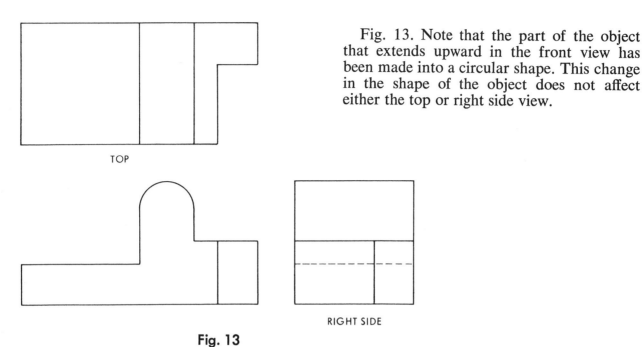

Fig. 13. Note that the part of the object that extends upward in the front view has been made into a circular shape. This change in the shape of the object does not affect either the top or right side view.

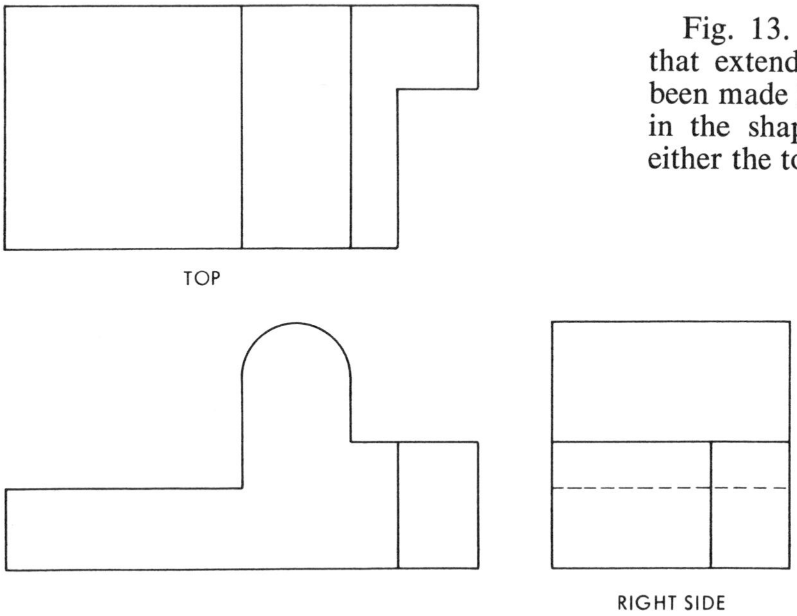

Fig. 13

Fig. 14. Note that if the object shown in Fig. 13 were viewed from the left side those lines that were *visible* in the right side view would become hidden (shown by dashed lines in Fig. 14). Also the line that was invisible in the right side view of Fig. 13 becomes visible in the left side view of Fig. 14.

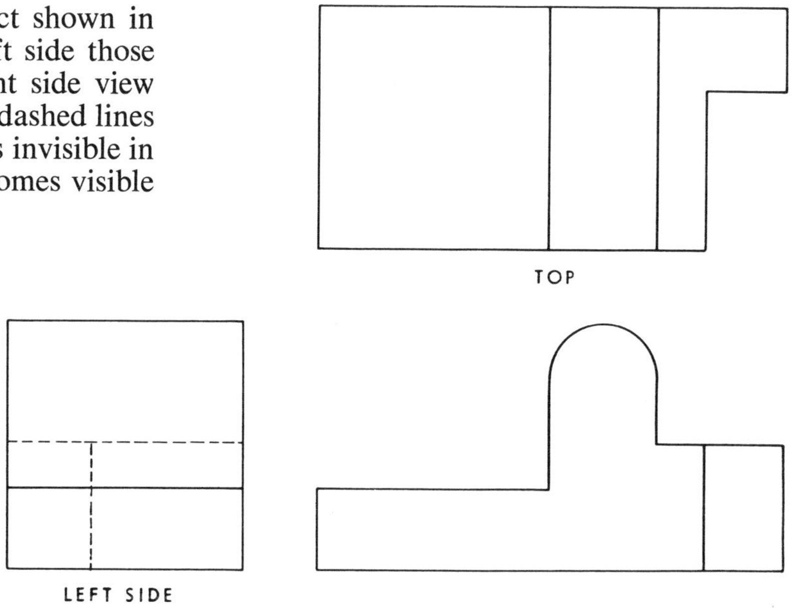

Fig. 14

VISUALIZATION

We have seen how a draftsman "takes apart" an object and represents it in views and how changes in the shape of an object affect those views. An understanding of the basic drafting principles is obviously essential in order to accurately and thoroughly read a blueprint. Equally important, also, to the machine tradesman is an understanding of how to "put together" or associate all of the views of a drawing and create a mental picture of the object.

ISOMETRIC SKETCHING

The ability to sketch pictorially from a blueprint is an invaluable aid in visualizing the object represented. Though the blueprint describes the object completely. Often it requires an effort of geometric imagination to visualize its appearance.

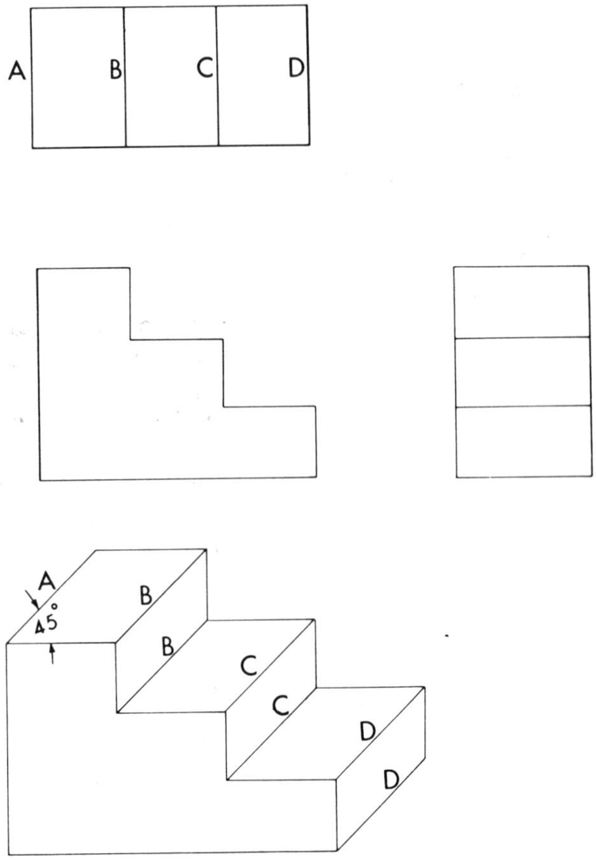

Fig. 15

Methods have been devised to give the pictorial effect that will aid the craftsman in visualizing an object. Basically, most methods are a form of axonometric projection whereby an object is positioned so that three surfaces are projected in one plane.

Figs. 15 and 16 illustrate poular methods of sketching used by craftsmen. In Fig. 15 the front surface of the object, as it is shown in the front view of the blueprint, is sketched; next, the vertical lines, as they appear in the top view, and horizontal lines as these appear in the side view, are sketched lightly at an angle approximately 45° starting from their corresponding edges of the sketch of the front view. These lines should be foreshortened approximately one-half of their actual length as shown in the top view. The horizontal lines of the top view should be drawn in their true length. The vertical lines in the front and side views will show in their true length.

Fig. 16 illustrates another method of pictorially representing a mechanical object. The sketching should start with the front surface of the object as it is shown in the front view. The horizontal lines of this view are drawn at angles 30° below horizontal. The vertical lines should be drawn, as shown in the front view, at their actual lengths. The lines drawn vertically in the top view should be drawn at their true length and at an angle 30° above horizontal. The horizontal lines of the top view are drawn at an agle 30° below horizontal. Horizontal lines in the side view are drawn at an angle 30° above horizontal and the vertical lines are drawn in a vertical position at their actual lengths. This is an isometric sketch, meaning the lengths are equal to blueprint dimensions.

It will be noted that the pictorial sketch made of the object appears much larger than the drawing. In order that a pictorial sketch show an object in its actual size, all lines would have to be foreshortened approximately 20 percent. This is not usually done since it requires a special scale.

Invisible surfaces are not generally shown in pictorial sketches.

Now test yourself by covering up the isometric sketch of the object and see if you can form an accurate mental picture of the object from the three views drawing. Remember

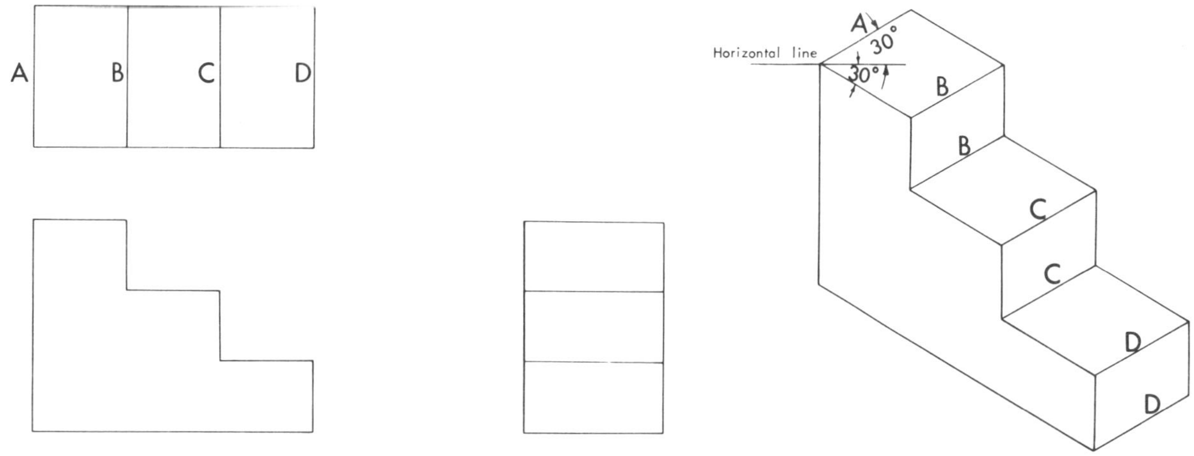

Fig. 16

that the proper procedure is to first visualize the general outline of the object and then progressively include all the details represented in the separate views. To develop the ability to visualize blueprints, try to mentally *construct* each new print in this book as you come to it. If you find it helpful to make experimental sketches, go ahead and do so. This lesson on isometric sketching should be considered only as a preliminary aid to visualization. Your goal should be to develop the ability to visualize the objects entirely mentally from the views and other information contained in notes and dimensions on the drawings. You will know you have clearly visualized a blueprint when you find yourself understanding the details with relative ease.

The student should do the following exercises before taking the Trade Competency Test, Chapter 2. Both of the following exercises are designed to provide practice in visualizing an object.

SELF STUDY SUGGESTION NO. 2

Exercise: Figs. 19 and 20 are three-view drawings, each accompanied by an incomplete isometric drawing of the object. After studying each three-view drawing, complete the isometric drawing in the ruled section by adding the missing lines, as demonstrated in Figs. 17 and 18. If you have difficulty review the material on isometric drawing. The correct answers to Figs. 19 and 20 are given at the end of the book.

DO NOT TEAR OUT

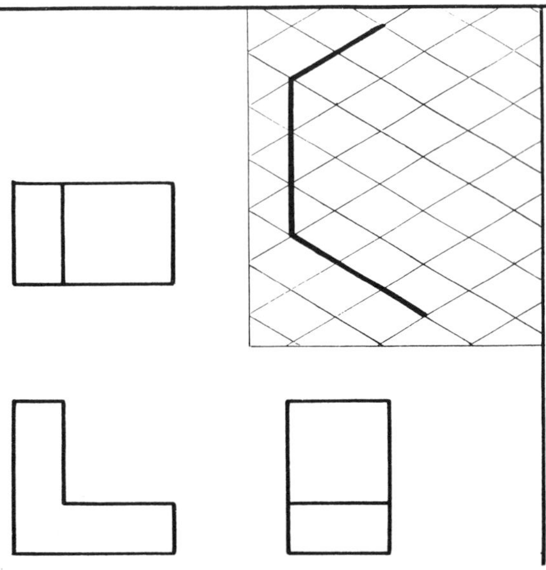

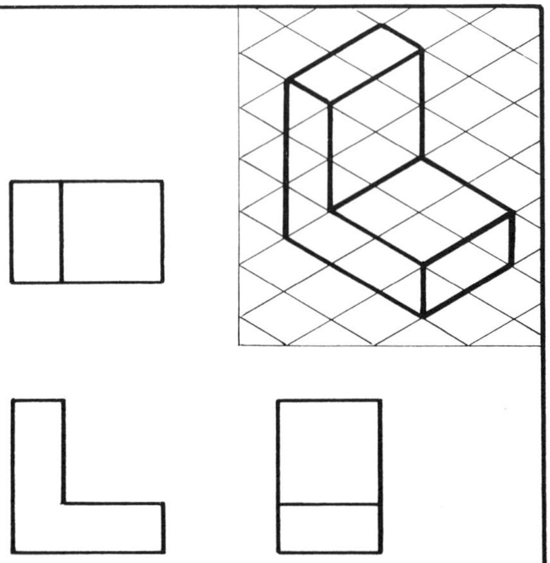

(FOR DEMONSTRATION ONLY)

COMPLETE THE ISOMETRIC VIEW

Fig. 17

ISOMETRIC VIEW COMPLETED

Fig. 18

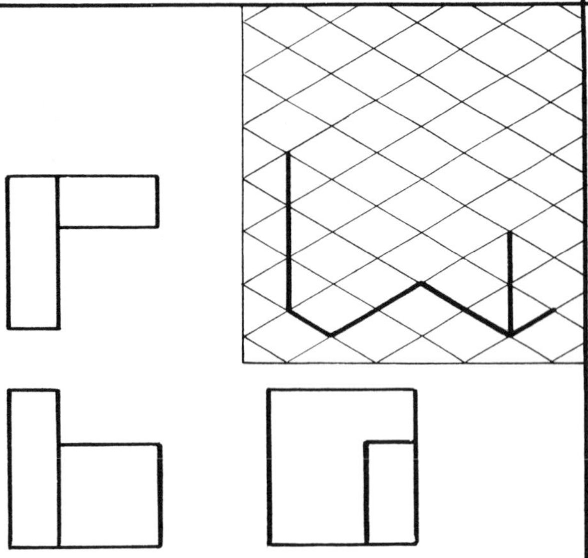

COMPLETE THE ISOMETRIC VIEW

Fig. 19

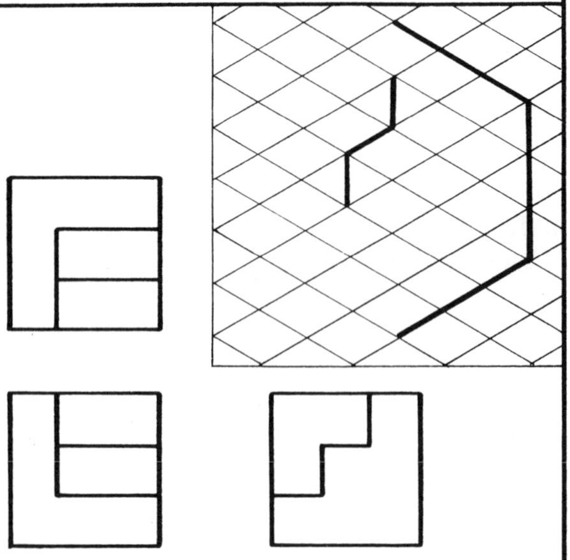

COMPLETE THE ISOMETRIC VIEW

Fig. 20

THIS PAGE FOR STUDENT NOTES

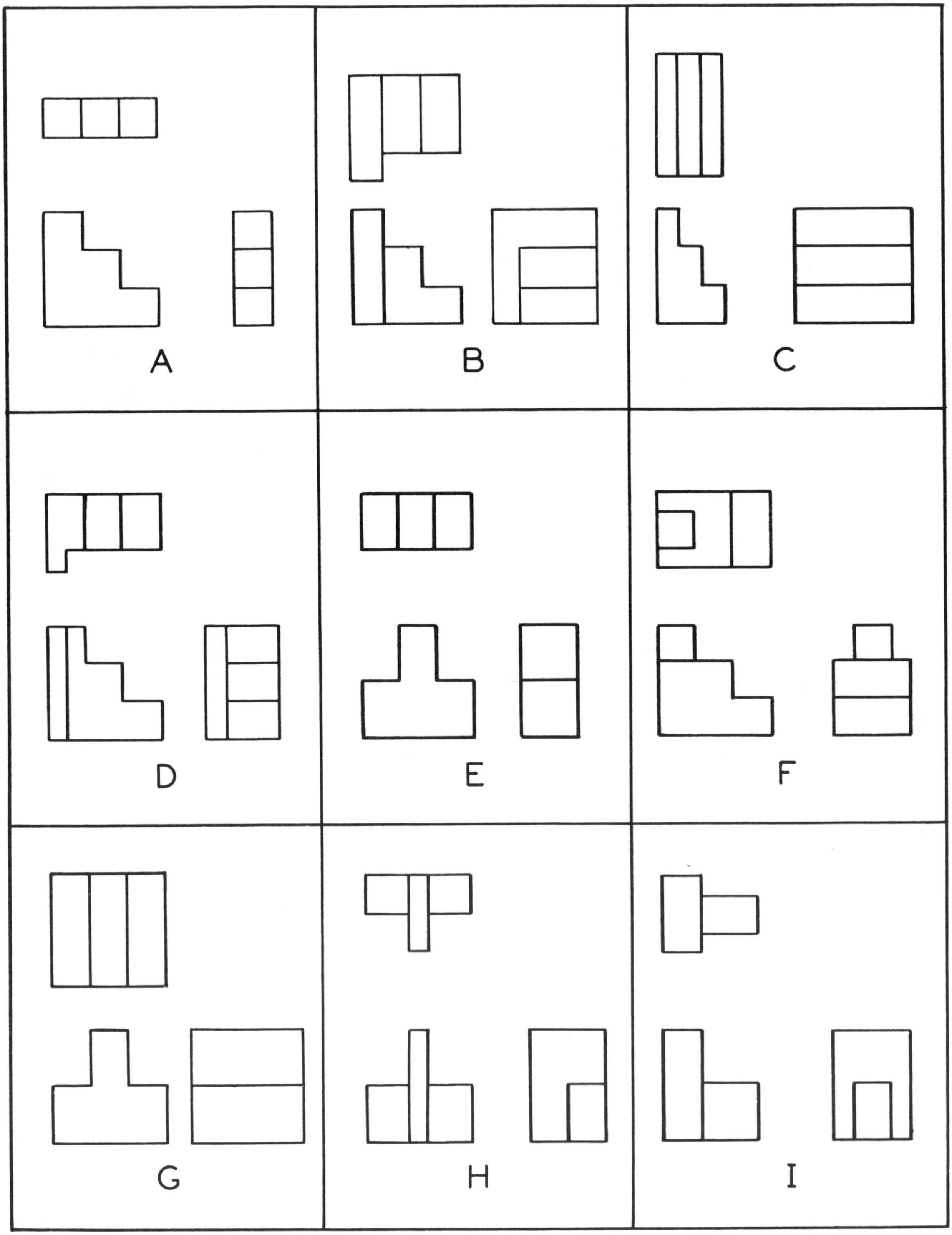

PLATE 1

TEAR OFF HERE

TRADE COMPETENCY TEST

Based on Plates 1-3, Chapter 2

Student's Name________________________ Instructor's Name________________________

1. **VISUALIZATION TEST.** Plate 1 shows nine three-view drawings of simple objects. Each of these drawings corresponds to one of the isometric drawings shown below. In the blank space below each isometric drawing you are to place the letter of the three-view drawing from Plate 1 which is correctly represented by the isometric drawing.

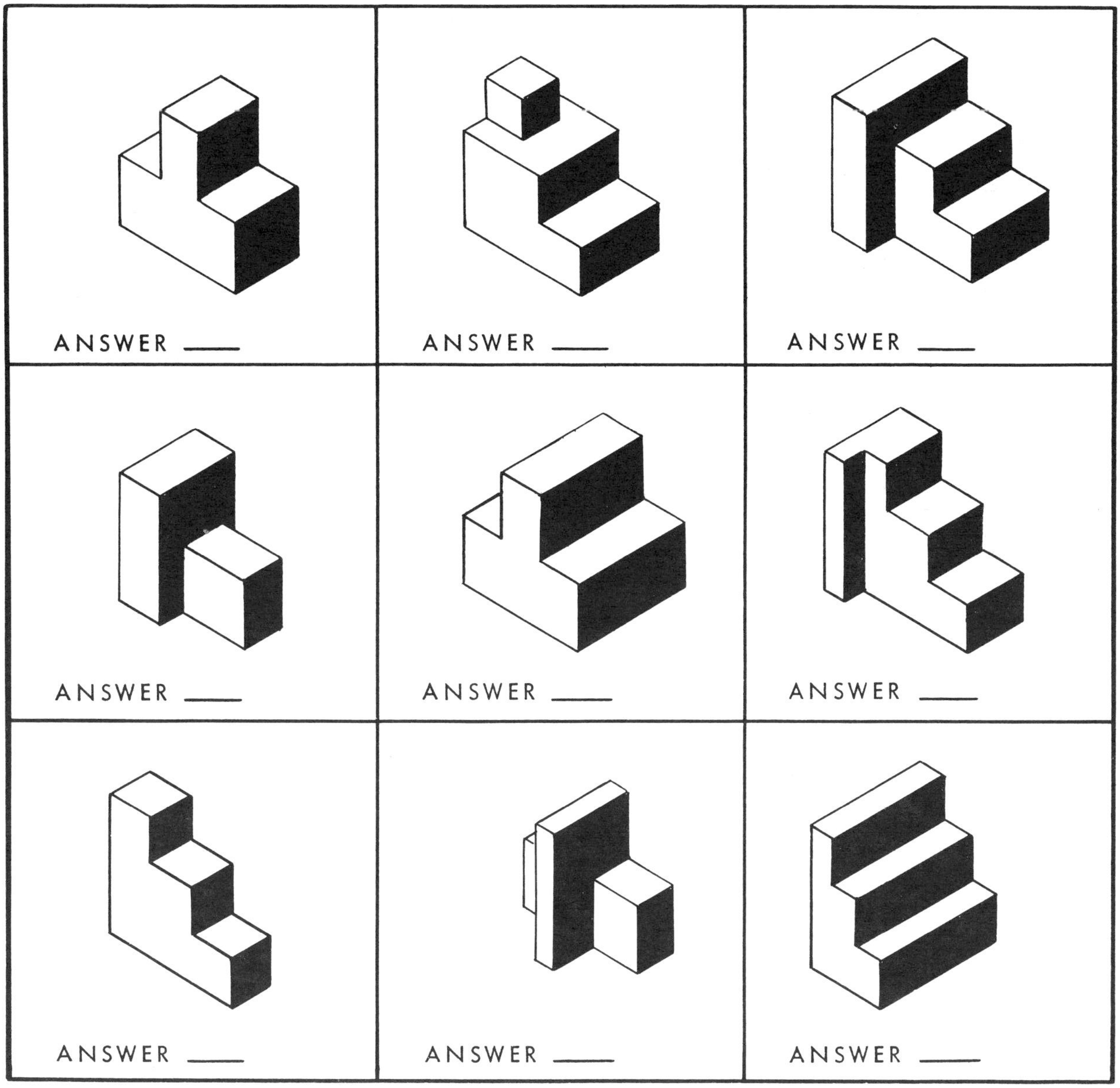

2. **OBJECT COMPLETION TEST.** Plates 2 and 3 consist of three-view drawings of simple objects. Each drawing is accompanied by an incomplete isometric drawing. You are to add the lines necessary to complete the isometric drawings so that they correctly represent the objects described by the three-view drawings.

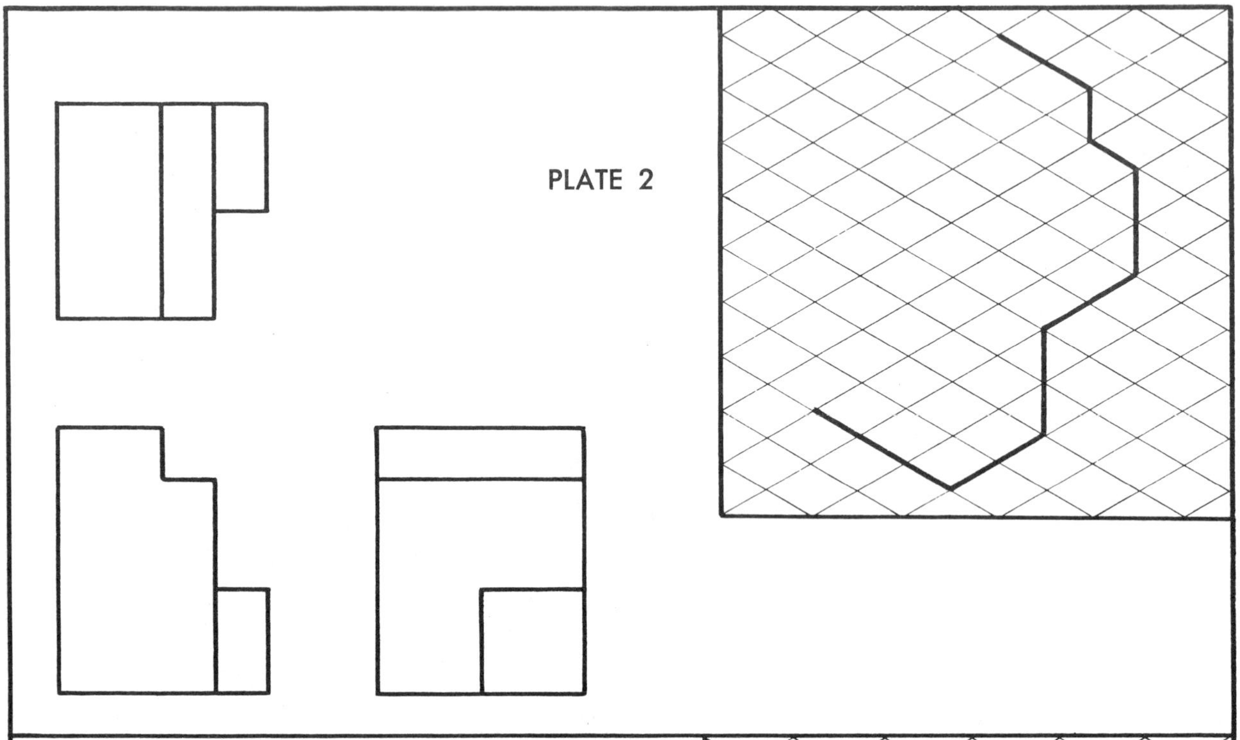

PLATE 2

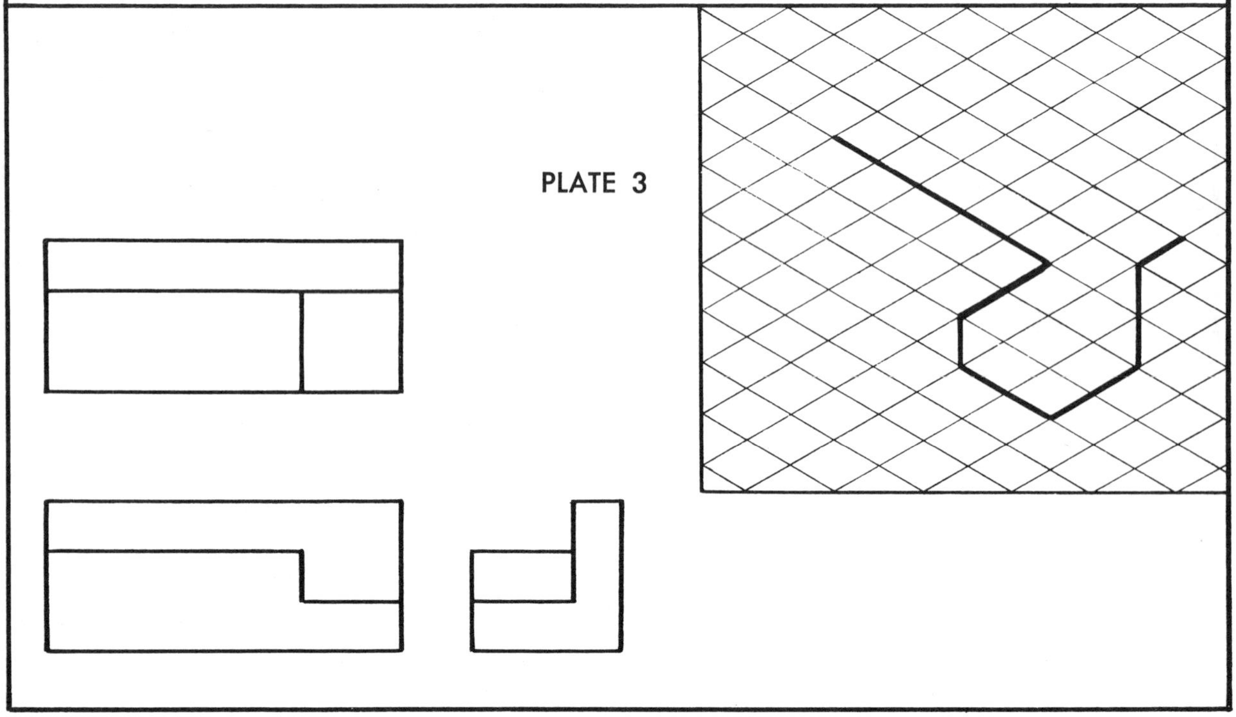

PLATE 3

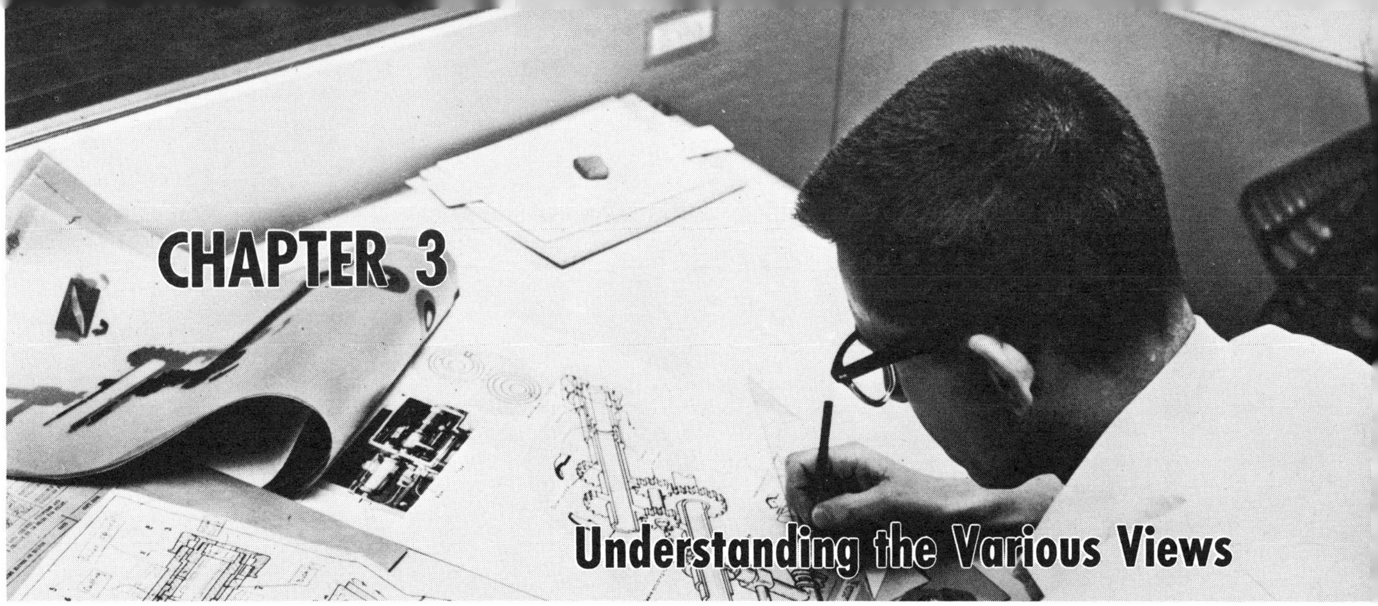

CHAPTER 3

Understanding the Various Views

Design News

We have seen in Chapter 2 that visualizing a blueprint consists of forming a mental picture of an object from the different views of it that are shown. From the examples in Chapter 2, it is obvious that the manner in which the drawings are arranged on the page is closely related to the structure or shape of the object.

Suppose, for instance, that we are given three views and that the views are arranged in the manner shown in Fig. 1.

It would be somewhat difficult to visualize this object from such a drawing because the different views are not arranged in order according to their *structural relationship* to each other; in other words, according to the way they are connected to each other.

ARRANGEMENT OF VIEWS

How are the views arranged on a blueprint so that they reveal their proper relationships or connections in the simplest manner?

Let us assume that we could mentally "unfold" a simple rectangular carton, step by step, in the manner illustrated in Fig. 2.

Assume that we will start with the side that shows the *length* and *height* of the carton, and call it the "Front."

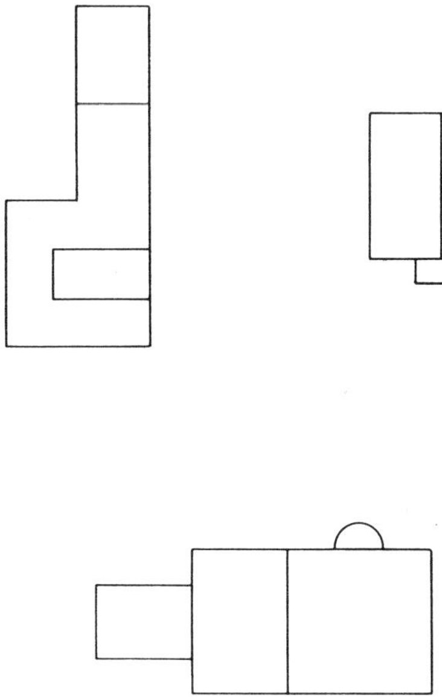

Fig. 1

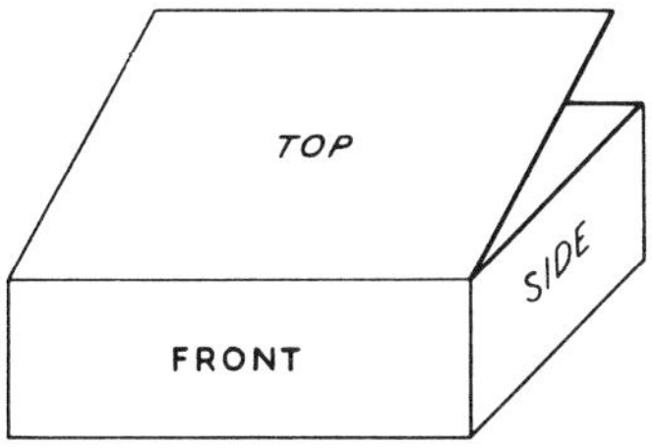

Fig. 2

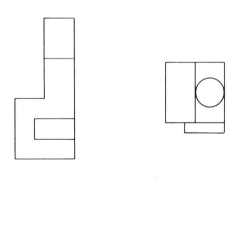

Proceeding in this manner, we begin with a view of the front alone, then progressively unfold the carton until it consists of a flat piece of cardboard, as shown in Fig. 3.

Remembering that a rectangular box has six sides, we may verify that the carton is completely unfolded by counting the number of sides or "views" contained in the completely unfolded piece in Fig. 3.

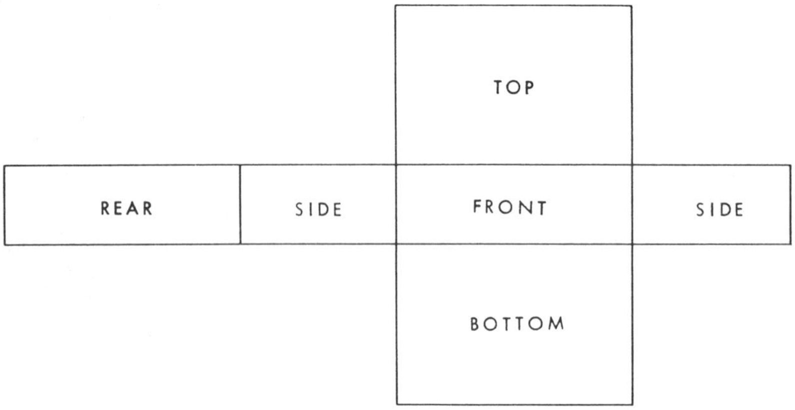

Fig. 3

As a further check, it should be possible to cut out the unfolded carton in Fig. 3, fold it along the edges, and reconstruct the carton.

It is clear then, that by unfolding the carton, we can see all the sides of it at once by means of views that are arranged so that they reveal the manner in which they are connected. By visualizing, or reconstructing the object in our minds, we can discover the true shape of the object.

In Fig. 4 we have cut along the edges of the unfolded carton and separated the pieces or views, keeping the same arrangement. This arrangement, besides being orderly, reveals and maintains the relationship that the views have to each other. Furthermore, the spaces between the views allow the insertion of dimensions which indicate the true size of the object.

This is the standard arrangement used

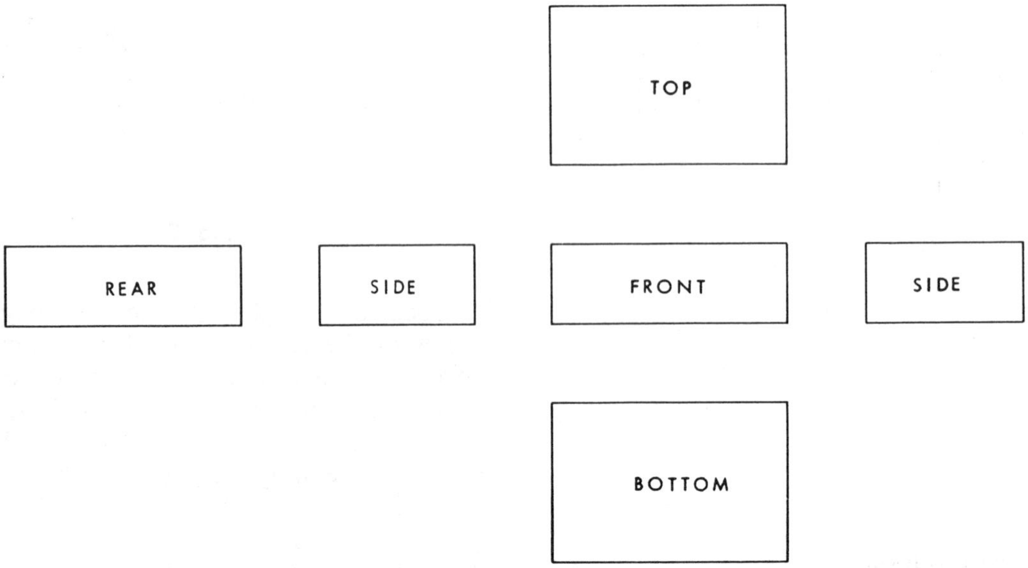

Fig. 4

throughout industry of the views of a blueprint and should be understood by anyone who works with blueprints.

You may have wondered why only three sides of the object were shown in Fig. 4 of Chapter 2, whereas we have just shown that a rectangular object is completely shown when all six sides are revealed, as in Fig. 3. Why were all six sides not shown?

We may answer this question by first returning to the views shown in Fig. 1. If these views were arranged in their proper order, you would have been able to visualize the object shown in the following isometric drawing (Fig. 5). What is the correct arrangement of the views in Fig. 1?

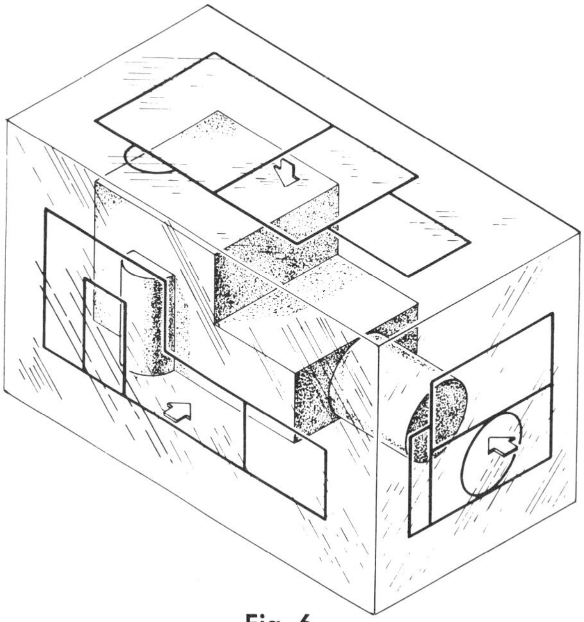

Fig. 6

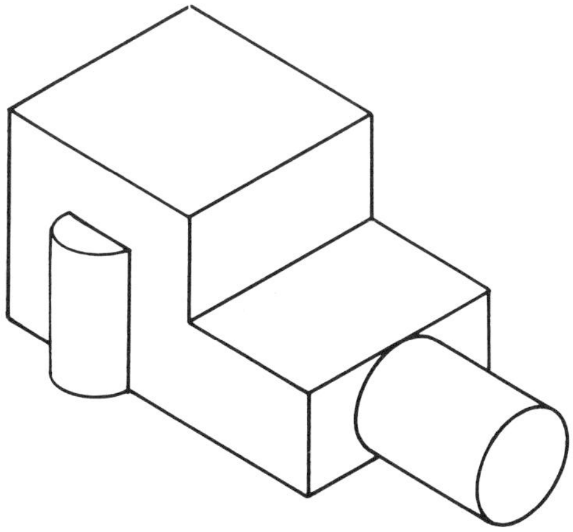

Fig. 5

Let us assume that a draftsman could carry out the procedure of mentally projecting and tracing the views of the front, top, and side of the object shown in Fig. 5. The result would be the "transparent box" shown in Fig. 6. The arrows in this illustration indicate the line of sight of an observer who is looking directly at the object so that its outlines appear as shown.

If our imaginary draftsman went further this time and traced views of the object on the *remaining three sides* of the box, he would have obtained *six views altogether,* each view being traced on one side of the box. If we imagined the box partially unfolded, it would look like Fig. 7.

It is convenient, in learning to read blueprints, to imagine a view of an object as if it were projected onto the side of a transparent box. The box may then be "unfolded" so that the different views are revealed at once in their proper arrangement. This is shown in Fig. 8 in which the box is completely unfolded and laid out flat. Notice that the arrangement of the views in Fig. 8 is the same as the standard arrangement of the sides of the carton in Fig. 3.

Studying these views in Fig. 8, we discover that the *outlines* of the front and rear views are identical, except that their positions are reversed, and that this is also true of both sides, and of the top and bottom.

The only differences we can discover between these pairs of corresponding or opposite views is that the bottom, left side, and rear views have hidden or invisible lines, and that these same lines are represented as solid outlines or *object lines* in the front, top, and right-side views. You may recall that hidden lines are used on certain views to represent edges or surfaces on the object that cannot be seen from that point of view. They are used to *locate* these hidden edges and surfaces and thus provide the reader with important information about the object.

In studying these views further, we discover, however, that the hidden lines do not tell us anything more than the corresponding

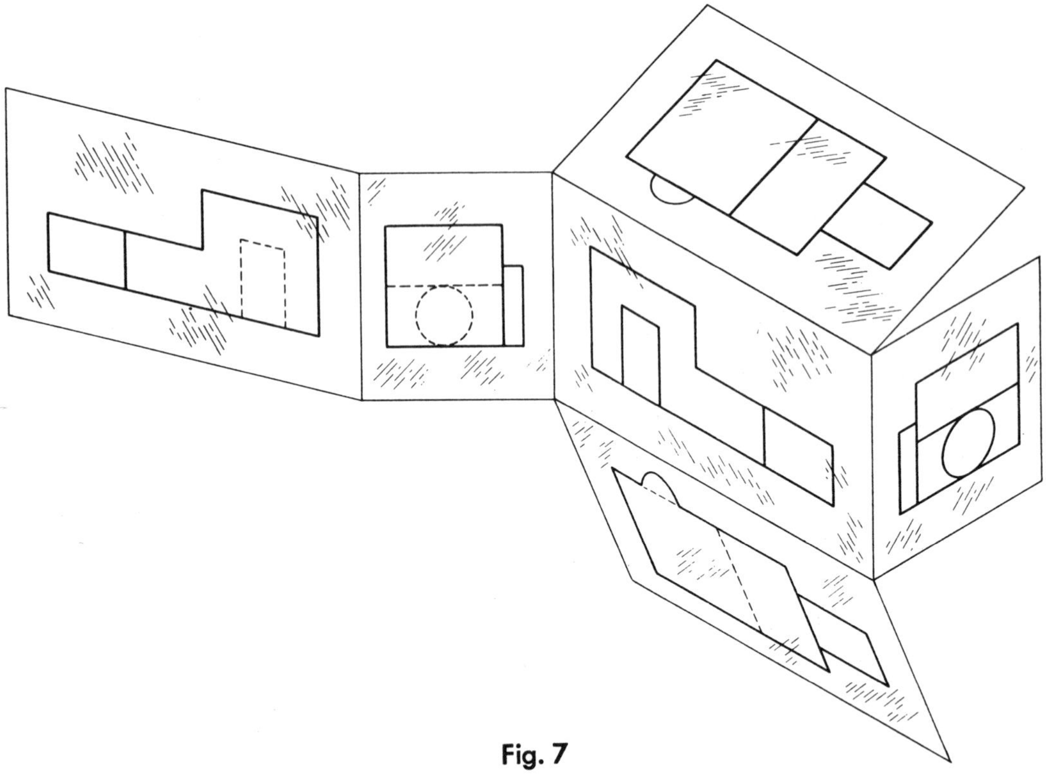

Fig. 7

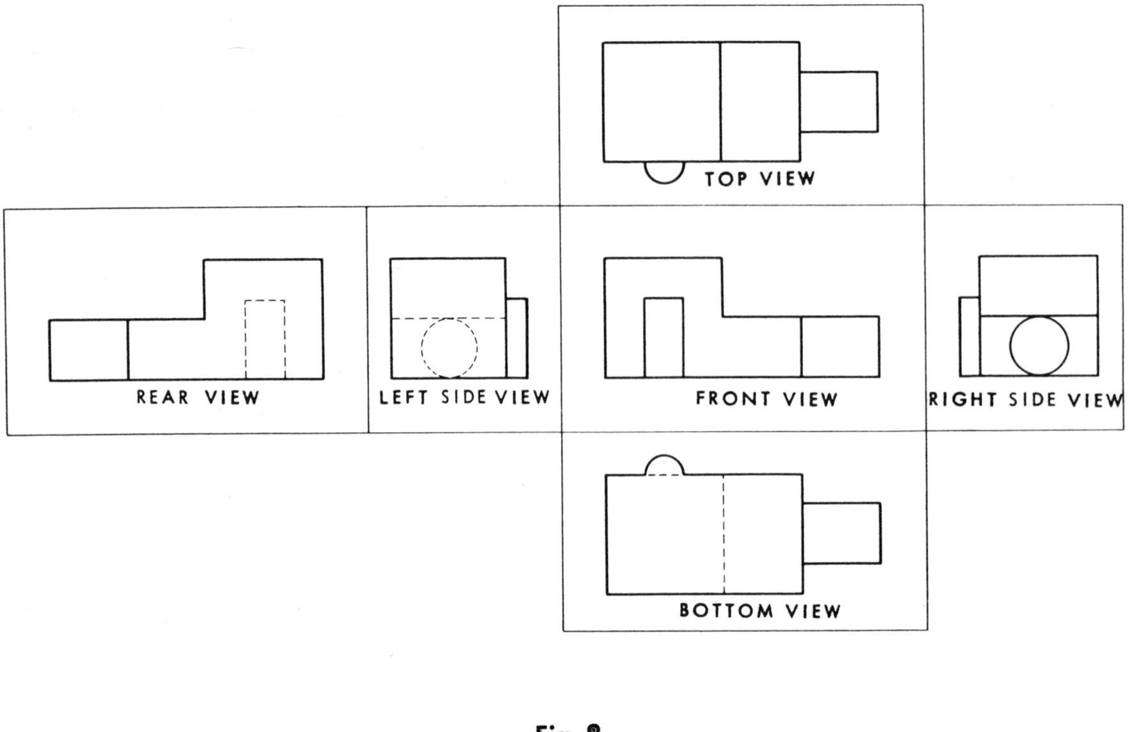
TOP VIEW
REAR VIEW
LEFT SIDE VIEW
FRONT VIEW
RIGHT SIDE VIEW
BOTTOM VIEW

Fig. 8

lines seen on the front, top, and right side views. For example, we know from studying the front, top, and right side views that there is a cylindrical projection on the right side of the object. The circular outline of this cylinder is shown as a dashed circle on the left side view. Since we already know this from the other views, the circular hidden line becomes unnecessary. The bottom and rear views, even though they contain hidden lines, do not tell us anything that the front and top views do not already indicate.

In other words, these hidden lines are the same lines that are already represented as outlines or object lines on the front, top, and right side views. When hidden lines convey the same information as object lines, it is generally preferable to employ the views that have object lines. Furthermore, the *overall outlines* of the rear, bottom, and left side views, being the same as the opposite views, are unnecessary.

It is a general practice, in drafting, to draw only as many views of an object as are necessary to give complete information about its shape and structure.

Since, as we have seen, the rear, bottom, and left side views do not convey any information that is not already shown by the front, top, and right side views, we may eliminate them. The object is completely represented by the three-view drawing shown in Fig. 9.

If you will compare the views shown in Fig. 9 with those shown in Fig. 1, you will notice that they are precisely the same views except that their arrangement is different. In Fig. 9 the views are arranged according to their *structural relationship* to each other; in other words, according to *the manner in which they are connected.*

It should be clear then, that even though the object in question has a total of *six sides,* it can be completely represented on a blueprint by only *three views.*

What we have demonstrated in the preceding paragraphs is essentially the same procedure that a draftsman goes through when he makes a working drawing of an object.

First, he studies or visualizes the object which he is about to draw.

Second, he analyzes, or mentally "takes apart" the object in order to determine which views of the object will be necessary to completely represent its true shape.

Third, he arranges the views in such a manner that they will reveal their relationships to each other and thus permit the blueprint reader to quickly and accurately visualize the object.

Generally, the views of a drawing will be arranged according to the standard arrangement regardless of the number of views in the drawing. Figs. 10 to 13 illustrate a number of ways in which views may be arranged according to the basic plan.

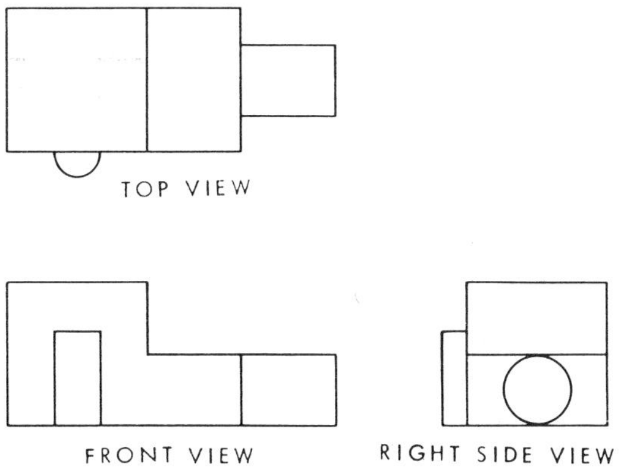

Fig. 9

SELECTION OF VIEWS

The Front View. Generally speaking, the front view is regarded as the principal view, and therefore the view around which the other views are arranged. The selection of the front view is determined by the shape or structure of the object. This involves two important factors:

1. The view of the object which reveals *its most important characteristics* is generally chosen for the front view. Since the largest dimension (length or height) of an object is an important characteristic, the front view in most cases includes this dimension.

2. Surface details are essential information. It is therefore logical to select a view which gives *the greatest amount of information* about the surface characteristic of the object. In any event, before drawing, the draftsman will look for the view that shows

the greatest number of signficant or important details and best reveals the shape.

In situations where no one view can clearly be called the front view, the draftsman will depend on his experience and judgment in selecting the views. It should be noted that the rules of selecting and arranging views cannot always be hard and fast. There are too many kinds of situations and problems in drafting and blueprint reading to attempt to always follow unbreakable rules.

The important thing for the blueprint reader to remember is that in the final drawing, the views should be *logically related* and should *completely and precisely describe the object*. He should also understand in general what determines the manner in which the object is represented on a blueprint.

The Top and Bottom Views. Once the front view has been selected, the top and bottom views are easily determined. In cases where only one of these views is necessary, the view conveying the most significant information is generally used. Whether it is the top or bottom view depends, of course, on its position in relation to the front view.

If the front view shows the *length* and the *height* of the object, the top and bottom views will show the *length* and the *width* (or depth) of the object, which will consist of the distance between the front and the back edges.

The Side Views. The side views are automatically determined once the front and top views have been selected. They are simply the views from the left or right side of the object from the point of view of a person who is looking directly at the front.

If only one side view is required, its location is, of course, dependent on whether it is the left or the right side view, as shown in Figs. 10 and 11.

Certain objects that include considerable detail are best shown by views drawn to the left and right of the front view. Fig. 12 shows this arrangement. It is considered good practice to omit hidden surfaces in either side view because this would be an unnecessary repetition of information.

Many objects are best shown in two views. If a right side view were shown of the object in Fig. 13, it would be identical to the front view and, therefore, would not convey any additional information.

The number of views to be drawn, as we have mentioned, is determined by the shape or structure of the object. In other words, the complexity of the object, the number of sides or surfaces, the amount of detail and the inner details if any, all have some influence in determining the number of views a drawing should have. In general, the draftsman will restrict the views to the minimum number required to completely represent the object, thereby making it more convenient for the

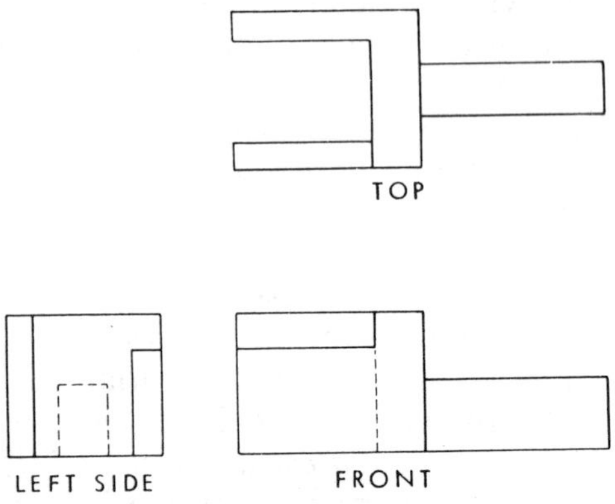

Fig. 10

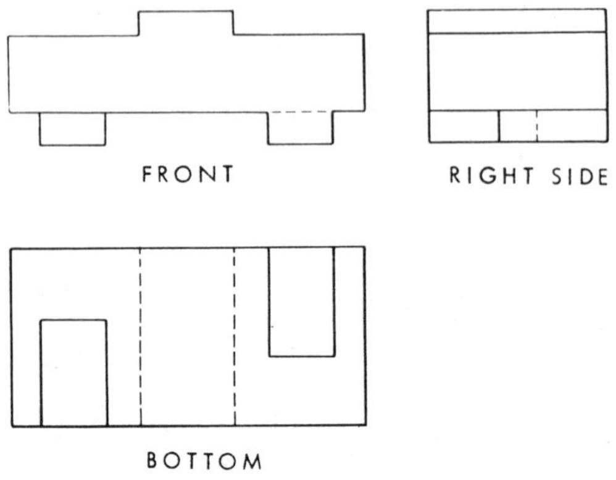

Fig. 11

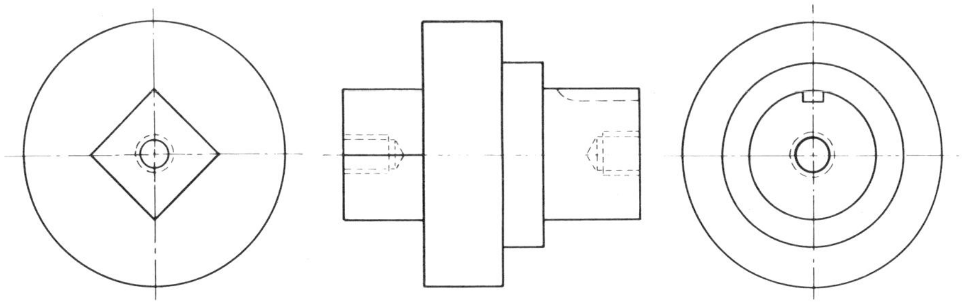

Fig. 12

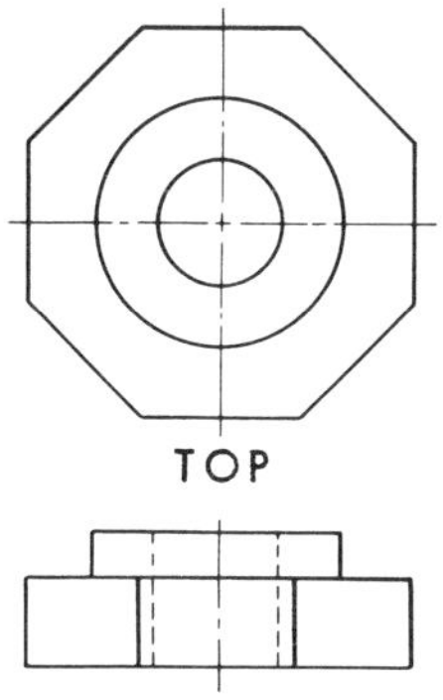

Fig. 13

blueprint reader to visualize and understand the blueprint.

SELF STUDY SUGGESTION NO. 3

Before doing Trade Competency Test, Chapter 3, you are advised to do the following study exercises. These exercises are designed to help you understand the relationship between a three-view drawing and the object which it represents. In doing these exercises, you should keep in mind the discussion in this chapter on the manner in which the views of an object may be imagined as being "projected" onto the sides of a transparent box.

Exercise 1. Figs. 16 and 17 each consists of an incomplete three-view drawing and a complete isometric drawing of the object. You are to examine the isometric drawings and then complete the three-view drawings by adding the necessary lines, as demonstrated in Figs. 14 and 15. (The correct answers are given at the end of the book.)

Exercise 2. Figs. 18 and 19 are provided to give the student further practice in understanding the relationship between views and the object they represent by actually completing drawings. The parts to be completed are specified in each drawing. (The correct answers are given at the end of the book.)

DO NOT TEAR OUT

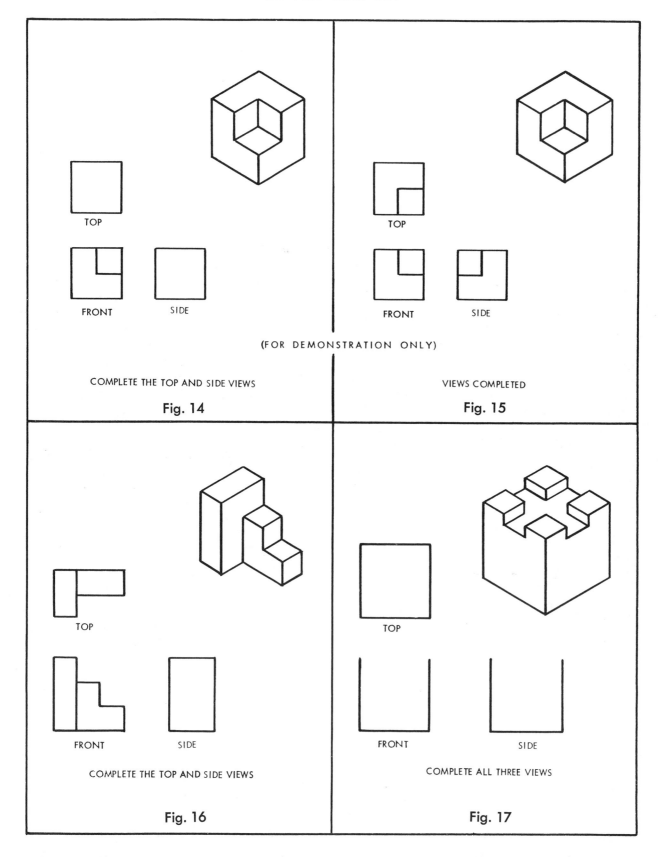

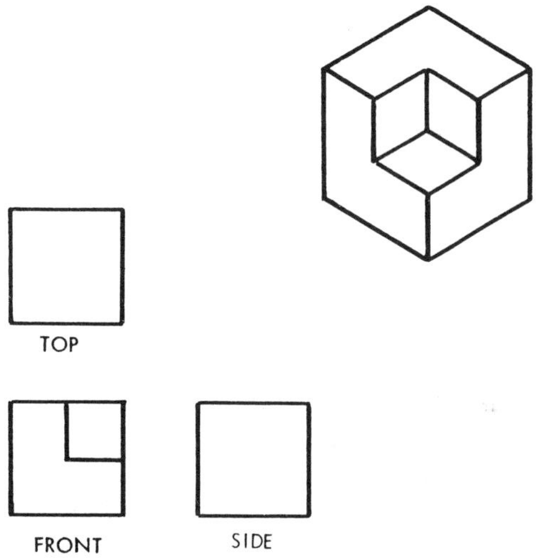

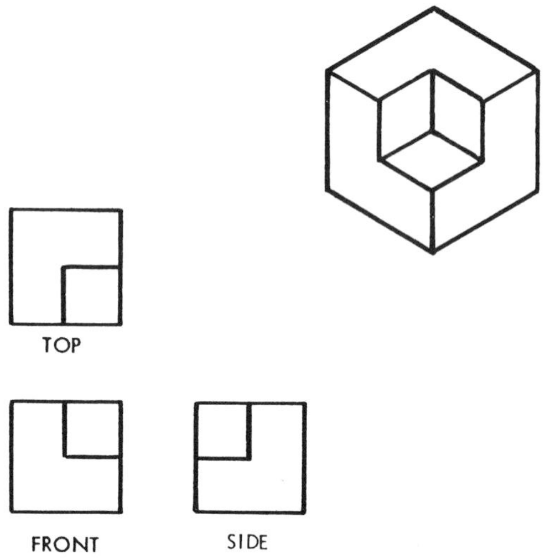

(FOR DEMONSTRATION ONLY)

COMPLETE THE TOP AND SIDE VIEWS

Fig. 14

VIEWS COMPLETED

Fig. 15

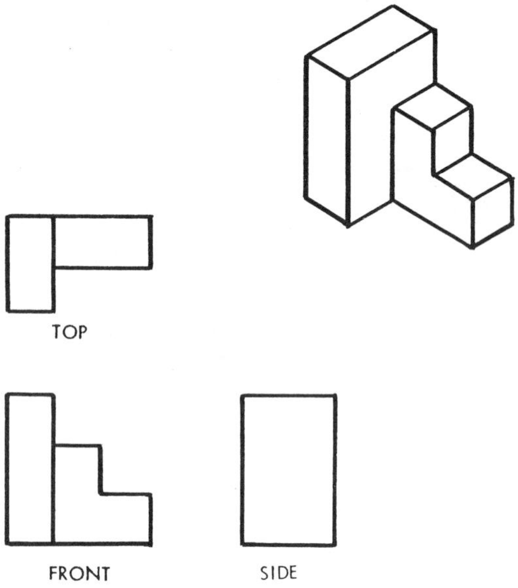

COMPLETE THE TOP AND SIDE VIEWS

Fig. 16

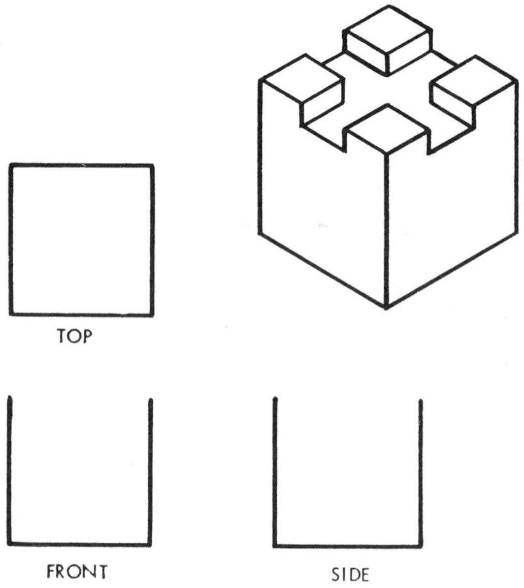

COMPLETE ALL THREE VIEWS

Fig. 17

DO NOT TEAR OUT

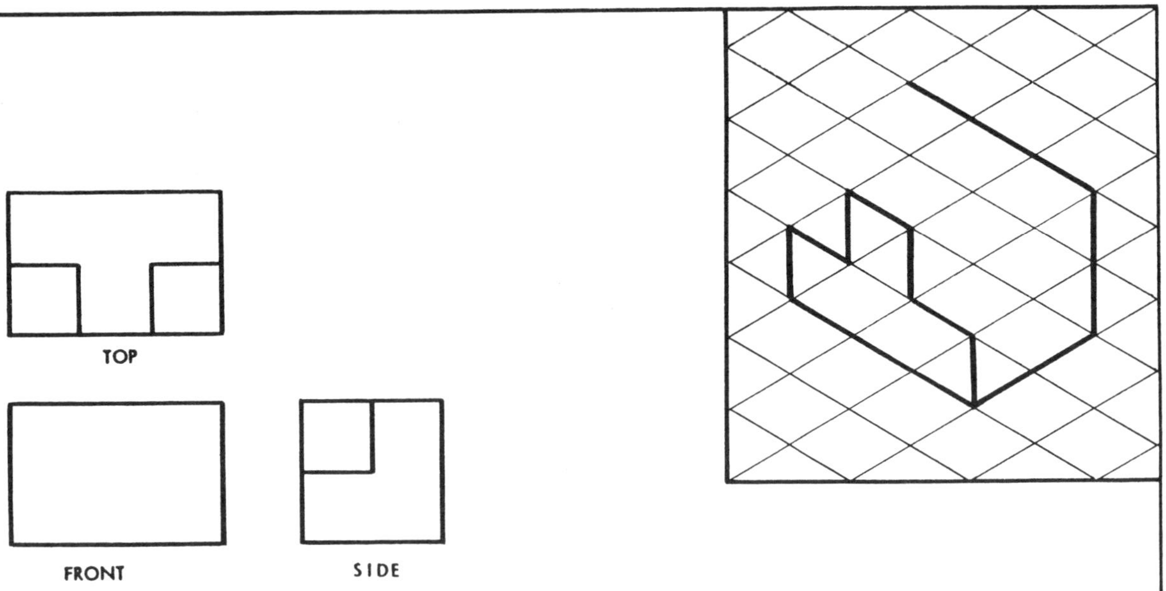

COMPLETE THE FRONT VIEW AND THE ISOMETRIC VIEW

Fig. 18

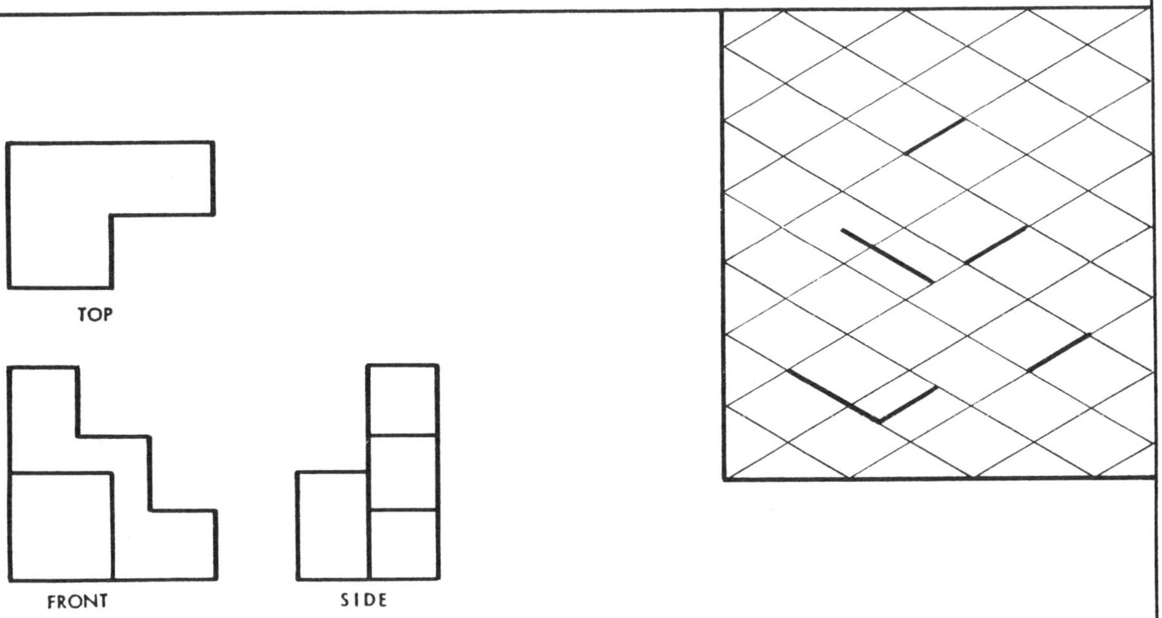

COMPLETE THE TOP VIEW AND THE ISOMETRIC VIEW

Fig. 19

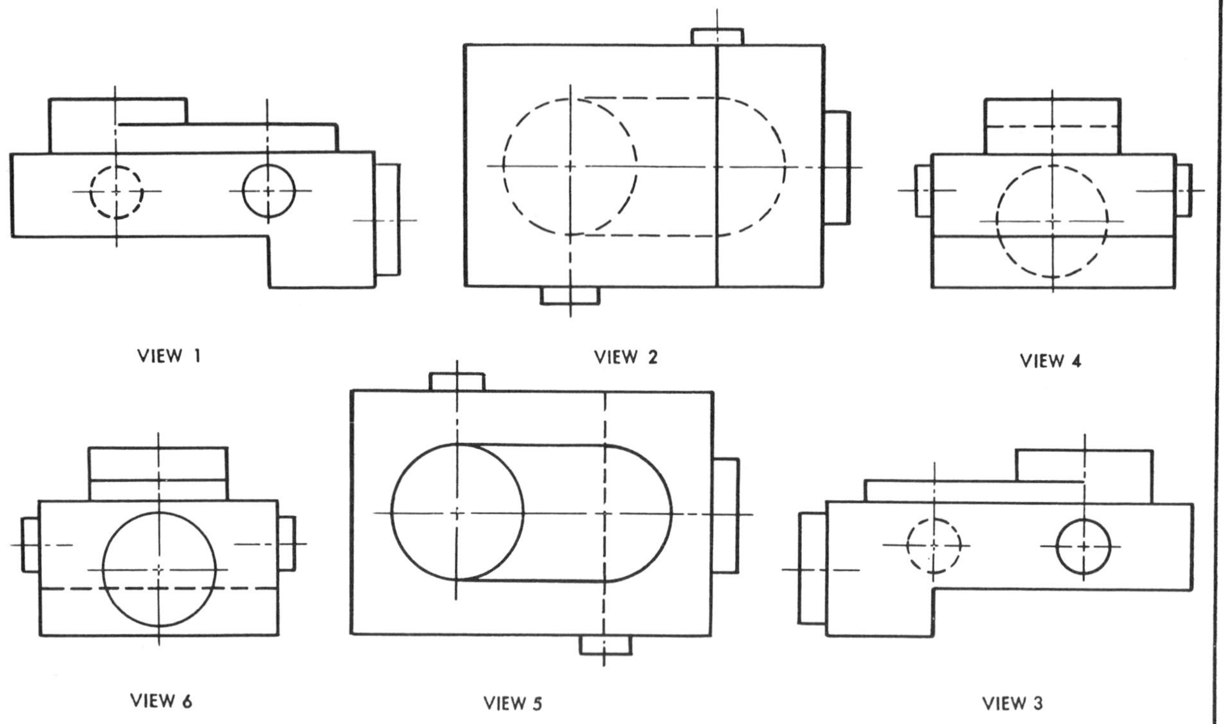

PLATE 1

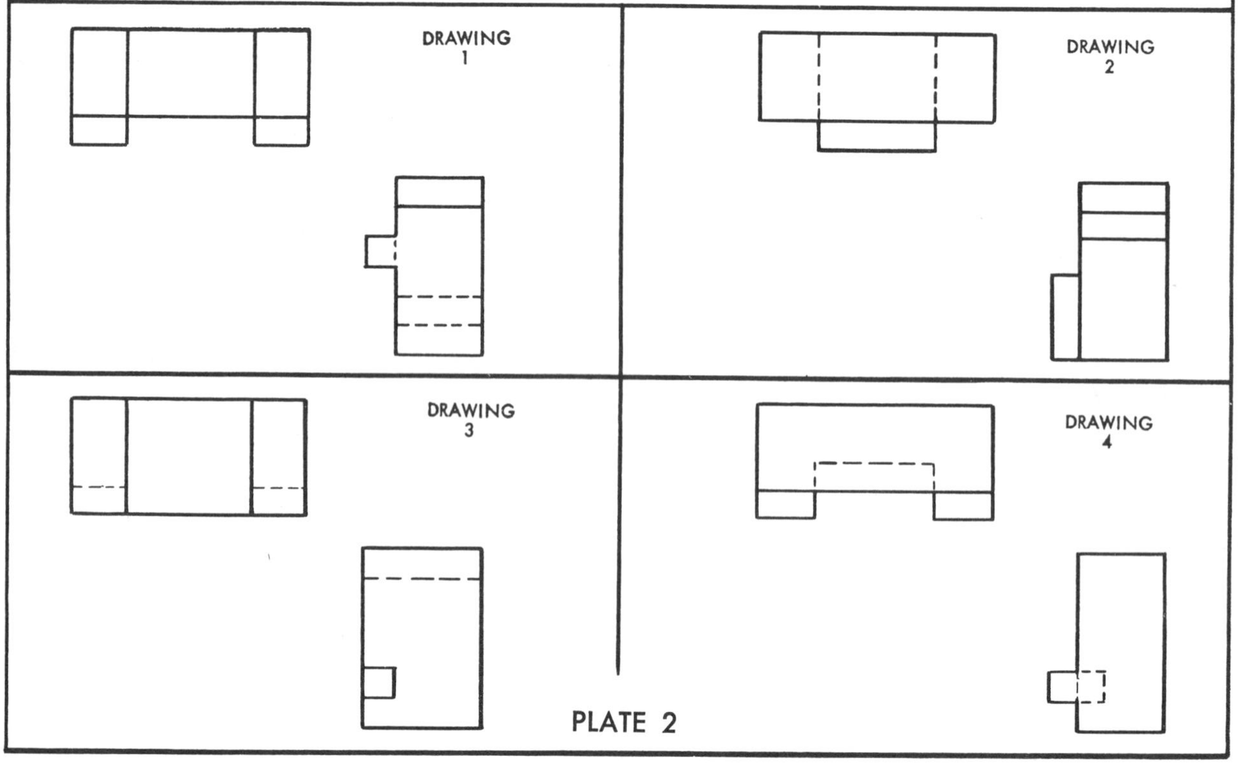

PLATE 2

TRADE COMPETENCY TEST

Based on Plates 1-4, Chapter 3

Student's Name____________________________Instructor's Name____________________________

1. **IDENTIFICATION TEST.** Plate 1 consists of six numbered views of a single object, but NOT arranged in the correct order. Below are six squares arranged and labeled according to the standard arrangement of views. In each square you are to place the number of the view which should correspond in position and name with the square. (View 1 is given as the Front View)

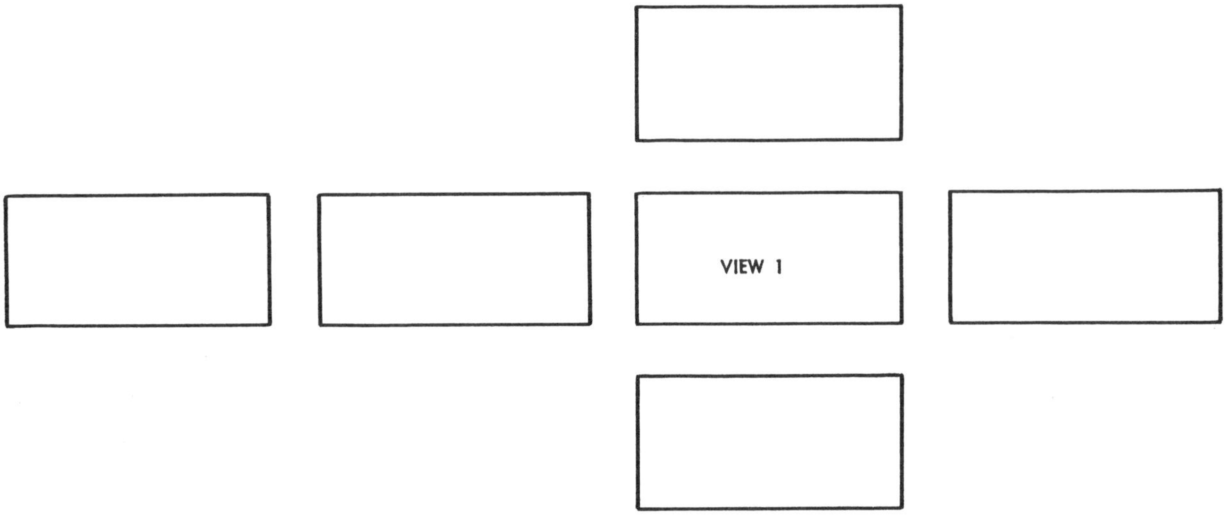

TEAR OFF HERE

2. **IDENTIFICATION TEST.** Plate 2 consists of the Top and Right Side Views of four different objects. Below are four Front Views, each of which completes one of the pairs of views in Plate 2. In the circle next to each Front View below, you are to place the number of the drawing from Plate 2 which corresponds to that Front View.

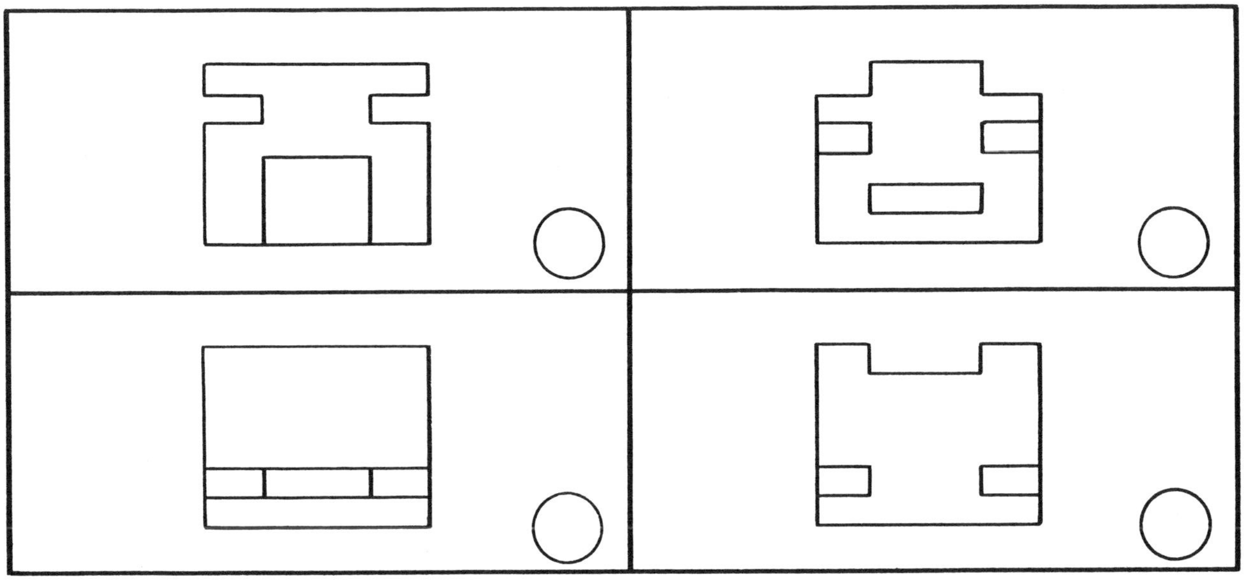

3. IDENTIFICATION TEST. Plate 3 consists of the Front and Right Side Views of four different objects. Below are four Top Views, each of which completes one of the pairs of views in Plate 3. In the circle next to each Top View below, you are to place the number of the drawing from Plate 3 which corresponds to that Top View.

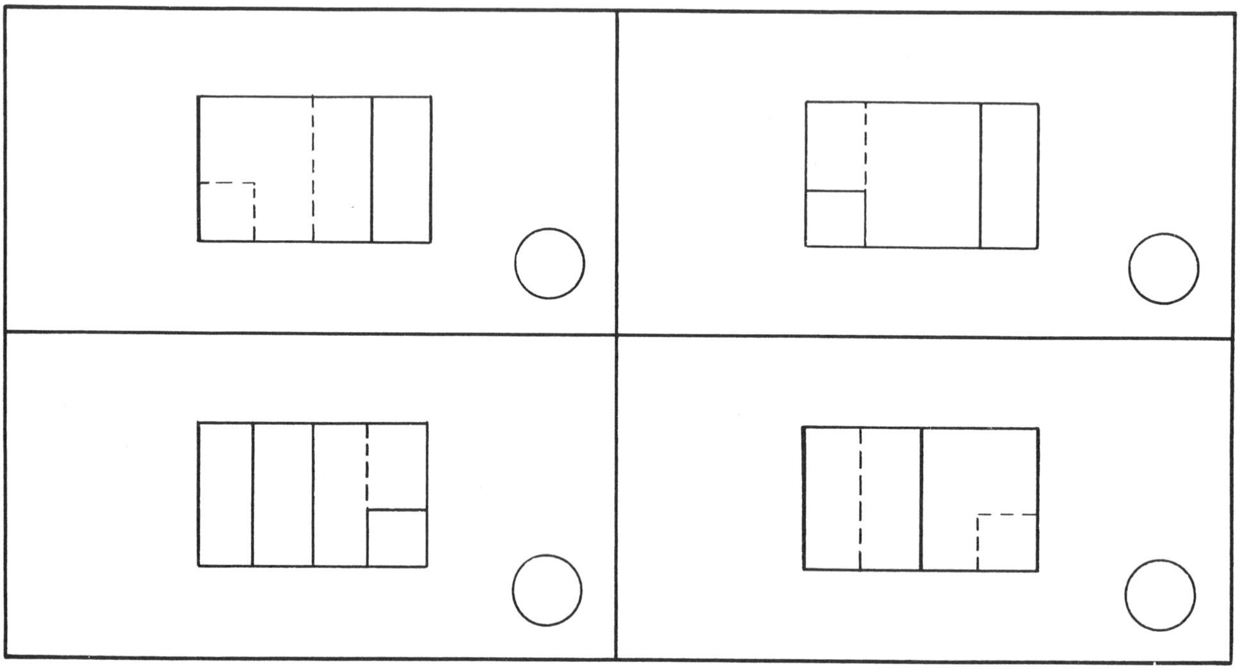

4. IDENTIFICATION TEST. Plate 4 consists of the Front and Top Views of four different objects. Below are four Right Side views, each of which completes one of the pairs of views in Plate 4. In the circle next to each Right Side View below, you are to place the number of the drawing from Plate 4 which corresponds to the Right Side View.

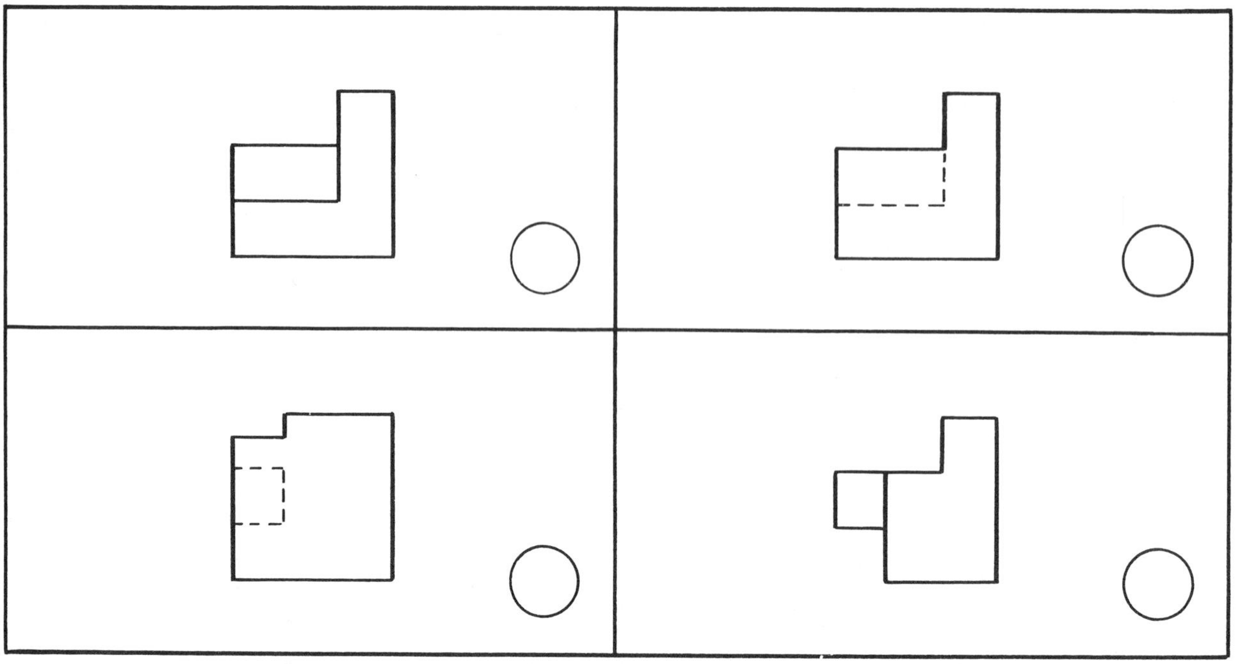

DRAWING 5

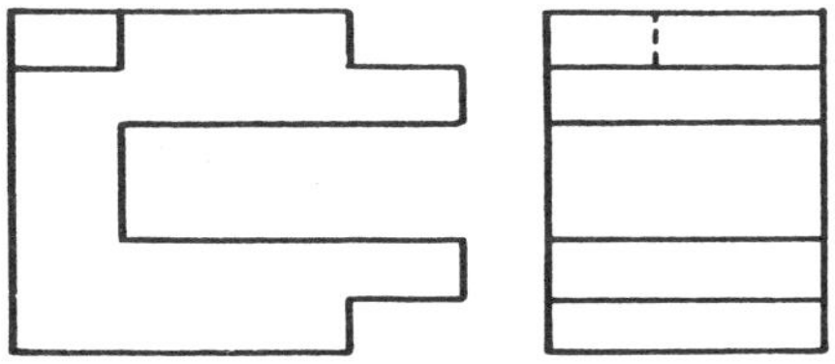

DRAWING 6

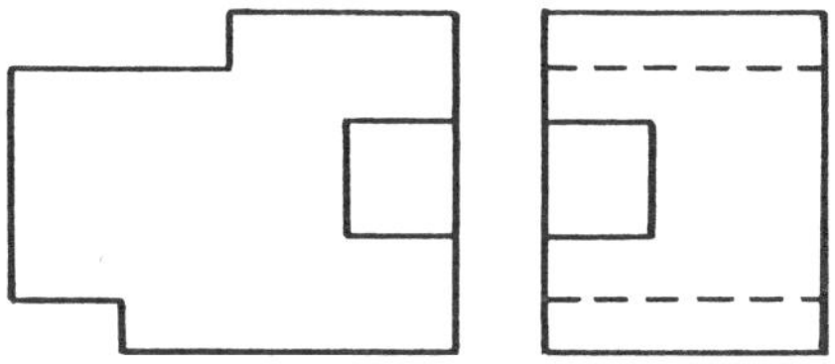

DRAWING 7

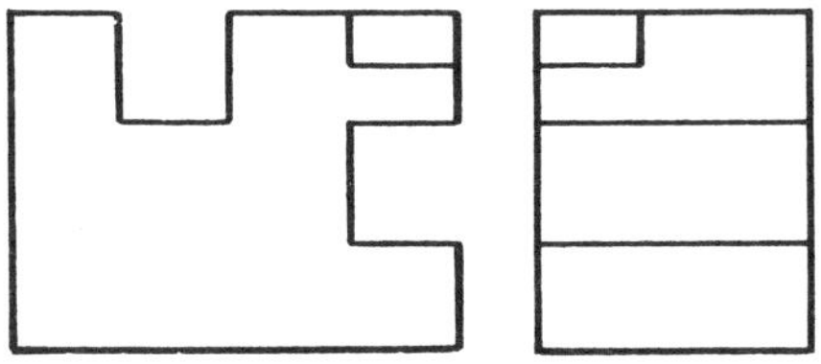

DRAWING 8

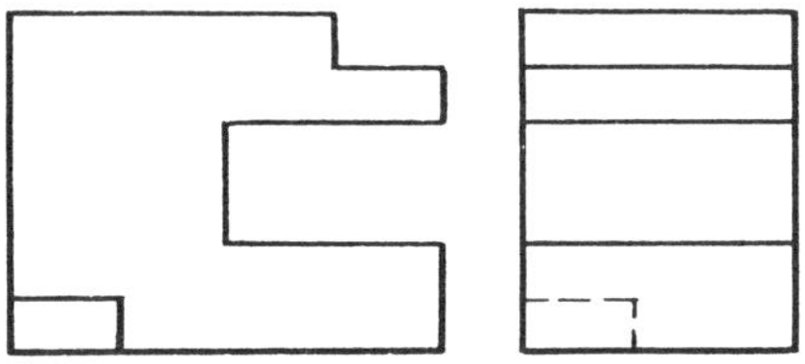

PLATE 3

DRAWING 9

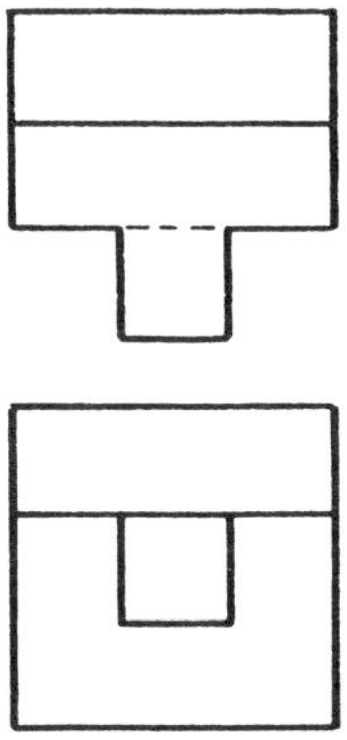

DRAWING 10

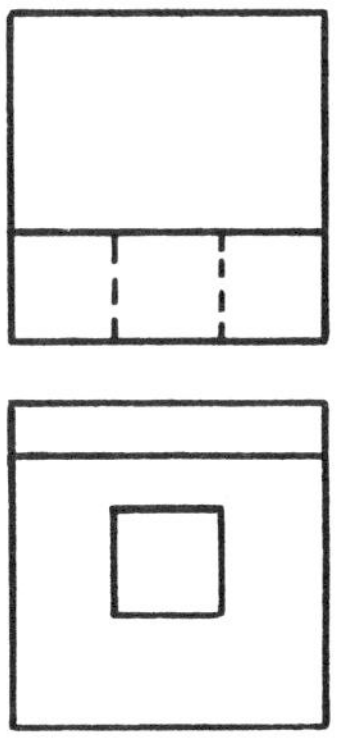

DRAWING 11

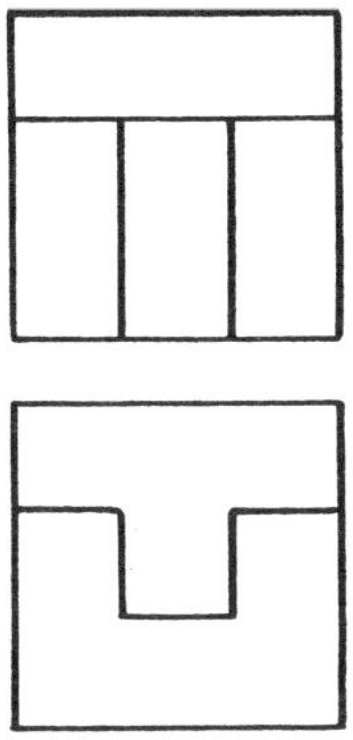

DRAWING 12

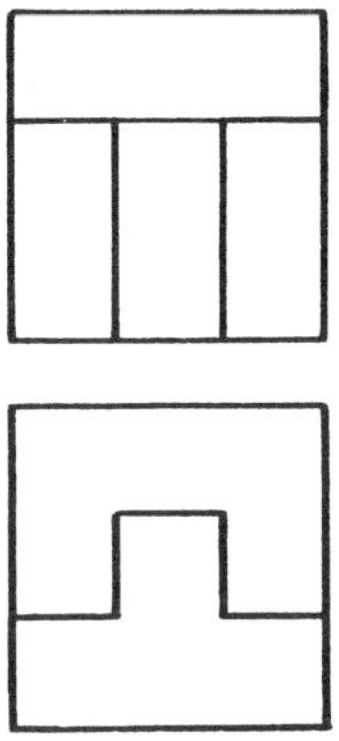

PLATE 4

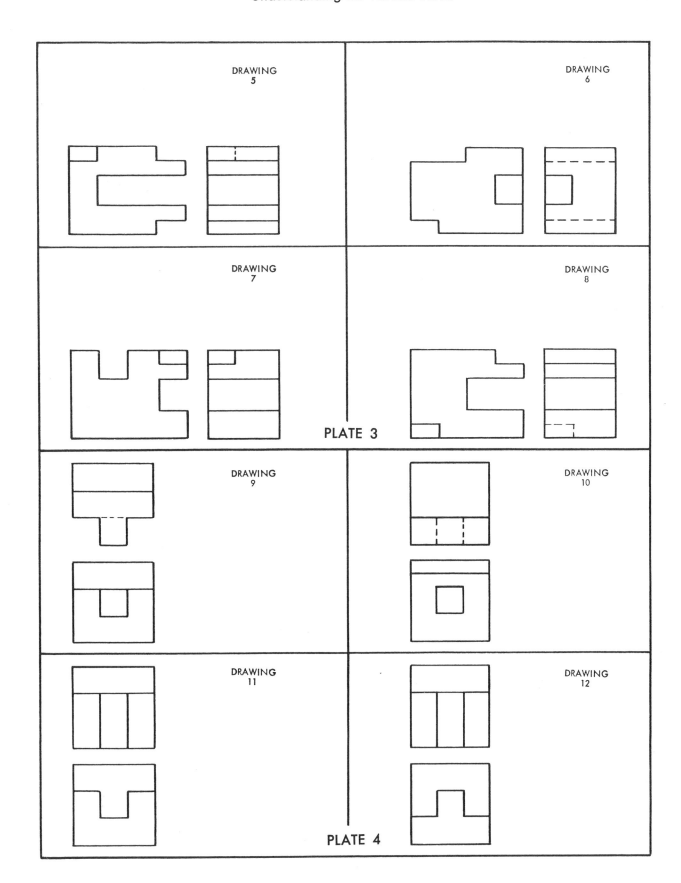

CHAPTER 4

Reading the Working Drawing

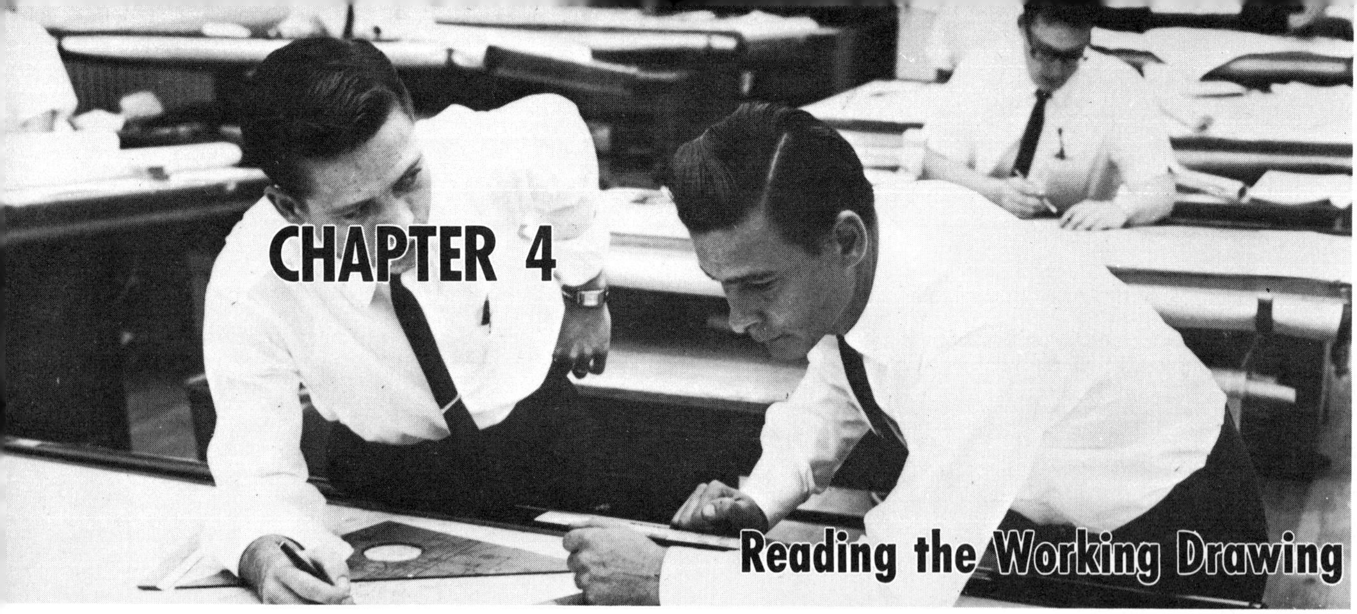

Ford Mo. Co.

The Working Drawing. So far, we have been concerned with the visualization of an object on the basis of the views of the object and the relationship between these views. Strictly speaking, however, a *working drawing* is a blueprint which indicates to us more than just the *shape* of the object. It must also convey complete and precise information concerning the *size* of the object and any other *instructions* necessary for the production of the object. In other words, the working drawing consists of the following parts:

1. *Lines* making up the views of the object
2. *Dimensions* indicating the size of the object
3. *Notes and symbols* providing additional information required for the production of the object.

Figure 1 of Chapter 1 is an example of a complete working drawing containing the parts mentioned above.

ALPHABET OF LINES

Just as words are made up of an alphabet of letters, the views of a blueprint are made up of an *alphabet of lines*. By combining different letters to form *words,* we can convey information in *writing*. By combining different lines to form views, the draftsman can convey information on a *blueprint*.

Figure 1 illustrates the most commonly used lines in the alphabet of lines. Notice that these lines differ from each other in two important ways:

1. In their thickness or *weight*
2. In their shape or construction (they may be continuous, broken, straight, or irregular, depending on their purpose).

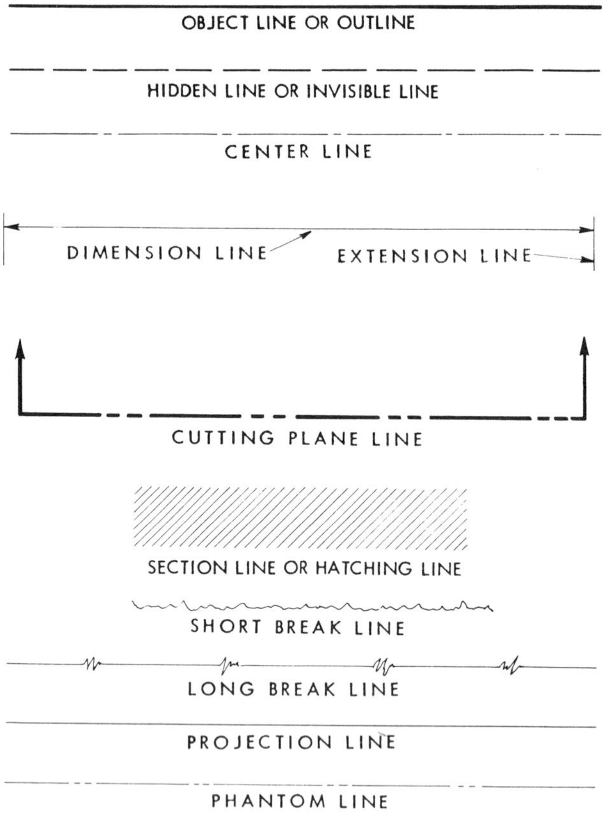

Fig. 1

Following are brief descriptions of the lines with an illustrative example of the usage of each type:

Center Lines. Views of objects that are symmetrical, especially views of circular parts, are centered by means of thin broken lines consisting of long and short dashes. They serve as reference lines for dimensions and, therefore, are usually the first lines that the draftsman makes for a working drawing. Before the draftsman draws the center lines, he must decide on the views and their arrangement. Fig. 2 consists entirely of center lines.

Object Lines. The *outline* or overall shape of an object is represented by heavy, continuous lines. Every edge or surface that is visible from a certain view is included in that view by means of an object line. In Fig. 3 the object lines have been drawn around the center lines in Fig. 2.

Hidden Lines. Surfaces or edges that are "hidden" from view are represented on a drawing by evenly spaced dashes that are slightly thinner than object lines. In the example shown in Fig. 4, the holes are "hidden" in the side view and are, therefore, represented as *hidden* or *invisible* lines. Since, as seen from the side view, the holes exactly coincide with each other, only one pair of hidden lines is shown. See Fig. 4.

Extension Lines. The dimensions of an object are shown by the use of extension and

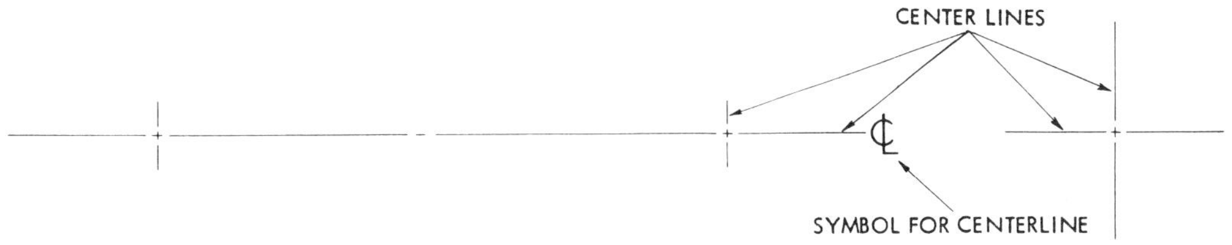

Fig. 2

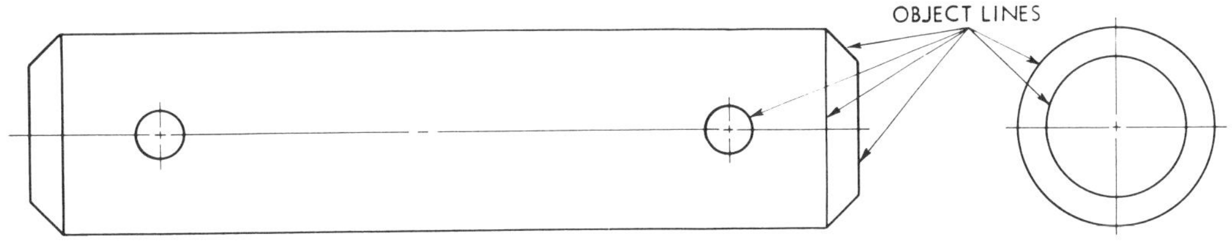

Fig. 3

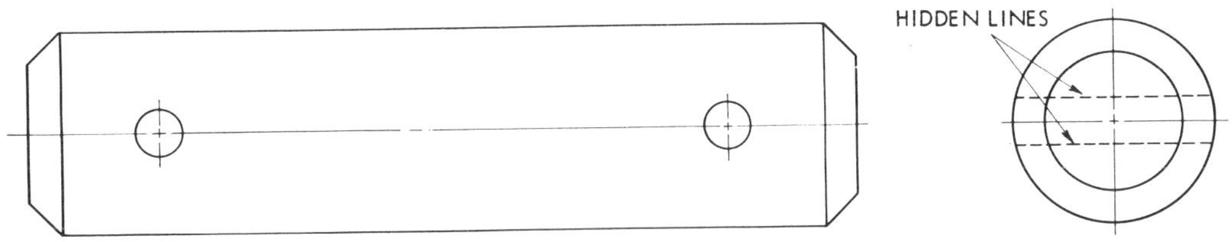

Fig. 4

dimension lines. Extension lines are thin, continuous lines which are extended from the object at the exact location between which the dimension lines are drawn. See Fig. 5.

Dimension Lines. Dimension lines, like extension lines, are thin and continuous, except that, where they meet the extension lines, they terminate with arrowheads. The dimension figure is placed in a break in the dimension line. See Fig. 6.

Break Lines. A view of a long object with uniform shape or cross section can be shortened by being "broken," thus permitting a saving in space on the drawing. Fig. 7 shows

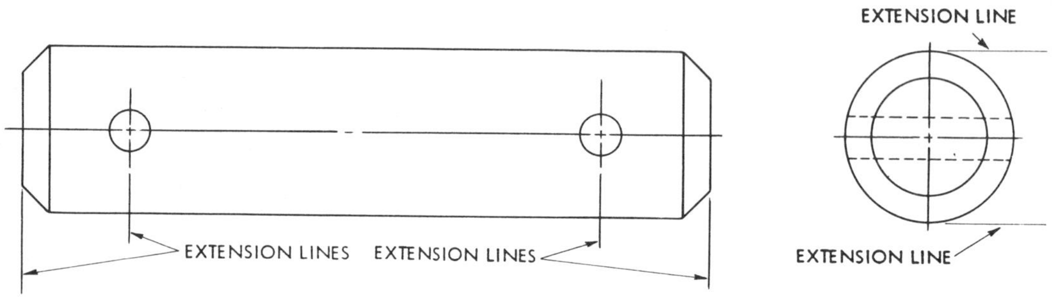

Fig. 5

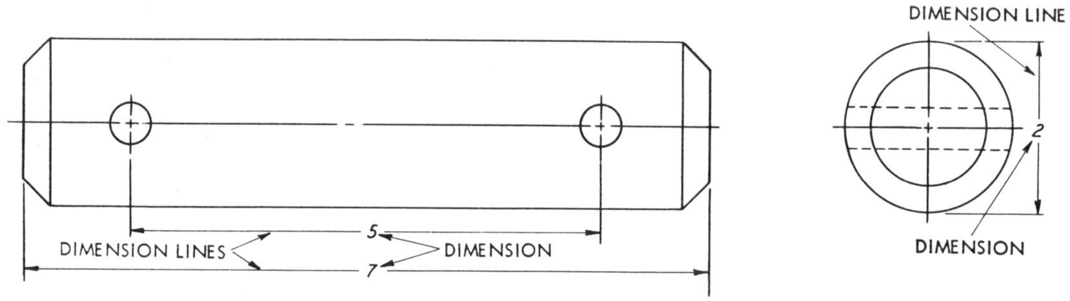

Fig. 6

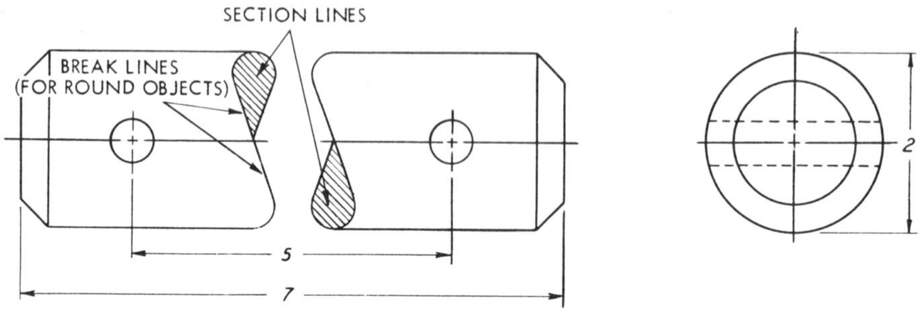

Fig. 7

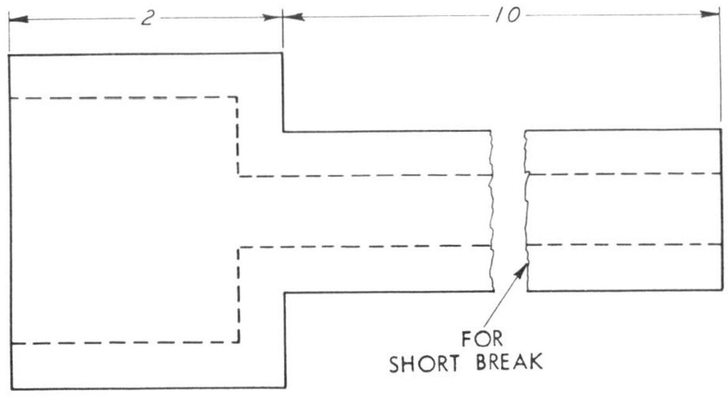

Fig. 8

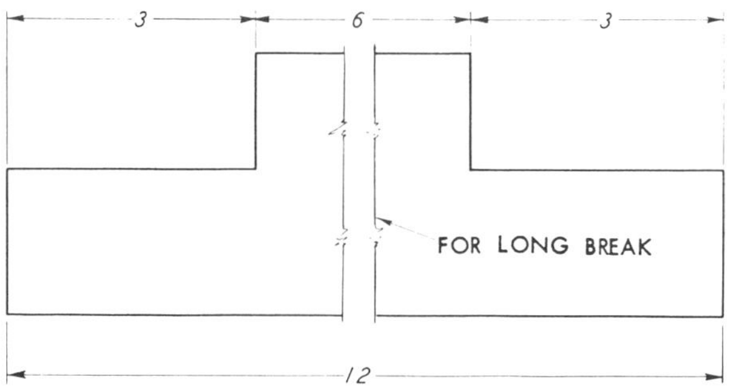

Fig. 9

the use of a round section break to shorten the drawing without losing any information or details.

Two other common methods of breaking employ the freehand line for short breaks on rectangular objects (Fig. 8), and the ruled zigzag line for long breaks on views of rectangular objects (Fig. 9). Notice in Fig. 9 that the specified length of the higher step is twice that of the lower steps even though in the drawing it is shorter.

Section Lines. When an object is represented in a drawing as if a part of it were "cut out" to reveal its inner structure, the section that is thus exposed is indicated by means of section lines. See Fig. 10. Different arrangements of section lines are used to indicate different kinds of materials. In Fig. 10 the section lines indicate that the exposed section is made of one type of material, cast iron. (Other material symbols will be discussed at the end of the chapter.)

Cutting Plane Line. A sectional view, as we have seen, is a method of revealing the inner structure of an object by "cutting through" the object and showing the exposed section.

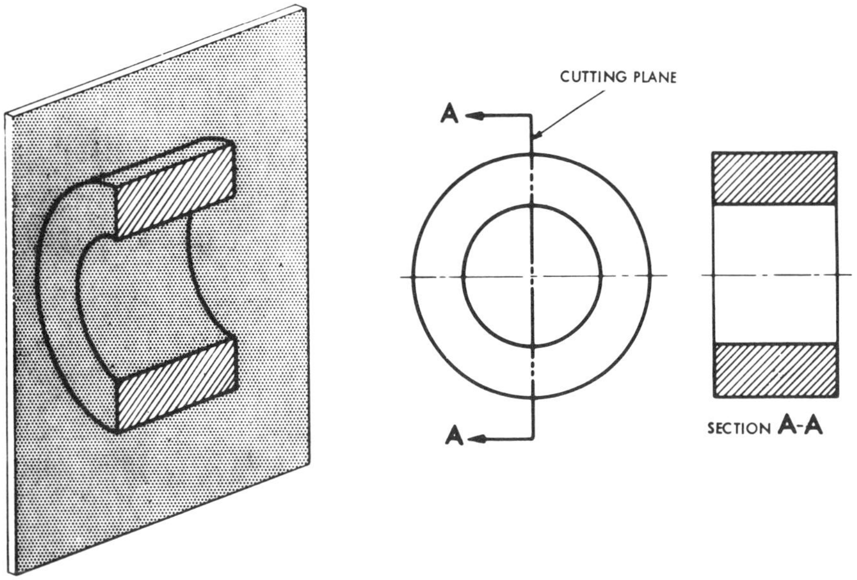

Fig. 10

A cutting-plane line is thick and is composed of alternating long dashes and pairs of short dashes. It is drawn across a view to show where the "cut" was made. See Fig. 10. The arrowheads indicate the direction in which the section is to be seen.

Projection Lines. As was pointed out in Chapter 2, projection lines do not usually appear on a finished drawing. The only exception to this is when an object is thought to be so complicated that projection lines would aid the reader in understanding the drawing.

Phantom Lines. Lines used to denote alternate positions of an object or used to denote machine surfaces are called phantom lines.

The use of phantom lines for the purpose of showing alternate position is described in Fig. 11. The phantom lines show the extent of travel of the actuating lever for a counter that records the number of strokes of a press.

Machine surfaces are sometimes shown on casting and forging drawings by use of phantom lines; conversely, in the same manner, they are used to show rough casting and forging surfaces on machine drawings.

At the left (side view) in Fig. 12, phantom lines show the rough shape of a casting before machining and to the right the phantom lines show the machine surfaces on the drawing of the casting.

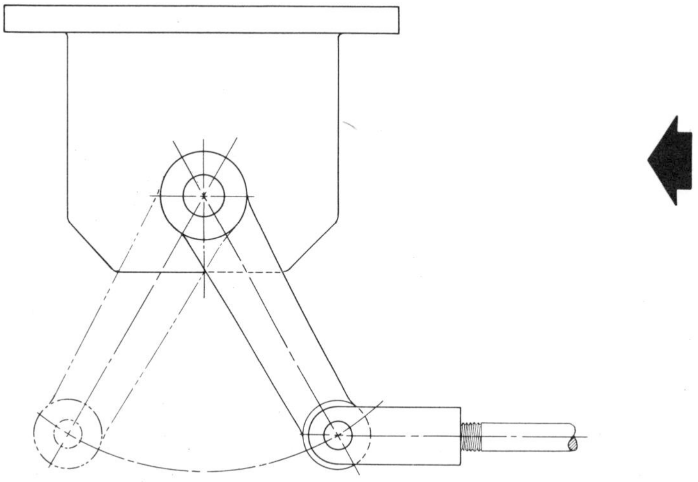

Fig. 11

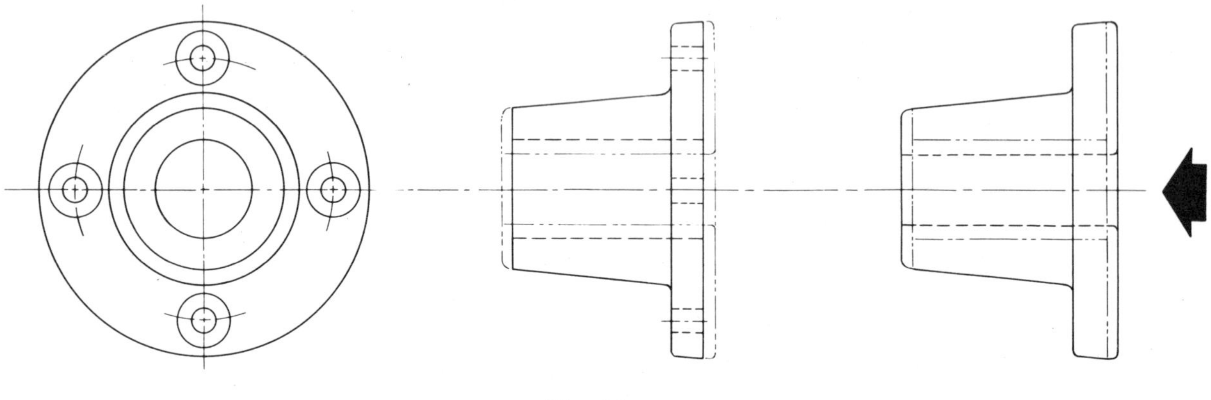

Fig. 12

DIMENSIONS

A machinist or other craftsman using a blueprint is guided in the production of an article or part by *views,* which indicate the *shape* of the object, and *dimensions,* which indicate the *size* of the object. There are three types of dimensions that are commonly used in blueprints: *fractional, decimal,* and *angular dimensions.*

Fractional Dimensions. Fractional dimensions are given on a blueprint in terms of *fractional* parts of an inch, as in Fig. 13A. Fractional dimensions are generally used on ordinary work not requiring a high degree of accuracy; that is, not requiring close tolerances. The *tolerance* of a dimension is the amount of variation allowed above or below the specified dimensions. For example, a part with a specified dimension of $4\frac{1}{8}''$ and a tolerance of $\pm \frac{1}{64}''$ would be acceptable if it measured anywhere between $4\frac{1}{8}'' + \frac{1}{64}'' = 4\frac{9}{64}''$, or $4\frac{1}{8}'' - \frac{1}{64}'' = 4\frac{7}{64}''$. The upper limit (in this case $4\frac{9}{64}''$) is called the *maximum* and specified as MAX. The lower limit (in this case $4\frac{7}{64}''$) is called the *minimum* and is specified as MIN.

Decimal Dimensions. Decimal dimensions are expressed on blueprints in hundredths of an inch (.01″), thousandths of an inch (.001″), ten-thousandths of an inch (.0001″) and, occasionally, in hundred-thousandths of an inch (.00001″).

The use of decimal dimensions in hundredths of an inch to replace fractional dimensions is becoming increasingly popular in a large segment of our mechanical industry. In other words, .31 is used instead of $\frac{5}{16}$; 4.50 instead of $4\frac{1}{2}$; and 4.13 in place of $4\frac{1}{8}$. See Fig. 13B.

The same tolerances that are allowed on fractional dimensions apply to decimal dimensions in hundredths of an inch. Decimal dimensions in thousandths are used on objects requiring greater precision with less tolerances. The tolerance specified in the example shown in Fig. 13C means that the finished piece may be anywhere between four-thousandths of an inch (.004″) greater and two-thousandths of an inch (.002″) smaller than the specified dimension. Thus, the piece is acceptable if it measures anywhere between 2.034″ and 2.028″.

Metric Dimensions. Because of increased trade in world markets, many United States companies make drawings showing metric as well as decimal-inch dimensions, converting measurements on the basis that one inch equals 25.4 millimeters. The usual practice is to show the millimeter equivalents in parentheses, following the dimension in inches. In doing so, a comma is used instead of a period to designate the decimal point, following European practice. (The subject of metric measurements is treated separately in Chapter 9.)

Limit Dimensions. These are the largest and smallest permissible dimensions. As in the example of limit dimensioning in Fig. 13D, the tolerance is the difference between the limits; in this case, a tolerance of .005″.

Angular Dimensions. The size of an angle is expressed on a blueprint in terms of degrees (°) which are further broken down into minutes or into fractional or decimal parts of a degree. You will recall that there are 360° to a circle, 60′ to a degree. An angle that measures 45° could be expressed in the manner shown in Fig. 13E. Angular tolerances may be given in a note on the drawing, or they may be shown in the dimension itself, as in

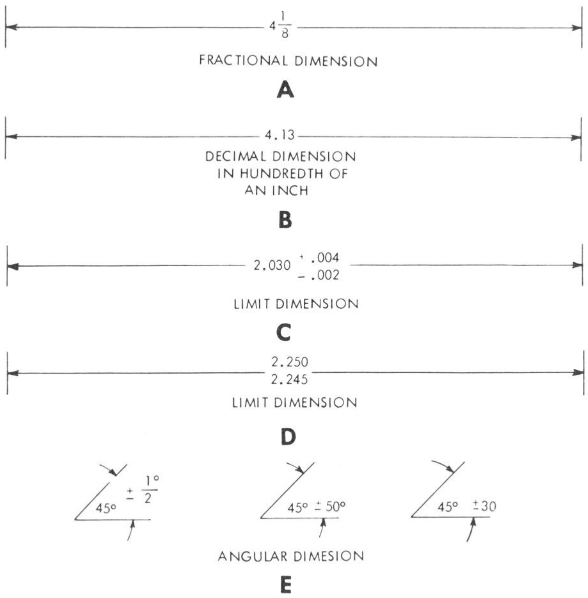

Fig. 13

the example shown in Fig. 13E. In this example, the finished angle may be as much as 30 minutes above or below the specified angle.

Geometric Dimensioning and Tolerancing. The recently introduced system of geometric dimensioning and tolerancing, based on concepts and principles to assure that parts will fit in assemblies and function as intended, is separately described and illustrated as Appendix A of this edition.

NOTES AND SYMBOLS

In a working drawing, it is often necessary to provide additional information which is not conveyed either by the views or the dimensions. Such information can be included by means of notes or symbols, depending on the nature of the information to be included.

Three common and important forms of supplementary information that the blueprint reader should be familiar with are: (1) finish marks, (2) thread symbols, and (3) materials symbols.

FINISH SYMBOLS

The term *finished surface* is applied to any surface that requires material to be removed from it in order to improve its size, geometry or smoothness. This can be done by such processes as planing, milling, turning, broaching or grinding. The method used depends on the contour, type of finish required, and kind of material.

Fig. 14 shows the accepted surface finish symbol applied to various surfaces so that it always appears in an upright position.

Present practice is to add numbers to the left of this basic symbol to designate surface roughness in *microinches* (MU in.) or millionths of an inch arithmetical average deviations from the centerline of the surface as measured by profilometer or surface analizer.

Also, numbers may be written above the horizontal extension of the symbol to designate maximum waviness height, in decimal inches, and maximum waviness width, in inches, placing the width notation to the right of the maximum height notation.

Additionally, the *lay,* or direction of predominant surface pattern, may be shown by a single letter or symbol placed below the horizontal extension of the surface finish symbol. See Fig. 15 showing relation of these numbers and symbols to surface characteristics.

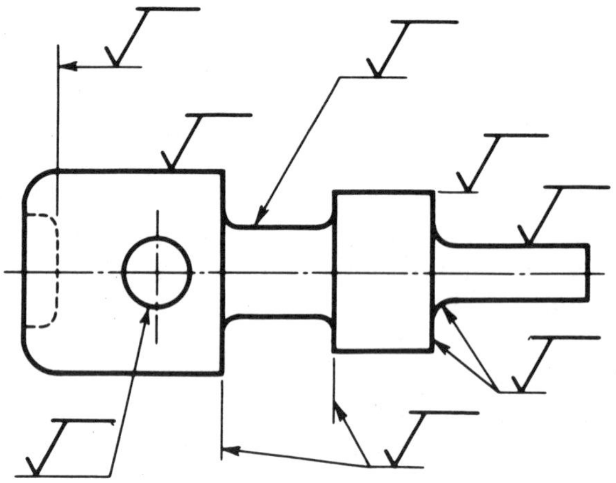

Fig. 14

Lay symbols are shown and explained in Fig. 16. Their importance is that each machining process produces a distinctive directive directional pattern which affects not only the frictional and load-bearing characteristics of the surface but also affects the technique of measuring the surface by instruments.

Earlier prints may still be found with finishes designated by a 60-degree "V" shape, alone or marked with a letter such as "M" for "machined" or "G" for "ground." Sometimes the letter "G," signifying "ground," was used alone, intersecting the object line. In still earlier practice, small *"f"* marks were drawn across the object line of surfaces to be finished by some sort of machining process. Fig. 17 shows these obsolete but sometimes still used symbols.

Earlier symbols were useful to the extent that they designated the need for a machining process to meet dimensional and smoothness requirements, but such specifications were only qualitative and no real measure of the surface condition.

THREAD REPRESENTATION

Because of their wide use, screw threads are represented on a drawing as described in

RELATION OF SYMBOLS TO SURFACE CHARACTERISTICS

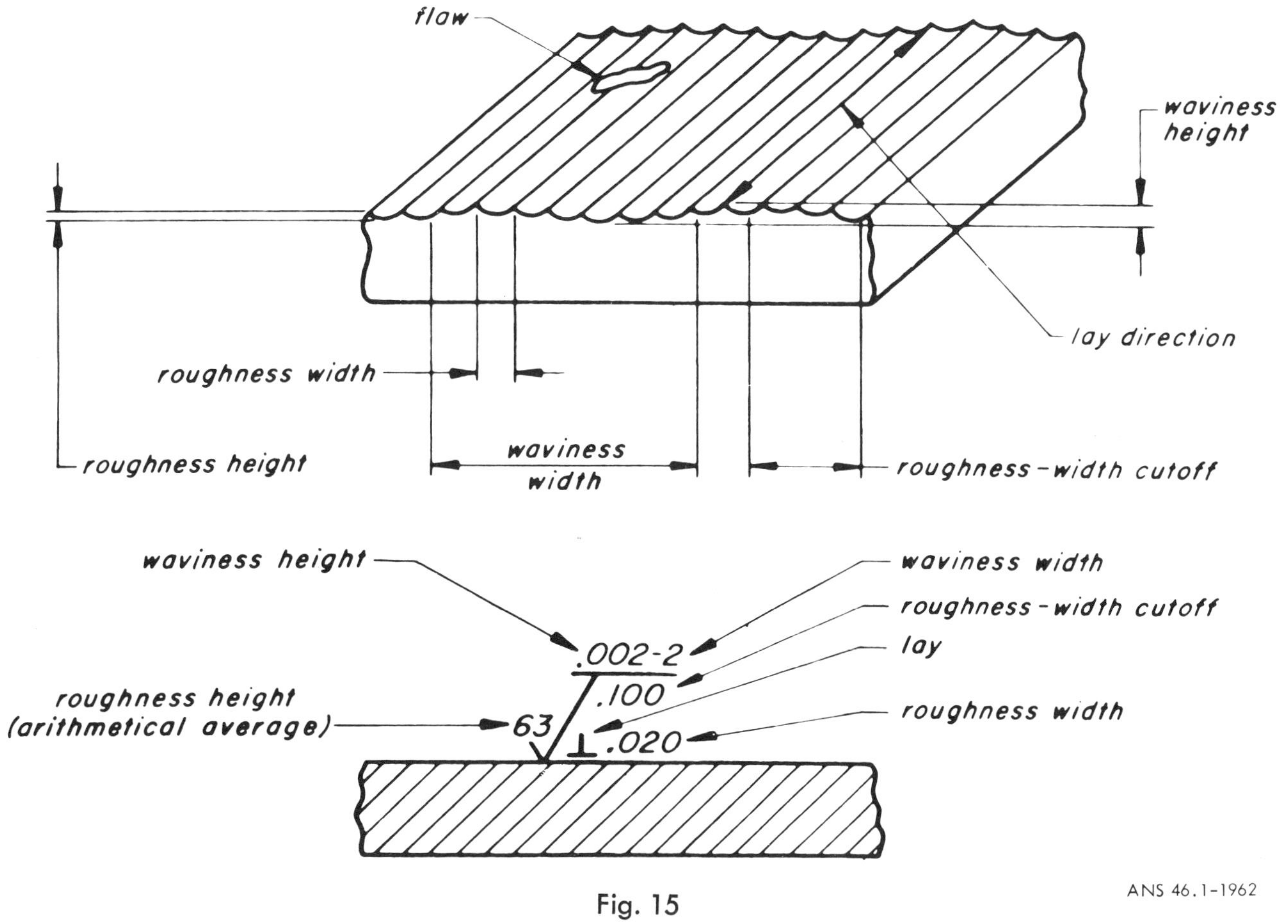

Fig. 15

Fig. 18. The simplified method has been adopted as general procedure by most engineering departments.

The American National Thread, most widely used until recently, has been replaced by the Unified National Thread Form. Though the basic thread dimensions are practically the same, some allowances were made for tool wear at the crest and root of the thread.

The Unified Thread series is specified on drawings as UNC (coarse thread), UNF (fine thread), and UNEF (extra fine thread). In addition to these there is also a uniform pitch series of 4, 6, 8, 12, 16, 20, 28, and 32 United National.

In the UNC, UNF, and UNEF thread series the number of threads per inch (pitch) varies with the diameter of the shaft or hole. However, in the 4, 6, 8, 12, 16, 20, 28 and 32 UN thread series, the number of threads per inch (pitch) is *constant* for different diameters. For example, in the 8 UN thread series, there will be 8 threads per inch for all diameters given.

Appendix B contains charts of standard thread dimensions and tap drill sizes for both the fine and coarse thread series.

Internal and External Threads. There are two basic types of screw threads.

1. *External threads.* Threads on the external surface of a cylindrical object.

LAY SYMBOLS		
LAY SYMBOL	DESIGNATION	EXAMPLE
‖	Lay parallel to the line representing the surface to which the symbol is applied.	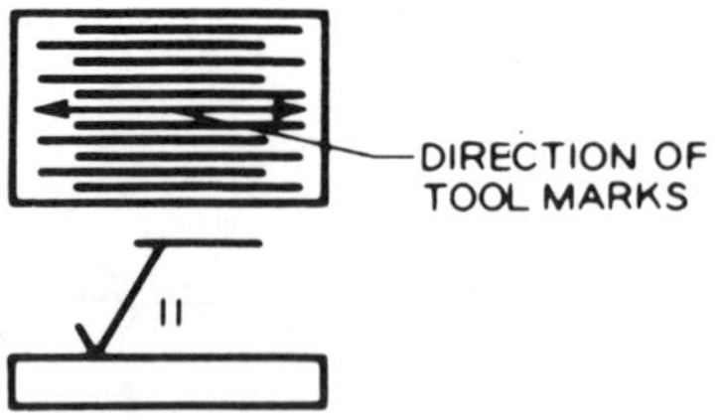
⊥	Lay perpendicular to the line representing the surface to which the symbol is applied.	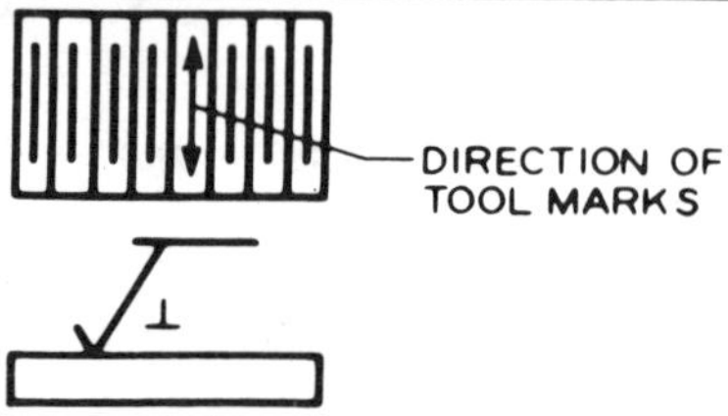
X	Lay angular in both directions to line representing the surface to which symbol is applied.	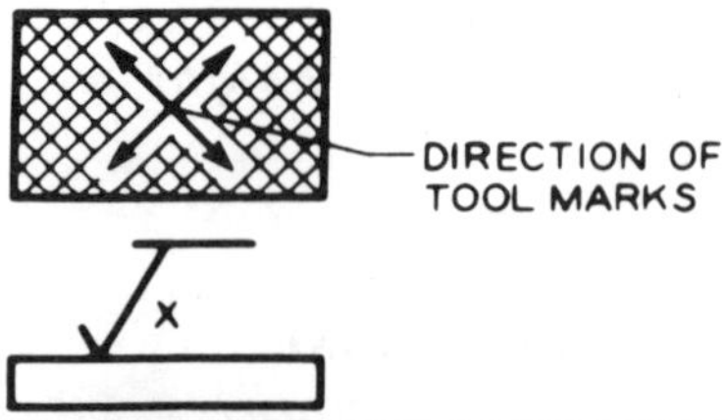
M	Lay multidirectional	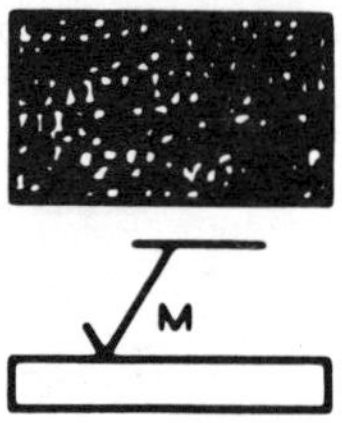
C	Lay approximately circular relative to the center of the surface to which the symbol is applied.	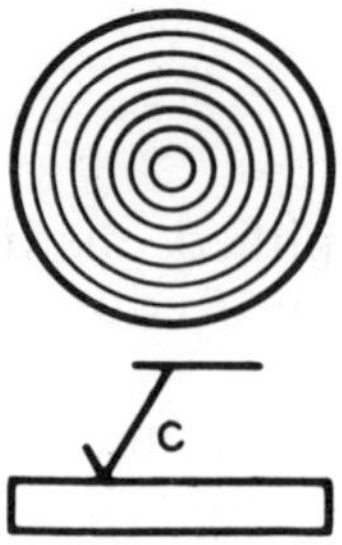
R	Lay approximately radial relative to the center of the surface to which the symbol is applied.	

Fig. 16

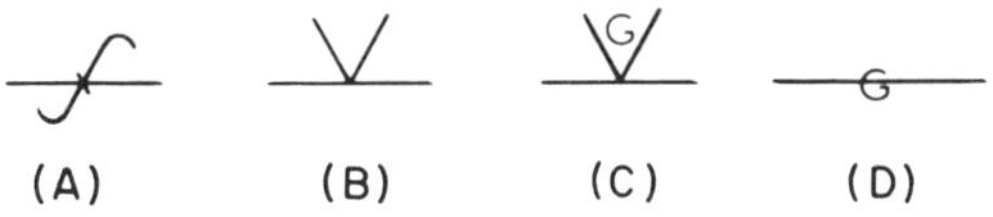
(A)
(B)
(C)
(D)

Fig. 17

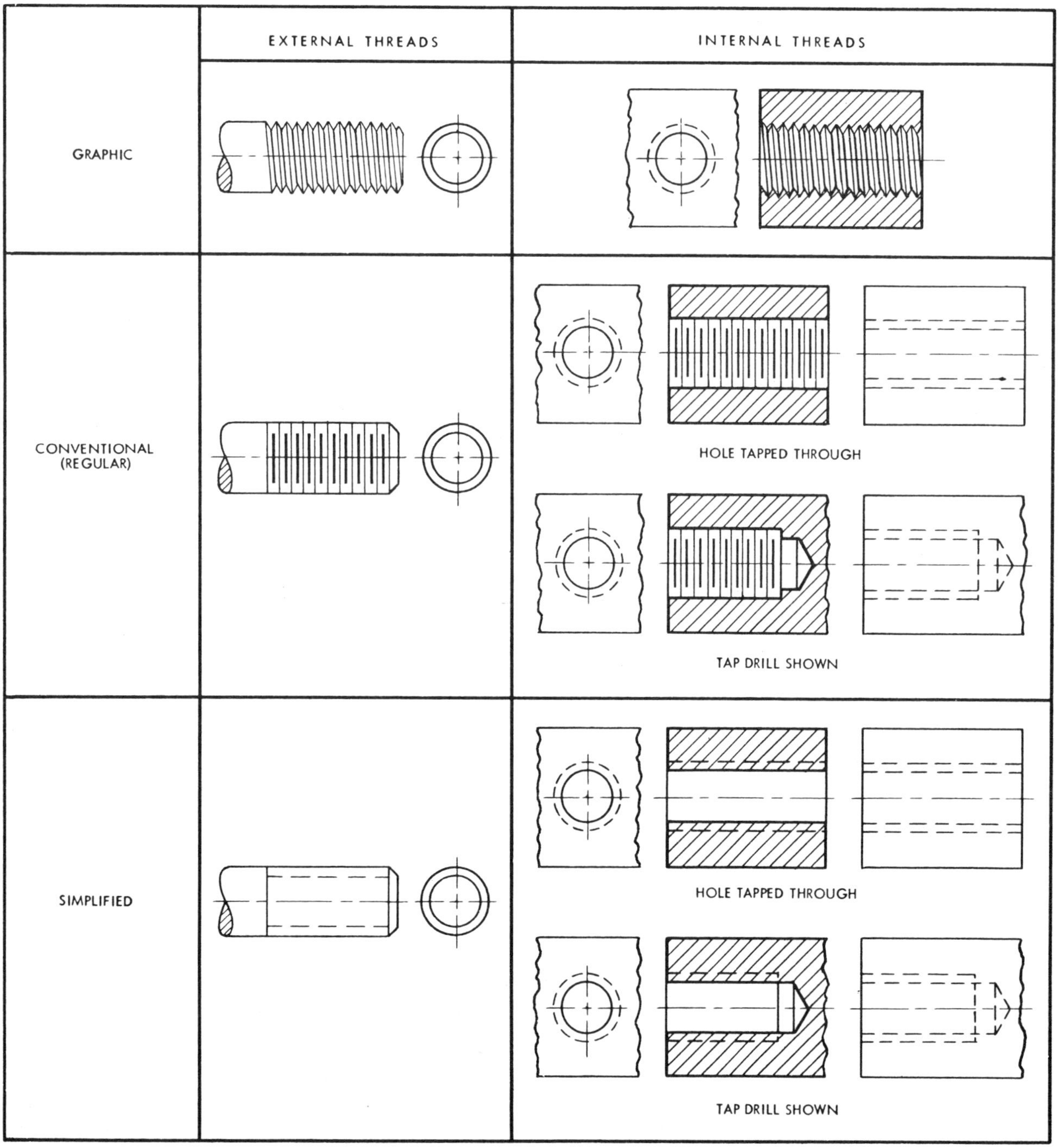
EXTERNAL THREADS
INTERNAL THREADS
GRAPHIC
CONVENTIONAL (REGULAR)
HOLE TAPPED THROUGH
TAP DRILL SHOWN
SIMPLIFIED
HOLE TAPPED THROUGH
TAP DRILL SHOWN

Fig. 18

2. *Internal threads.* Threads on the internal cylindrical surface of an object; *tapped threads.*

Fig. 18 illustrates these two types of threads and shows different methods of representing them on blueprints.

Both external and internal threads may be represented on a drawing in three different ways:

1. *Graphically.* Screw threads may be drawn graphically, that is, as they appear to the eyes. This method is not a standard practice and is seldom used. See Fig. 18.

2. *Conventionally.* Screw threads may be drawn conventionally; that is, according to adopted standards of representation. This method is also known as the *regular* method of thread representation. See Fig. 18.

3. *Simplified.* A thread symbol may be simplified by the omission of some of its details. (This is the most commonly used method.) See Fig. 18.

Thread Dimensions. Whatever the method of representation chosen, the thread should be specified by means of a standard notation which conveys certain items of information, such as illustrated in Fig. 19. The most important of these identifying features are:

1. The diameter of the threads
2. The number of threads per inch
3. The form of the thread (Unified National)
4. The series of the thread (coarse, fine, 8-pitch, etc.)
5. The class of fit, or the degree of looseness or tightness required between matching threads. There are three classes of fit, class 1 being the loosest, class 3 being the most accurate.
6. Each class is further designated as A or B. "A" designates external thread; "B" designates internal thread.

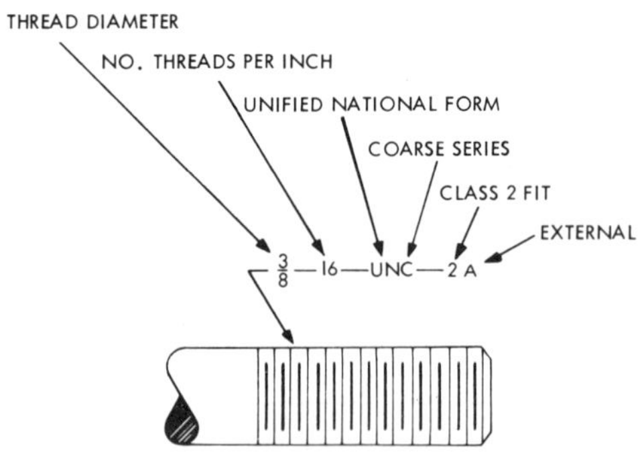

Fig. 19

In the example shown in Fig. 19, the notation means:

1. ⅜″ diameter threads
2. 16 threads per inch
3. Unified National form
4. Unified National Coarse series
5. Class 2 fit
6. A (external thread)

To illustrate the difference in the allowance between the three classes of Unified Screw Threads, the tolerances for a ⅜-16-UNC external thread are shown. See Table 1.

SECTION OR CROSS HATCHING LINES

In section views, it is useful and convenient to distinguish the sectioned or interior parts of the object from the surfaces. This is done

TABLE 1.

Class	Allowance	Major Diameter		Pitch Diameter		Minor Diameter
		Max	Min	Max	Min	
1A	0.0013	0.3737	0.3595	0.3331	0.3266	0.2970
2A	0.0013	0.3737	0.3643	0.3331	0.3287	0.2970
3A	0.0000	0.3750	0.3656	0.3344	0.3311	0.2983

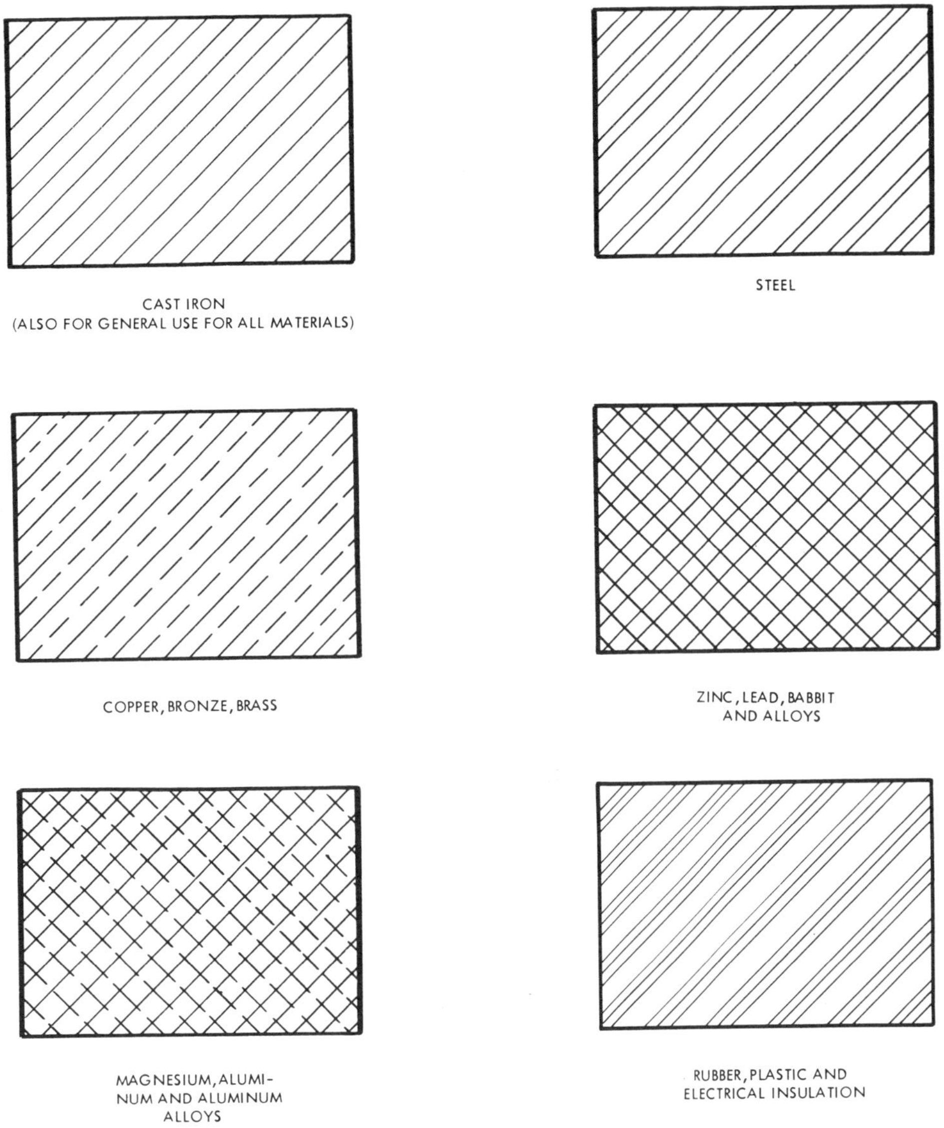

Fig. 20

by the use of conventional symbols indicating the kind of material out of which the section is made. Some of the more common conventional material symbols and the materials they represent are shown in Fig. 20.

This device is also useful in clarifying sections where more than one kind of material is used, as in Fig. 21.

It should be noted that material symbols represent the *kind of material*. The exact composition, such as an alloy or a specific material, must be specified on the drawing by means of a note.

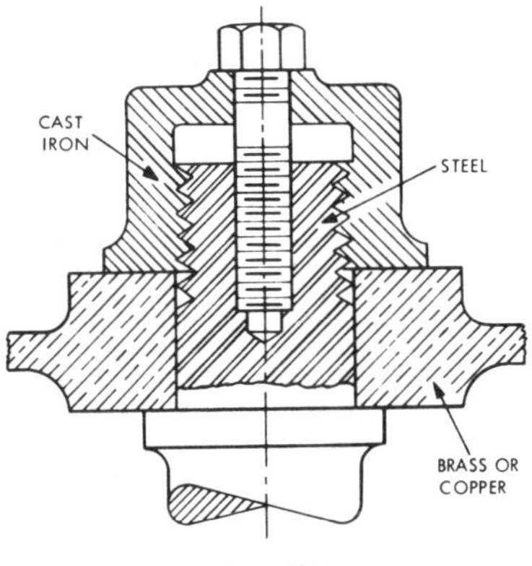

Fig. 21

SELF STUDY SUGGESTION NO. 4

Before doing Trade Competency Test, Chapter 4, you are advised to do the following study exercises designed to help you understand the various types of lines used in working drawings.

Exercise 1. Fig. 22 shows a sketch of a Tool Room Arbor Press. This sketch illustrates most of the lines included in the alphabet of lines. Fill in the correct name of each line indicated. The answers will be found in Answers to Self Study Suggestions near the end of the book.

Exercise 2. With the aid of the Decimal Equivalent Chart shown in Appendix C give the fractional equivalent of the decimal dimensions shown.

DO NOT TEAR OUT

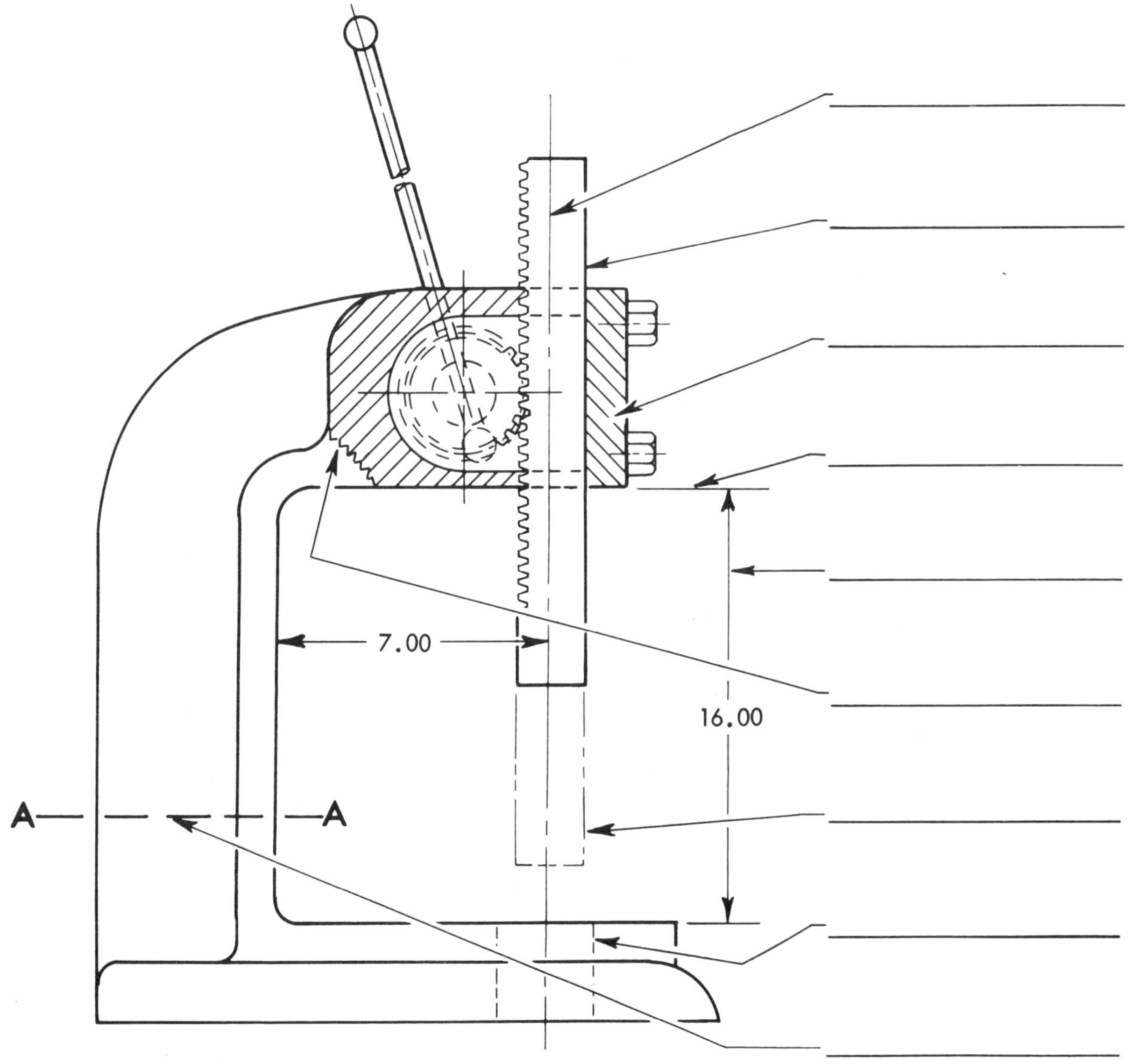

3.06 = $3\frac{1}{16}$ 2.50 = ____   .69 = ____

.28125 = ____   3.750 = ____   .13 = ____

.31 = ____   .781 = ____   1.875 = ____

Fig. 22

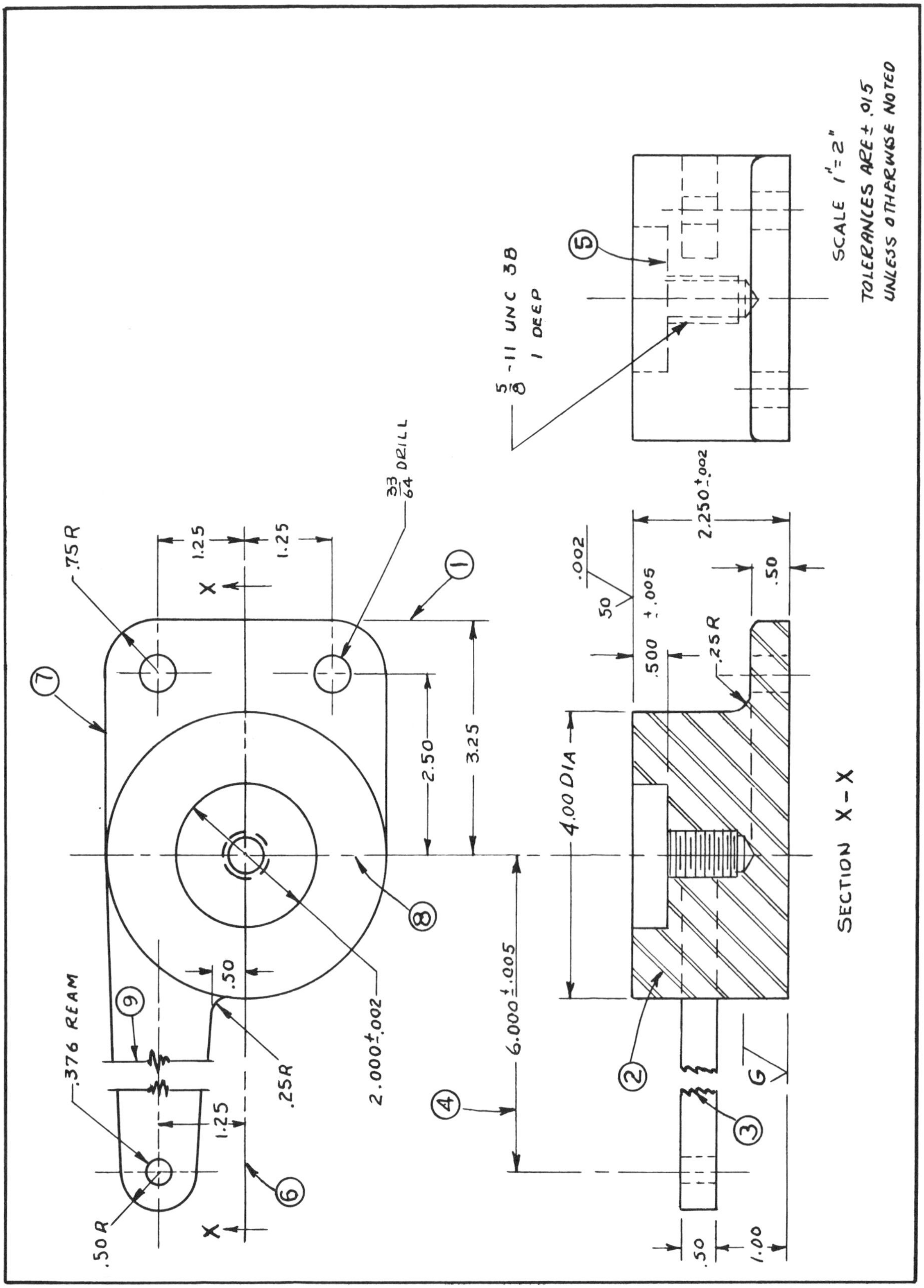
.75R
1.25
1.25
X
$\frac{33}{64}$ DRILL
$\frac{5}{8}$-11 UNC 3B
1 DEEP
SCALE 1" = 2"
TOLERANCES ARE ± .015
UNLESS OTHERWISE NOTED
2.250 ± .002
.002
50
.500 ± .005
.50
.25R
2.50
3.25
4.00 DIA
SECTION X – X
.376 REAM
.50
.25R
2.000 ± .002
6.000 ± .005
1.25
G
.50R
X
.50
1.00

PLATE 1

TRADE COMPETENCY TEST
Based on Plate No. 1, Chapter 4

Student's Name____________________________Instructor's Name____________________________

1. IDENTIFICATION TEST. Plate 1 is a working drawing in which the different types of lines which are used are labeled. In the space after each number below you are to place the name of type of line used in Plate 1 which corresponds to that number.

LABEL	NAME OF TYPE OF LINE	SCORE
1.		
2.		
3.		
4.		
5.		
6.		
7.		
8.		
9.		
	TOTAL	

TEAR OFF HERE

2. MULTIPLE CHOICE TEST. The following questions are based on Plate 1. After each question, there are labeled solutions which may be used to answer the question. In the answer column you are to place the letter of the solution which correctly answers the question.

Example:

What is the decimal equivalent of 1/16?

A. .0625 C. .1250
B. .9375 D. .1875

1. What is the upper limit of the width of the object?

A. 4.000″ C. 4.050″
B. 4.015″ D. 4:500″

2. What is the lower limit of the diameter of the 2.0″ hole?

A. 1.998″ C. 2.448″
B. 2.002″ D. 2.502″

	ANSWER	SCORE
	A	
1.		
2.		

3. What is the lower limit of the diameter of the circular projection on the top of the object?

A. 4.00″
B. 3.950″
C. 3.985″
D. 3.995″

4. What is the specified length of the object?

A. 9.255″
B. 9.500″
C. 9.750″
D. 11.000″

5. How many tapped holes are required in the making of the object?

A. 2
B. 3
C. 4
D. 1

6. Which of the following specifications is indicated for the threads?

A. 8 external threads per inch
B. 11 internal threads per inch
C. 11 external threads per inch
D. 8 internal threads per inch

7. What is the class of thread fit specified for the tapped hole?

A. Class 2B
B. Class 1B
C. Class 3B

8. From what type of rough blank should this part be made?

A. Casting
B. Bar stock
C. Stamping

9. What type of finish is specified for the lower surface of the object?

A. Coarse
B. Rough
C. Smooth
D. Ground

10. How are the threads represented in the hole in the section view?

A. In Simplified form
B. In Graphic form
C. In Conventional form

11. Of what material is the object to be made?

A. Cast iron
B. Steel
C. Brass
D. Copper

	ANSWER	SCORE
3.		
4.		
5.		
6.		
7.		
8.		
9.		
10.		
11.		
	TOTAL	

CHAPTER 5

Understanding Supplementary Information

General Motors Corp.

We have already considered the basic elements in the working drawing. In this chapter we will be concerned with some supplementary kinds of information which are often used in working drawings. These concern scaling, holes, fillets and rounds, diameters, title blocks, bills of materials, and change blocks.

SCALING

When the size of an object allows, the draftsman generally will make the drawing full or actual size. In many cases, however, the object is so large that it would be inconvenient or impossible to draw it full or actual size. In these cases the drawing is *reduced* to some convenient size.

On the other hand, there are cases where the original object is so small that a full size representation would be inconvenient from the point of view of both the draftsman and the blueprint reader. The solution in these cases is to *enlarge* the drawing of the object to some convenient size.

A drawing in which the original object is represented as being *reduced or enlarged* is known as a *scale drawing*. The ratio or proportion between the original and the scale drawing is known as the *scale*.

For example, if a block actually 8″ square measures 4″ square on the drawing, the drawing is *half as large* as the object. The drawing is then said to be drawn on a scale of ½; that is, the ratio between the *drawing size* and the *actual size* is 1 to 2, or ½. On a drawing, the scale is indicated in the following manner: Scale: 1″ = 2″. This simply means that 1″ on the drawing equals 2″ on the object.

If the same block which is actually 8″ square is represented on a drawing as being 16″ square, the drawing is then twice as large as the original object and is then said to be drawn to a scale of 2 to 1, or *double scale*.

Whatever the scale of a drawing may be, the dimensional indications do not change, since they always refer to the actual size of the object.

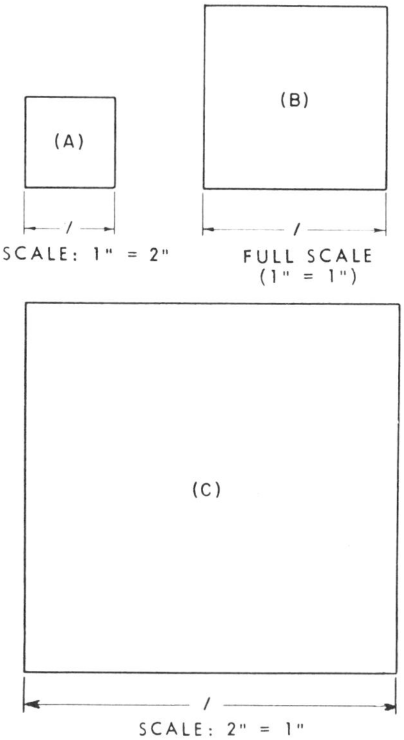

Fig. 1

Fig. 1 shows the same square drawn to three different scales: (A) ½ scale, (B) full scale, (C) double scale. *Note that the dimensional indications are the same throughout.*

In some cases the scale of the object is not specifically indicated on the drawing, since the dimensions should convey all the necessary information concerning size.

However, if different views of the same object are drawn to different scales, the draftsman should indicate the different scales to prevent confusion or error.

In any case, the machinist or craftsman should not attempt to take a dimension by direct measurement from the drawing sheet since processed blueprints may shrink or stretch enough to make such a practice erratic. *Always use the dimensions as given.*

HOLES, FILLETS, AND ROUNDS

Efficient and accurate blueprint reading depends to a great extent on the ability to interpret and understand conventional methods of conveying information. Some of the most common of these conventions concern such machining operations as *drilling, countersinking, counterboring,* etc., and such common details as *fillets and rounds.*

Drilled, Reamed, and Bored Holes. The primary distinction between *drilled, reamed,* and *bored holes* is that drilled holes require less accuracy and smoothness of finish than reamed and bored holes. Reaming is the operation of finishing a drilled hole to exact size with a reamer. Boring consists of enlarging a drilled hole to specific size with a boring tool in a lathe or milling machine.

All of these types of holes are represented identically on a drawing. The operation is specified by a standard method of notation, as in Fig. 2. The notation contains the required information in the following order: (1) diameter of the hole, (2) operation to be performed, (3) number of holes to be drilled, if more than one.

Counterbored Holes. A *counterbored hole* is one in which a second hole of specified depth and diameter is machined concentrically, usually to accommodate a socket or fillister head machine screw, as in Fig. 3.

The dimensions for counterbored holes include: (1) the diameter of the drilled hole, (2) the diameter of the counterbore, (3) the depth of the counterbore, and (4) the number of counterbored holes to be made, if more than one.

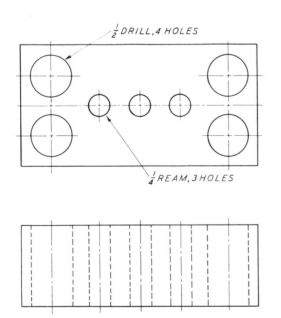

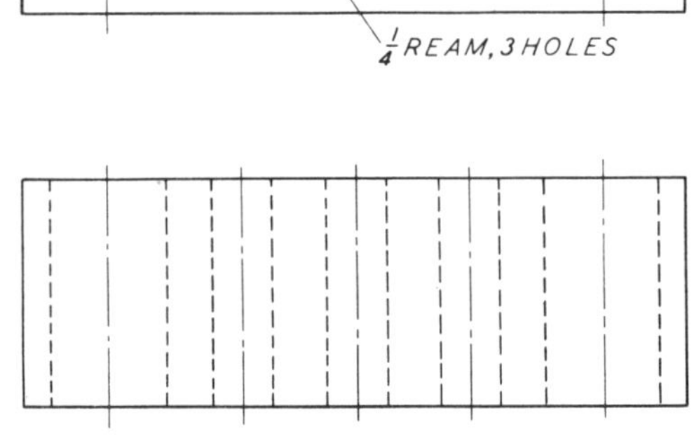

Fig. 2

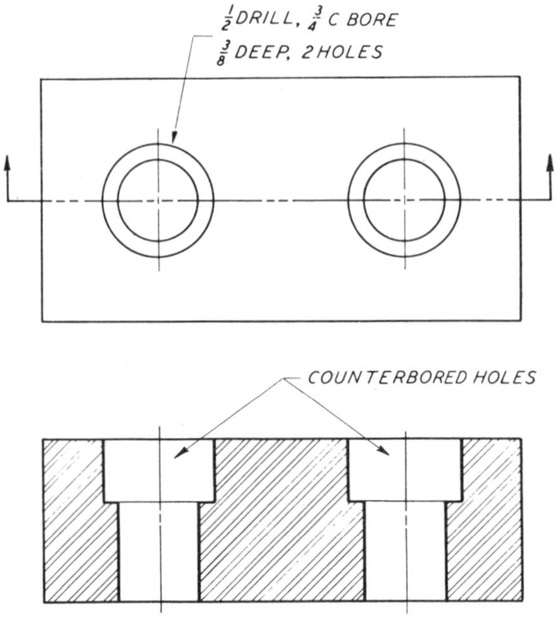

Fig. 3

Countersunk Holes. Holes are often *countersunk* to permit seating of flat head (FH) machine screws so that the top of the screw is flush with the surface of the object, as in Fig. 4. The information conveyed by the dimension should contain: (1) the diameter of the countersunk hole or the type of screw or bolt for which the countersink is to be made and the number of holes to be countersunk is usually given.

Spotfacing. A spotfaced surface is a smooth, flat, circular surface machined on a part in order to provide a flat seat for the head of a nut, bolt, cap screw, etc. A spotfacing specification may include only the diameter of the spot, or it may include the type of screw for which the spotfacing is to be made, as in Fig. 5.

Fillets and Rounds. Fillets and rounds are specified in terms of their radius. The proper notation consists of the dimension followed by the letter "R", as in Fig. 6. A *fillet* is a rounded corner made by the meeting of two inner surfaces on an object. A *round* is a rounded corner made by the meeting of two outside surfaces on an object.

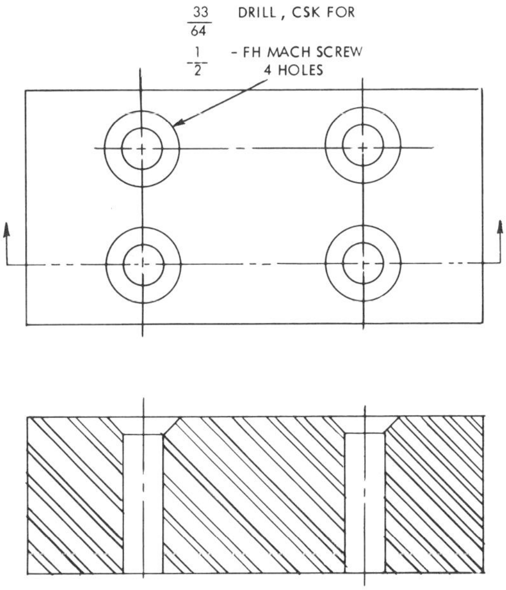

Fig. 4

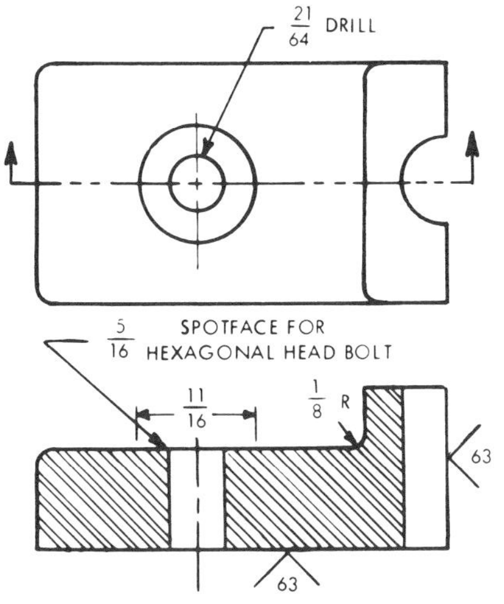

Fig. 5

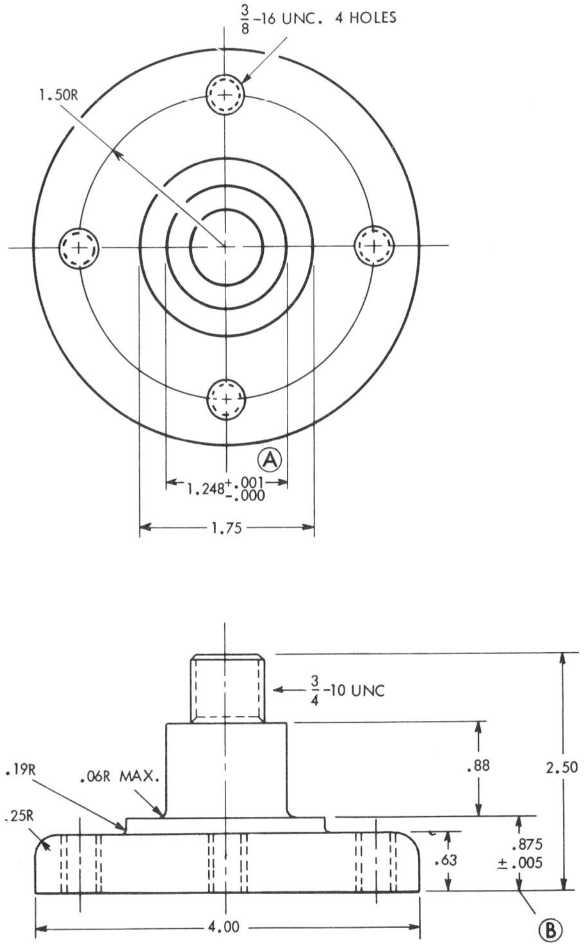

DETAIL	ALTERATION	DATE	BY
(A)	WAS 1.248 +.002 −.000		
(B)	WAS .873 ±.005		

Fig. 6

Diameters. The dimension for a circular part or surface should be indicated in terms of the diameter rather than the radius of the circle as shown in Fig. 6.

Fastener Specifications. Fastening devices may be classified into two main groups. One group includes fasteners which are intended to join parts permanently together. Permanent fasteners are rivets, nails, and wood screws. (Welding can also be considered as falling into this classification, since any welding process will fasten sections together permanently.)

The second group of fasteners are those which permit parts to be disassembled whenever necessary. In this category are such devices as bolts, cap screws, machine screws, set screws, keys and pins.

Usually fasteners are specified as to size and kind. Some also spell out what the material is or such specific information as type of head or finish.

Rivets. Rivets are available in various head shapes such as cone, button, truss, countersunk, pan and flat. The sizes of rivets are usually indicated by the diameter and length of the stem.

Tapping Screws. Tapping screws are used in assemblies of sheet metal parts, plastics and soft castings. They form their own mating threads as they are driven into the material.

Tapping screws are manufactured with a plain steel finish or with a plated finish of brass, zinc, nickel or cadmium. They are available with slotted or Phillips driving recesses in eight basic head shapes: flat, oval, round, fillister, truss, pan, hex, and hex washer. These screws come in a variety of types and sizes.

The sizes of tapping screws are designated by the length and wire gage number. On a drawing, tapping screws are indicated by a note as

⅜—NO. 4 TYPE A-RH TAPPING
SCREW—NICKEL FINISH

Bolts. Basically there are two series of square and hexagon head bolts known as *regular* and *heavy*. These bolts are also classified as *finished, semifinished,* and *unfinished.*

On a drawing, the specifications of a bolt should include diameter, number of threads per inch, series, class, type of finish, type of head, name and length, as

⅜—16 UNC—2*A* x 2½
SEMI-FIN HEX HD BOLT

Cap Screws. These range in diameter from ¼ to 1¼ inches and are made with five types of heads: hexagon, round, flat, fillister, and socket. See Fig. 7.

The specifications of a cap screw should be given as

½—13 UNC—2*A* x 1
ROUND HD CAP SCR

Machine Screws. Machine screws are similar to cap screws except that they are smaller and are used chiefly on small work having thin sections. Below ¼″ size, machine screws

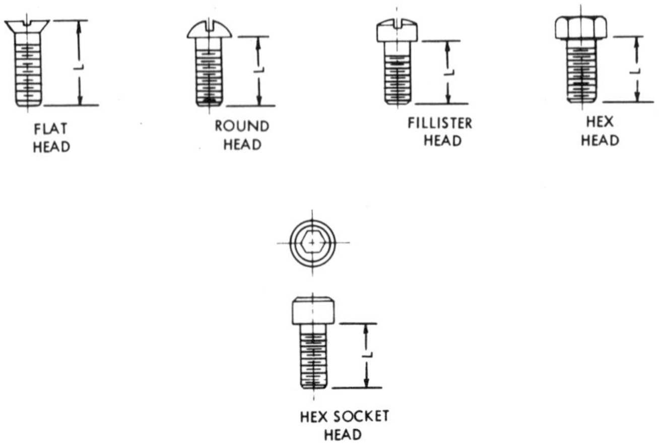

Fig. 7

are specified by numbers from 2 to 12. Above ¼″ the size is indicated by diameter. The threads run the entire length of the stem of screws 2 inches and under in length. On a drawing machine screws are shown as

No. 10—24 UNC—2*A* x ⅜
RD HD MACH SCR

Set Screws. The function of a set screw is to prevent rotary motion between two parts, such as the hub of a pulley and shaft, as shown in Fig. 8. It is also used to make slight adjustments between mating parts.

Set screws are available in the following types: headless slotted, hexagon socket, fluted socket, square head, and safety or socket head. Each of these can be obtained with these points: cup point, flat point, oval point, cone point, full dog point and half dog point. (See Fig. 8, right.) Specifications for set screws should include diameter, number of threads per inch, series, class of fit, type of head, type of point, and length. Example:

¼—20 UNC—2*A* x ½
SLOTTED CONE PT SET SCR

Keys. Keys are used to prevent parts attached to shafts, wheels, cranks, etc., from rotating. Fig. 9 illustrates one of the more common types of key: the Pratt and Whitney key. Also commonly used are square keys, flat keys, gib-head keys, and Woodruff keys.

On drawings, flat, taper, square, and gib-head keys are specified by a note giving the width, height, and length. Pratt and Whitney keys are shown by a number. Woodruff keys are also specified by a number with the last two digits representing the nominal diameter in eighths of an inch, and the preceding digits indicating the width in thirty-seconds of an

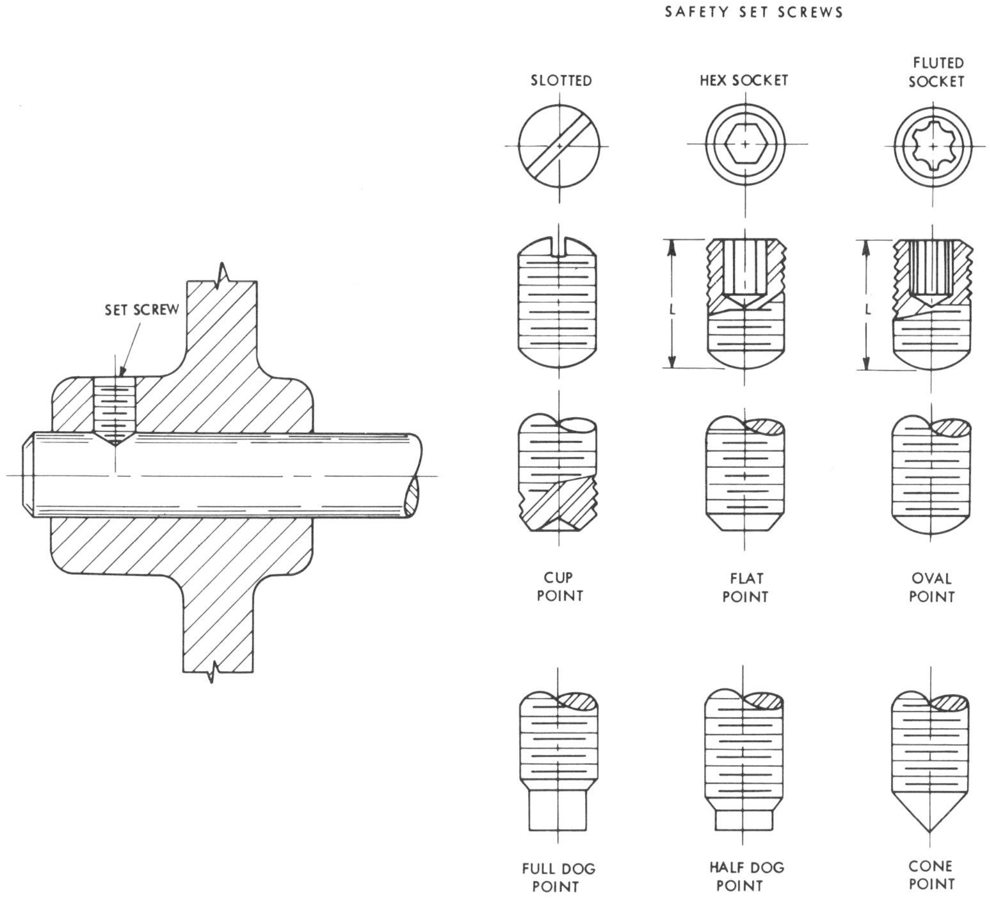

Fig. 8

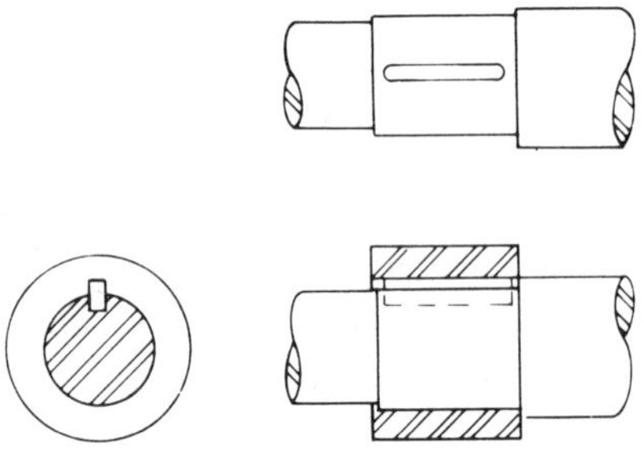

Fig. 9

inch. The information for keys should be listed as

$\frac{3}{16}$ x $1\frac{1}{4}$ SQUARE KEY
$\frac{1}{4}$ x $\frac{3}{16}$ x $1\frac{1}{4}$ FLAT KEY
No. 12 PRATT & WHITNEY KEY
No. 304 WOODRUFF KEY

ALTERATIONS AND REVISIONS

Dimensions on a drawing are sometimes revised without requiring the change of any other part of the drawing. Instead of being completely redrawn, the drawing may be brought up to date by means of notes indicating the change. Often these notes are placed in a *change block* and may consist of a detail number, a description of the change, the effective date of the change, and the initials of the person authorizing or making the change. Fig. 6 shows an example of the use of a change block. (.875 ± .005 for example *was* .873 ± .002).

TITLE BLOCK AND BILL OF MATERIALS

In addition to the elements that make up the working drawings proper, a blueprint should have some means of identification which includes the name of the object and other supplementary data, such as the number of the print, the producing agent or company, the scale of the drawing, etc. Usually this identifying information is placed at the lower right-hand corner of the drawing and is known as the *title block* or *title strip.*

There is no standard form for the title block, and the kinds of information contained in it may also vary. Most companies use drawing sheets with printed title blocks in which the appropriate data is filled in by the draftsman, as in Fig. 10.

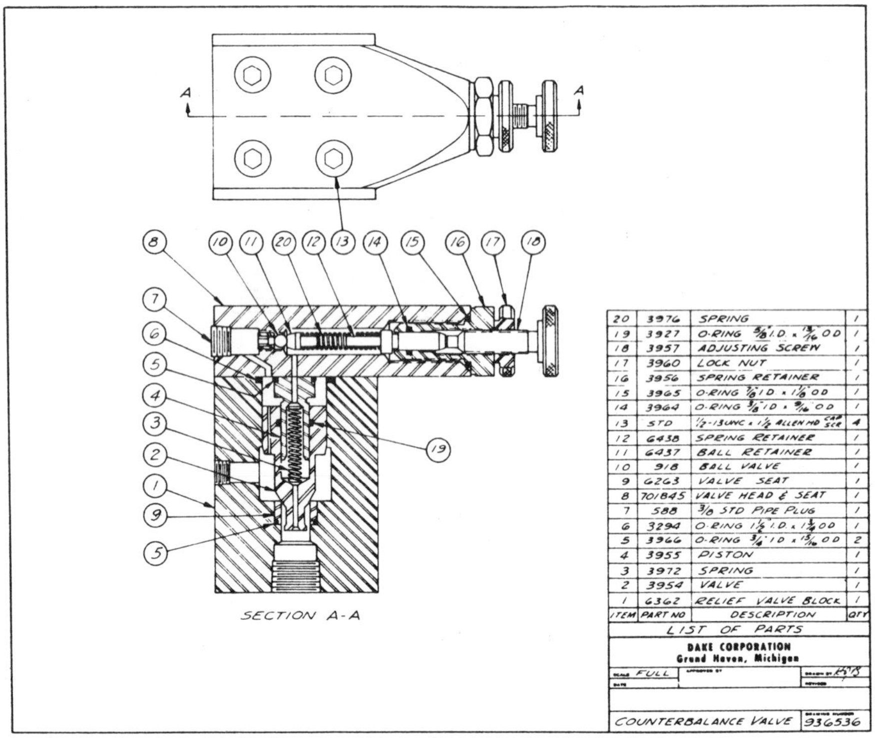

ITEM	PART NO	DESCRIPTION	QTY
20	3976	SPRING	1
19	3927	O-RING 5/8 I.D. x 13/16 OD	1
18	3957	ADJUSTING SCREW	1
17	3960	LOCK NUT	1
16	3956	SPRING RETAINER	1
15	3965	O-RING 7/8 ID x 1 1/8 OD	1
14	3964	O-RING 3/8 ID x 9/16 OD	1
13	STD	1/2-13UNC x 1 1/2 ALLEN HD CAP SCR	4
12	6438	SPRING RETAINER	1
11	6437	BALL RETAINER	1
10	918	BALL VALVE	1
9	6263	VALVE SEAT	1
8	701845	VALVE HEAD & SEAT	1
7	588	3/8 STD PIPE PLUG	1
6	3294	O-RING 1 1/2 I.D. x 1 3/4 OD	1
5	3966	O-RING 3/4 ID x 15/16 OD	2
4	3955	PISTON	1
3	3972	SPRING	1
2	3954	VALVE	1
1	6362	RELIEF VALVE BLOCK	1

LIST OF PARTS

Fig. 10

Bill of Materials. When objects consist of several or many parts, a *bill of materials* is prepared, listing each detail of the assembly. This bill of materials, also referred to as "list of parts" or "materials list," gives information regarding the construction of the assembly.

In the case of standard parts it describes them by name and gives the size and quantity required, as shown under item 13 on the list of parts for the Counterbalance Valve, Fig. 10. Part numbers that refer to drawings are given on those parts that are of special design. Frequently a bill of materials will give specific information as to the type of material and size for parts of special design. This procedure, which is particularly suited to tool, die and fixture design work, enables the tool order department to requisition the complete material requirements for the design from the bill of materials. This is illustrated in the lesson for Print 137.

SELF STUDY SUGGESTION NO. 5

The following study exercise should be performed before you take Trade Competency Test, Chapter 5.

Exercise: In Figs. 11 and 12 you are given two complete isometric drawings and two views of each subject. You are to sketch the missing views of each object in order to obtain three-view drawings which fully describe the object. To achieve this, you should first determine how the given view or views are related to the object. The needed views should then be sketched in relation to the given views. (Answers are given at the end of the book.)

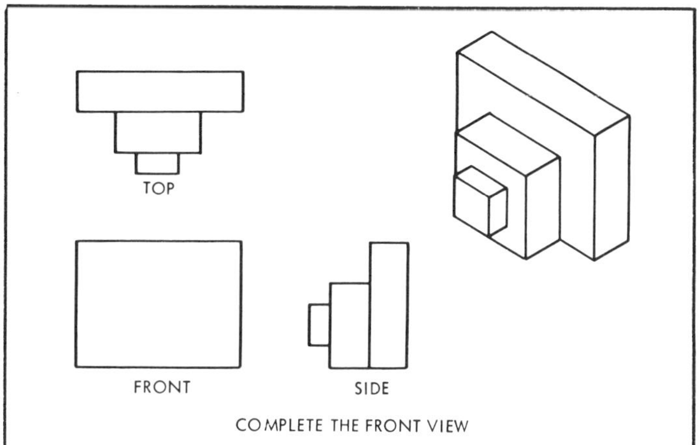

Fig. 11

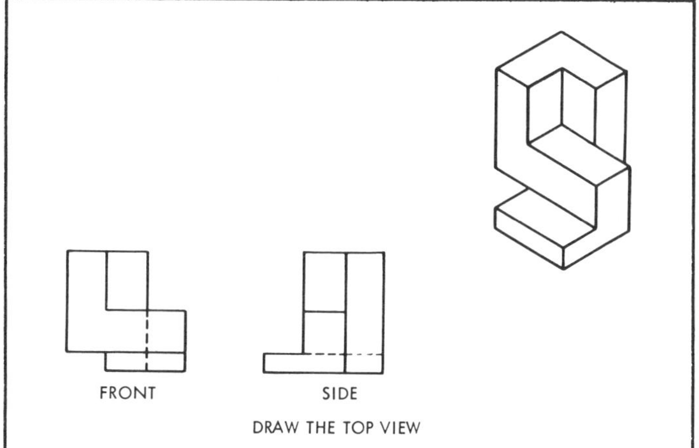

Fig. 12

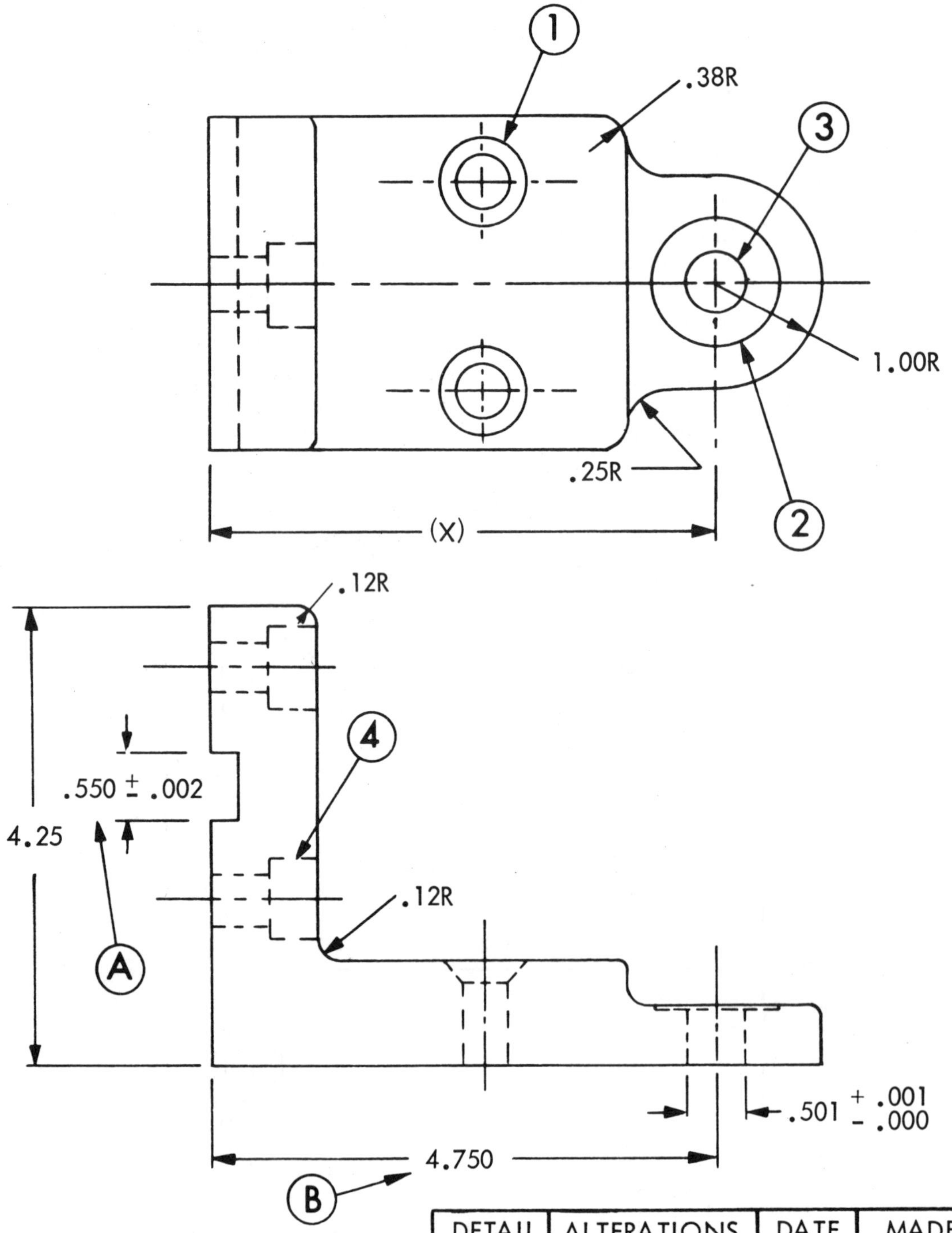

DETAIL	ALTERATIONS	DATE	MADE BY
Ⓐ	WAS .500	1/7	
Ⓑ	WAS 4.70		

TOLERANCES UNLESS NOTE
ARE TO BE ±.003

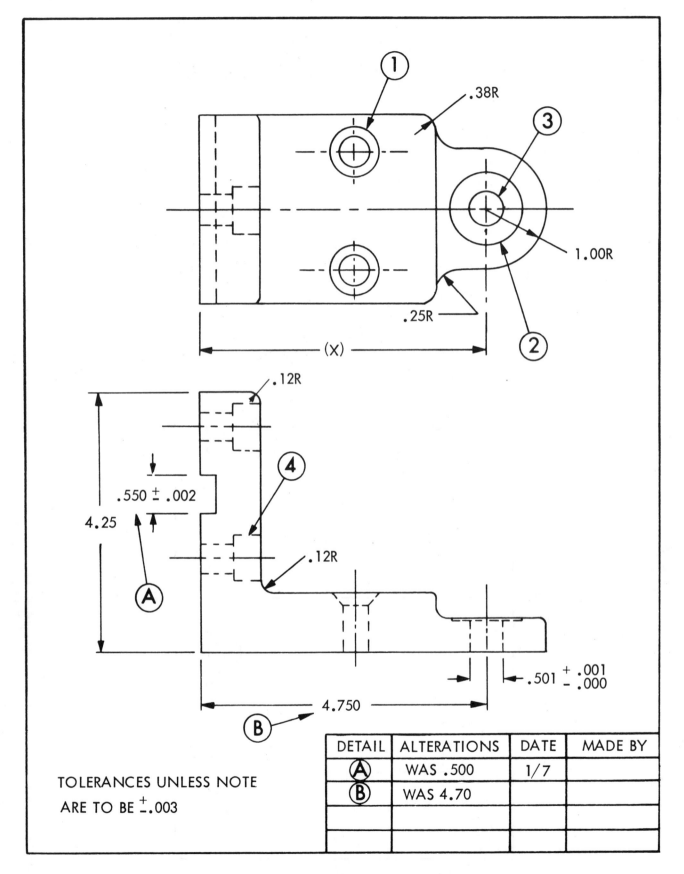

Plate I

TRADE COMPETENCY TEST
Based on Plate No. 1, Chapter 5

Student's Name______________________________Instructor's Name______________________________

MULTIPLE CHOICE TEST. After each of the following incomplete statements and questions, there are four labeled solutions which may be used to complete the statement or answer the question. In the answer column you are to place the letter of the solution which *correctly* completes the statement or answers the question.

Example:

What is the decimal equivalent of 5/8?

A. .580 C. .525
B. .625 D. .875

TEAR OFF HERE

1. If the object were drawn to a scale of 1″ = 2″, what would be the actual length of (X)?

   A. 2.375 C. 9.50
   B. 4.75 D. 9.75

2. If the object were drawn to a scale of 2″ = 1″, what would be the actual length of (X)?

   A. 9.75 C. 9.50
   B. 4.75 D. 2.375

3. Which of the detail numbers in Plate 1 refer to a counterbored hole?

   A. 1 C. 3
   B. 2 D. 4

4. Which of the detail numbers in Plate 1 refer to a spotfaced surface?

   A. 1 C. 3
   B. 2 D. 4

5. Which of the detail numbers in Plate 1 refer to a countersunk hole?

   A. 1 C. 3
   B. 2 D. 4

6. Which of the detail numbers in Plate 1 refer to a reamed hole?

   A. 1 C. 3
   B. 2 D. 4

7. For what reason is it necessary to ream this hole?

   A. To make it round C. Close tolerance
   B. Because material is hard D. To locate it accurately

	ANSWER	SCORE
	B	
1.		
2.		
3.		
4.		
5.		
6.		
7.		

8. What is the MAX dimension indicated in detail A ?

A. 0.547
B. 0.548
C. 0.552
D. 0.5502

9. What is the MIN dimension allowed by detail B ?

A. 4.750
B. 4.747
C. 4.7503
D. 4.753

10. Assume that two specimens of the object are produced. What would be the maximum difference in the overall height of the object if both were within tolerance?

A. .004
B. .005
C. .003
D. .006

11. What is the minimum diameter of detail 3 ?

A. .500
B. .5049
C. .501
D. .5009

12. What are the dimensions of the fillet radii given?

A. 1.00
B. .38
C. .25
D. .12

13. Which of the following specifications for counterboring follow the usual method of notation?

A. 4 holes, ½ drill, ¾ C′ bore, ½ deep
B. ½ drill, ½ deep, ¾ C′ bore, 4 holes
C. ½ drill, ¾ C′ bore, ½ deep, 2 holes
D. ¾ C′ bore, ½ deep, ½ drill, 4 holes

14. Prior to January 7, what was dimension A?

A. .500
B. .470
C. .550
D. .545

15. What is the purpose of spotfacing?

A. To provide clearance for socket head wrench
B. To provide a recess for a bolt or pin
C. To provide a flat, smooth surface for a nut or screw head
D. To provide a smooth, overall finish on a surface

	ANSWER	SCORE
8.		
9.		
10.		
11.		
12.		
13.		
14.		
15.		
TOTAL		

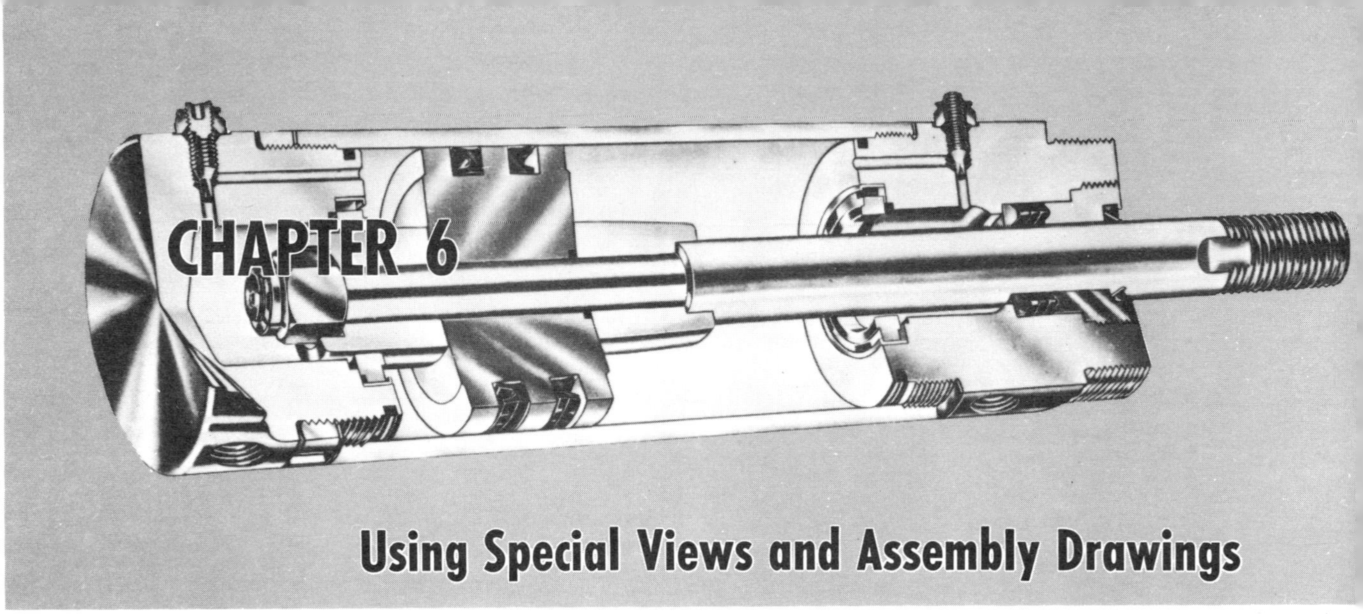

CHAPTER 6

Using Special Views and Assembly Drawings

Tomkins-Johnson Co.

In addition to having a basic knowledge of the fundamental parts of the working drawing, the Machine Tradesman should be familiar with certain techniques of drafting which convey kinds of information which the ordinary methods of drafting do not. Among the most common of these techniques are the various types of sectional views, auxiliary views, assembly drawings, and the various methods of representing gears.

SECTION VIEWS

Because of the complexity of certain objects, especially those that consist of several or many parts, working drawings often include *sectional views*. Sectional views, as we have seen, are views of an object whose inner construction is revealed by passing a *cutting plane* through the object. See Fig. 10, Chapter 4.

The various types of sectional views are all based on the same principle and differ largely according to the kind of information they convey. Following are descriptions and illustrative examples of the most common types of sectional views.

Full Section. When the cutting plane passes entirely through an object, the exposed section is called a *full section*. The example shown in Fig. 10 of Chapter 4 is a full section. Notice that the cutting plane not only passes through the *entire object* but also through its largest dimension, the diameter.

The effectiveness of a section in aiding blueprint reading will be clearly demonstrated if we first study the following regular exterior view of a chuck (Fig. 1, top).

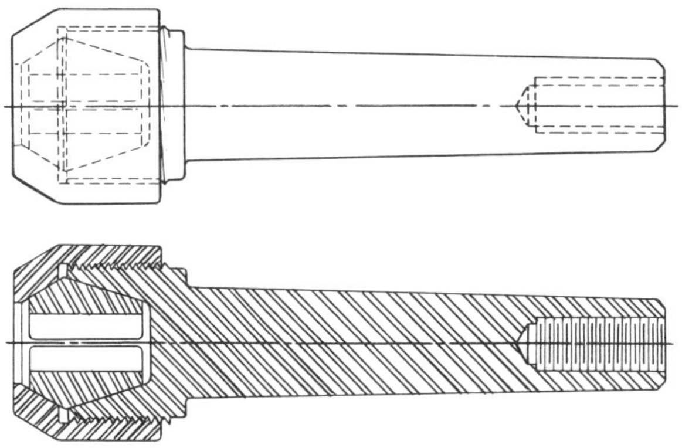

Fig. 1

Obviously the large number of hidden lines makes it rather difficult to clearly visualize the inner construction of the object. Now compare Fig. 1 (top) with a sectional view of the same part shown in Fig. 1 (bottom). The lines that were hidden are now exposed and are drawn as visible *object lines*.

It should be clear, then, that the chief advantage of a section view is that it aids blueprint reading by avoiding the use of hidden lines and by revealing clearly the inner details of an object.

Offset Section. An *offset section* is similar to a full section except that, instead of being a straight unbroken line, the cutting plane line is broken or *offset* in order to reveal inner details which are not in line with each other on the object. Fig. 2 shows an example of an offset front sectional view.

Half Section. In certain situations, it is sufficient to show one-half of a view as a section, leaving the other half as an exterior view, as in Fig. 3. This occurs when the sec-

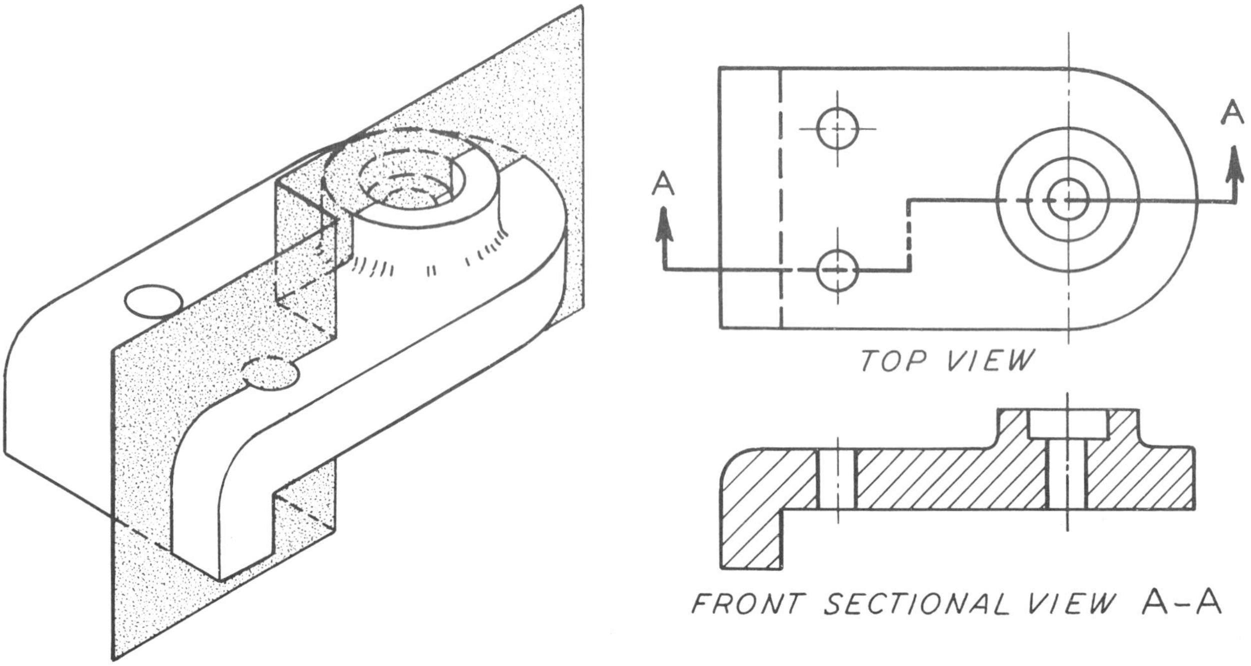

Fig. 2

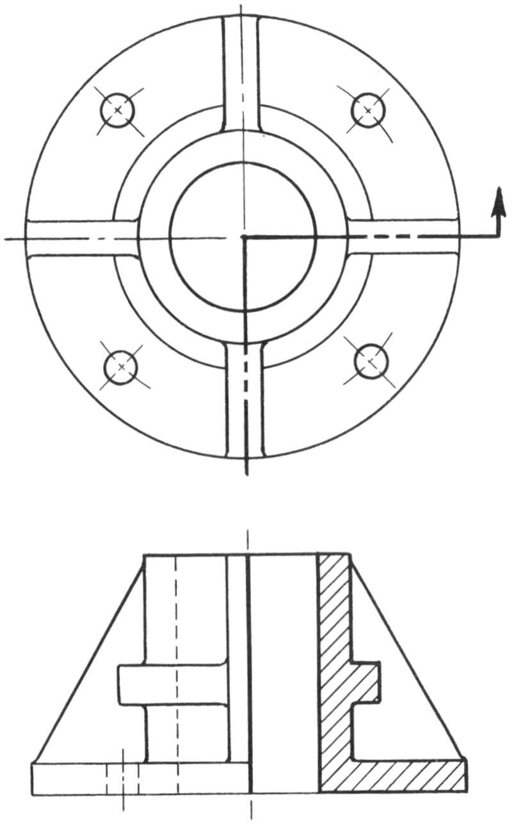

Fig. 3

tioned half is uniform or identical in structure with the non-sectioned half.

Notice in Fig. 3 that the inner construction is represented by a section on the right side of the front view and by the hidden lines on the left side of the same view. Notice also in the top view that the cutting plane line runs through the angular rib, and the rib is not cross-hatched in the section view. This is the usual treatment of ribs of this nature.

Broken-out Section. The inner construction of an object may sometimes be revealed by the removal of just a small section. This is achieved by what is known as a *broken-out section,* and usually shows the boundary as an irregular outline, as in Fig. 4.

Partial Section. A section which shows only a portion of a view of an object, as in Print No. 108, Chapter 8, is known as a *partial section.* Notice that section AA cuts across only that portion of the object which would otherwise be represented by hidden lines.

Revolved Section. In certain types of objects which have a more or less uniform shape, such as spokes, bars, and ribs, sectional

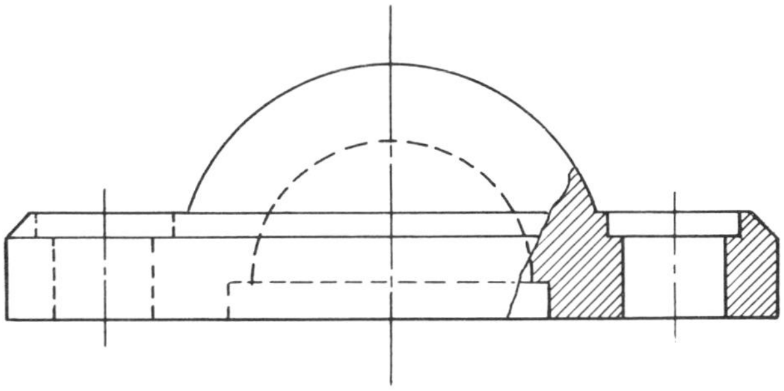

Fig. 4

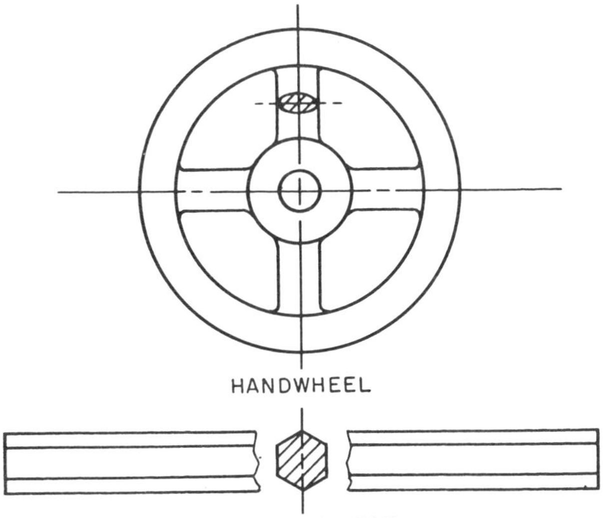

Fig. 5

views may be shown by turning or *revolving* the section in order to reveal its true shape. This is achieved by turning the cutting plane 90° on its axis, as illustrated in Fig. 5.

The effect of a revolved section is to reveal the inner construction on the object where the cutting plane passes through it, but at a 90° angle to the view on which it is shown.

Removed Section. A *removed section* is nothing more than a revolved section which has been removed from its point of origin on a view to some other place on the drawing, as in Fig. 6. This avoids confusion when several sections are made on one view, as in Print No. 111, Chapter 8.

A removed section may also be used to advantage when it is desired to enlarge a revolved section in order to clarify certain

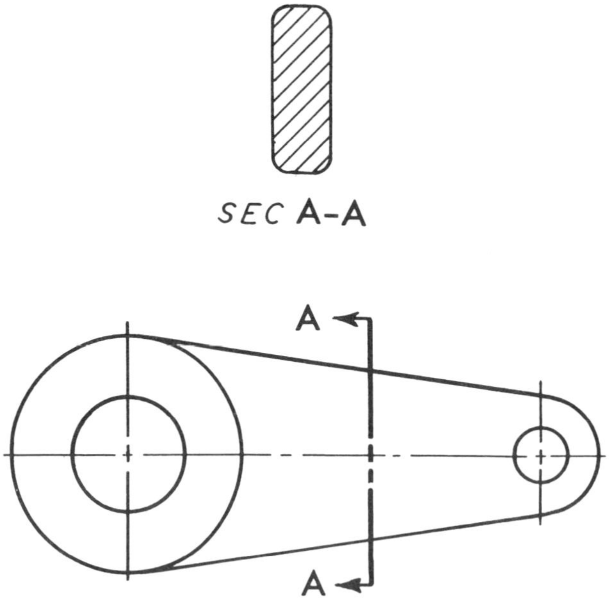

Fig. 6

details. Removed sections are also known as *detail sections.*

SELF STUDY SUGGESTION NO. 6

Exercise: Two complete views of two objects are provided in Figs. 9 and 10 with only the *outline* of the third view *given.* After studying the *given* views, you are to complete the unfinished view as a *section view,* as demonstrated in Figs. 7 and 8. The section view should be based on the cutting plane indicated. Remember that the section lines are used to indicate the material out of which the part is to be made.

AUXILIARY VIEWS

Objects may have slanted surfaces which cannot be adequately represented by ordinary exterior views. The true shape of such an object can be represented only by an *auxiliary view,* which is nothing more than the outline of a *slanted surface* as it would appear to an observer looking directly at it. Fig. 11 shows how a slanted surface (marked A) appears on a side view, and as it appears on an auxiliary view.

The auxiliary view is obtained by *projecting* the slanted surface on a plane *parallel* to it, as in Fig. 11. Notice that the circular hole and the circular edge which appear dis-

DO NOT TEAR OUT

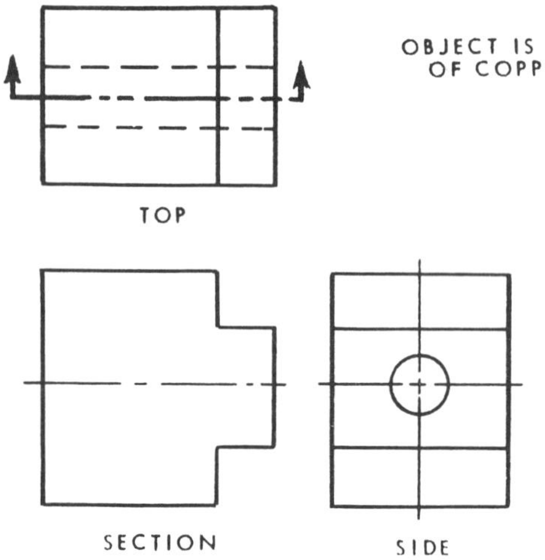

OBJECT IS MADE OF COPPER

COMPLETE THE SECTION VIEW

Fig. 7

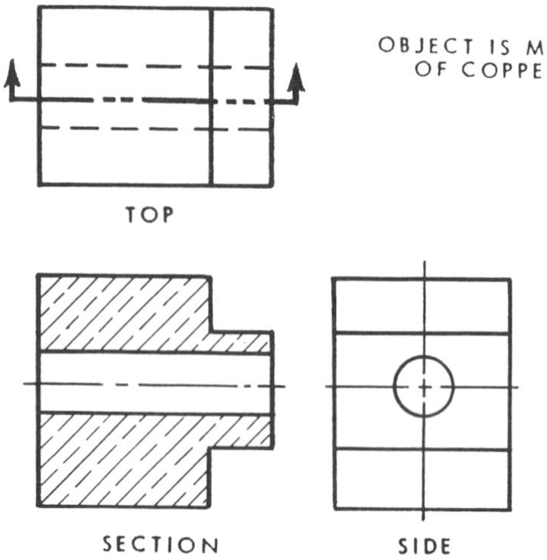

OBJECT IS MADE OF COPPER

(FOR DEMONSTRATION ONLY)

SECTION VIEW COMPLETED

Fig. 8

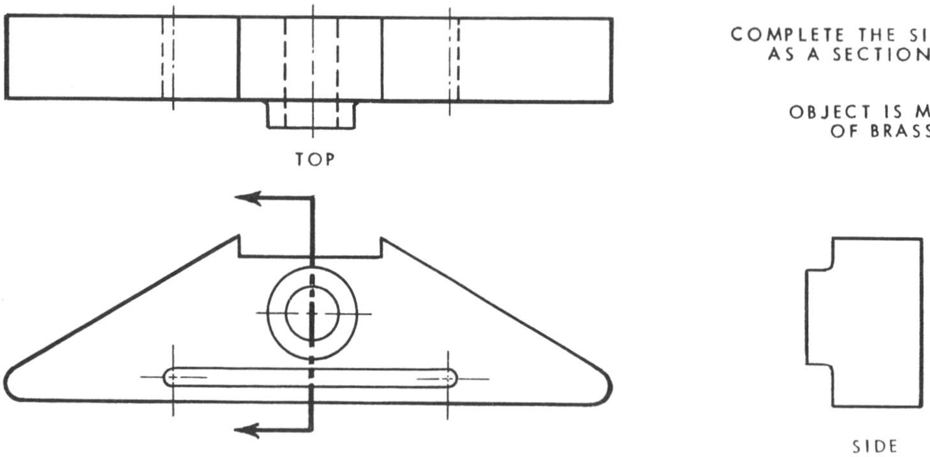

COMPLETE THE SIDE VIEW AS A SECTION VIEW

OBJECT IS MADE OF BRASS

FRONT

Fig. 9

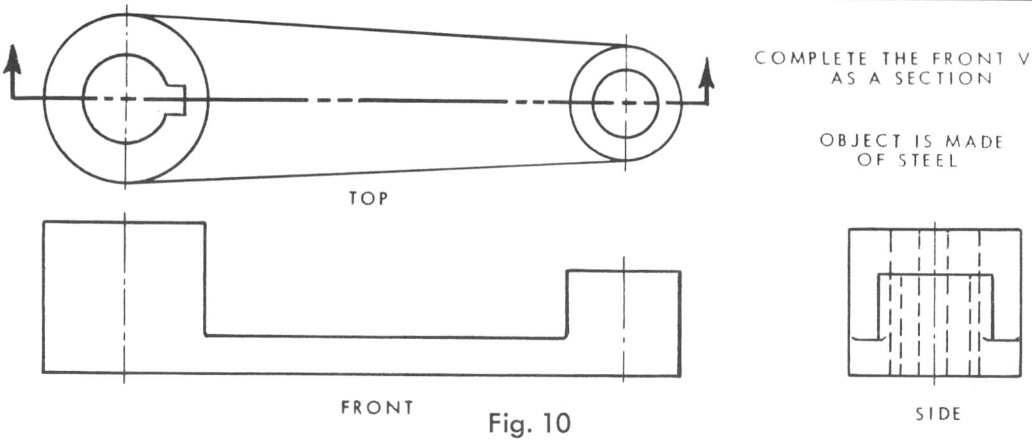

COMPLETE THE FRONT VIEW AS A SECTION

OBJECT IS MADE OF STEEL

Fig. 10

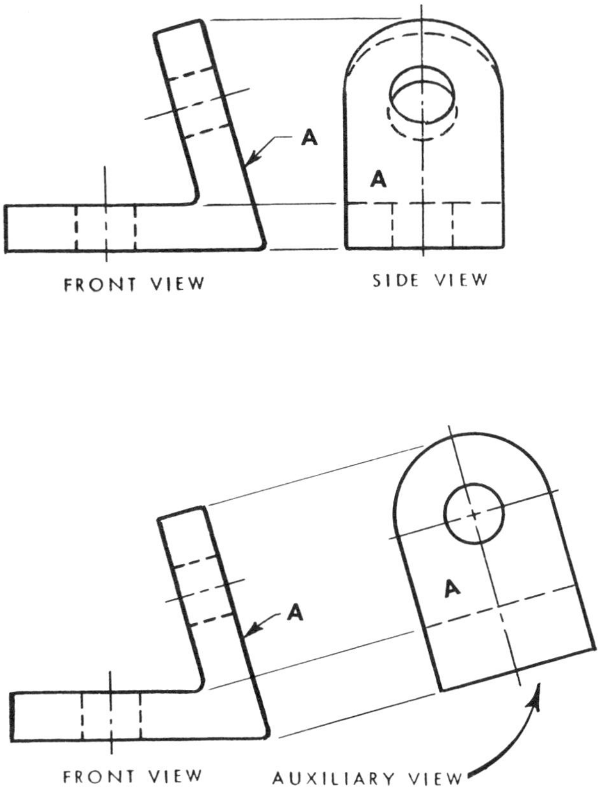

Fig. 11

torted on the end view assume their *true outlines* on the auxiliary view.

ASSEMBLY DRAWINGS

A blueprint which indicates the correct position and location of units in a mechanism is known as an assembly drawing. An *assembly drawing* differs from a working drawing in that the working drawing conveys the information necessary for the *production* of the individual parts, whereas the main purpose of the assembly drawings is to aid the engineer in design work and to show the craftsman how the individual parts should be *assembled* into the final unit.

Among the principal types of assembly drawing are the *unit assembly drawing* and the *working assembly or detail assembly drawing*.

Unit Assembly Drawing. An assembly drawing which shows a complicated object *already assembled in its final form,* with indications of the proper positions of the parts, is called a *unit assembly drawing*. Fig. 10, Chapter 5, is an example of a unit assembly drawing. Note that dimensions are omitted in this drawing. Since the unit consists of parts for which dimensioned working drawings were used in their production, the assembly drawing need show only their correct position on the assembled unit.

Unit assembly drawings may, however, include the overall dimensions of the assembled unit, the distances between centers, and certain other dimensions that might aid in assembly.

Working Assembly Drawing. A working assembly drawing simply combines the main features of a working drawing and a unit assembly drawing. In other words, it provides all the information necessary for the production of the individual parts of an object and also shows how the individual parts fit together to form the final assembled unit. See Fig. 12.

The working assembly drawing is especially useful for objects which have relatively few and simple parts, as in the example shown in Fig. 12. By showing the position and function of each part on the unit, it provides the machinist with a guide as to the importance of the various dimensions of the individual parts. For instance, in Fig. 12, the part indicated by detail No. 4 has to be more accurately machined than the part indicated by No. 1 since the former has to be precisely fitted to the part indicated by No. 2. The machinist therefore knows that he has to machine the dimensions of the plunger more accurately than some *outside* dimension, such as indicated by details 1 and 2.

GEARS

A basic knowledge of gears is essential to the machine tradesman both because of the universal use of gears in industry, and because of the existence of standard practices for representing gears on blueprints.

The methods of representing gears are based on the following factors:

1. The different types of gears and their main features.

2. The different parts of gears and their terminology.

3. The use of formulas for computing the dimensions of the parts of a gear.

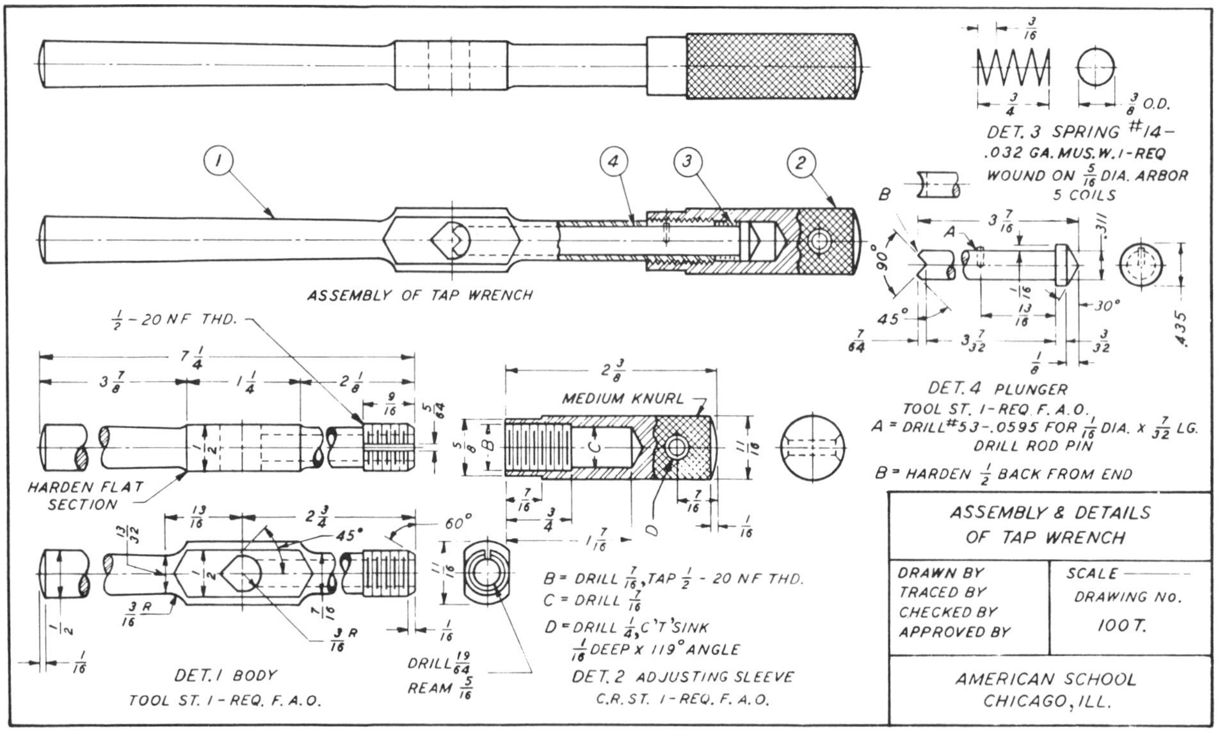

Fig. 12

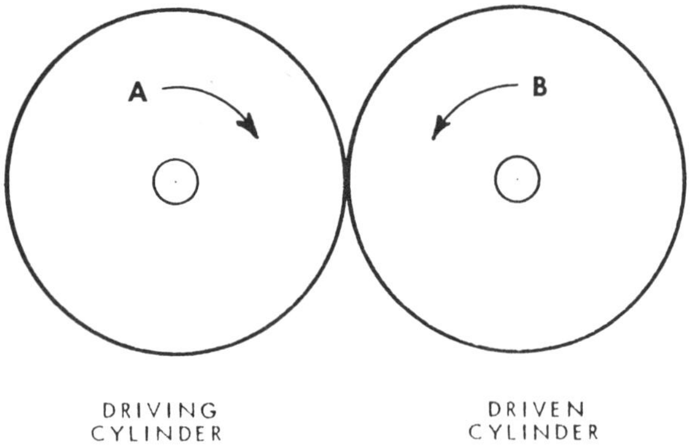

Fig. 13

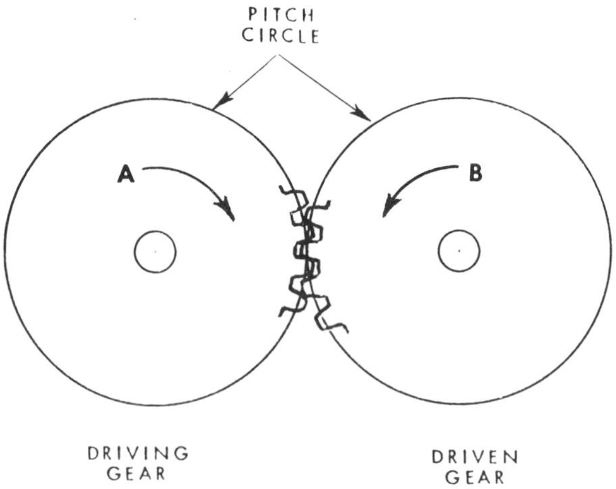

Fig. 14

The basis of all gearing consists of the *imparting of rotary motion by one part to another*. Fig. 13 shows how one cylinder can impart motion to another one through friction. In the situation illustrated in Fig. 13, a great deal of motion would be lost by *slippage* between the cylinders, thus reducing the effectiveness with which cylinder A moves cylinder B.

The slippage can be overcome by the use of teeth on one cylinder interlocking with *teeth* on the second cylinder. The interlocking teeth provide "grab" through which *one gear can impart motion effectively to another gear*. Fig. 14 illustrates the manner in which two "cylinders" on which teeth are cut can perform the mechanical operation known as *gearing*. All the different methods of gearing are based on the principle of efficiently imparting rotary motion from one part to another part by means of *interlocking teeth*.

Spur Gear. When two gears are meshed in the manner shown in Fig. 14, the *imaginary circle* formed by the points of contact between the gears is called the *pitch circle*. Notice that the pitch circle lies about halfway between the tops and the bottoms or roots of the gear teeth. With these facts in mind we may now consider the definition of the parts of a spur gear, as illustrated in Fig. 15.

DEFINITIONS OF GEAR PARTS

Pitch Circle. The imaginary circle of a gear which could be moved by friction alone if it were a simple cylinder. On a gear this imaginary circle is located *approximately halfway* between the tops and the roots of the teeth.

Pitch Diameter. The diameter of the pitch circle.

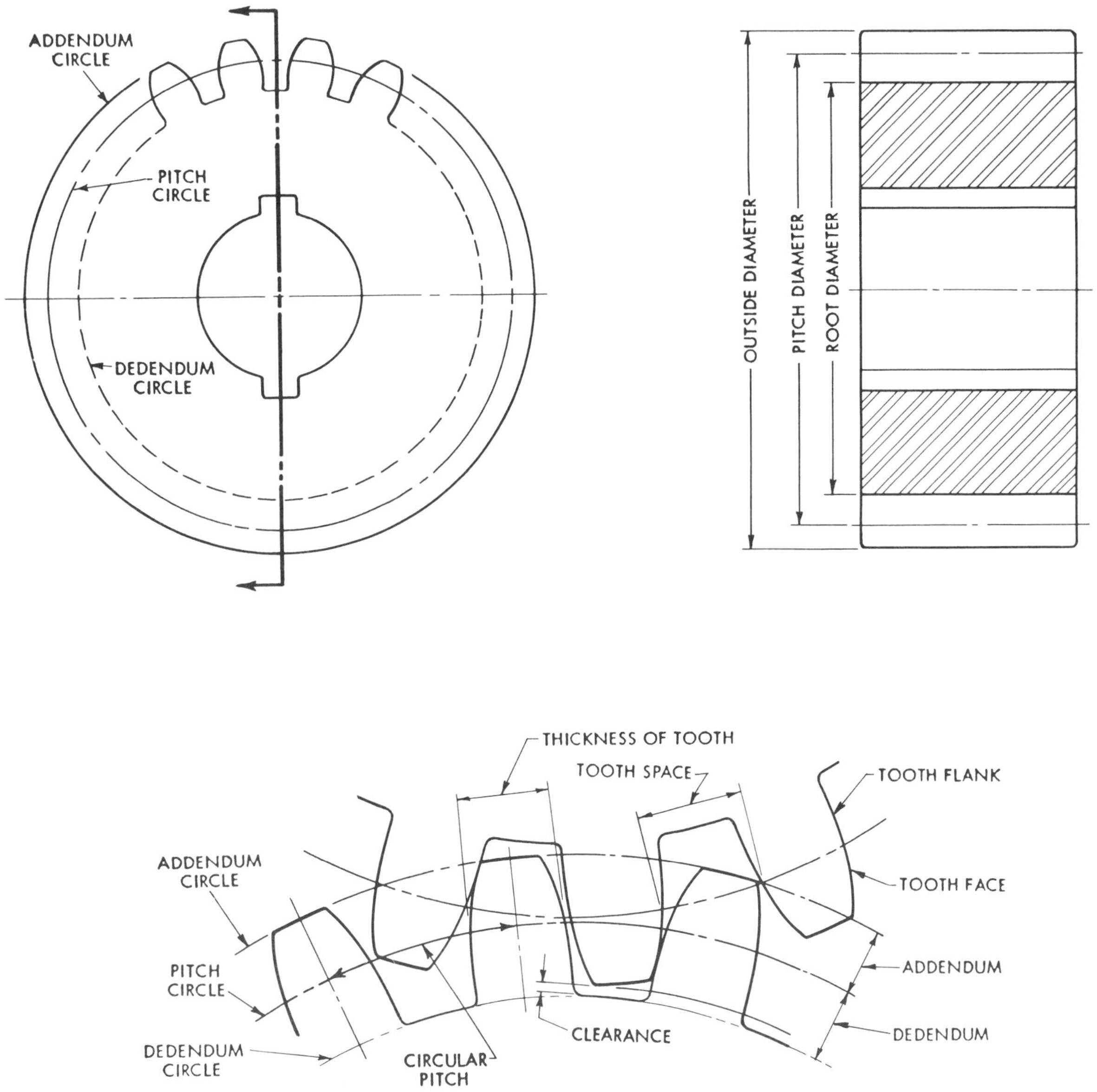

Fig. 15

Addendum. That portion of the gear teeth which is "added" to the pitch diameter.

Addendum Circle. The circle whose diameter equals the pitch diameter *plus twice* the addendum; also the circle formed by the tops of the gear teeth.

Outside Diameter. The diameter of the addendum circle; also the diameter of the gear blank.

Dedendum. That portion of the gear teeth which is "deducted" from the pitch diameter.

Dedendum Circle. The circle whose diameter is equal to the pitch diameter *minus twice* the dedendum; also the circle formed by the roots of the gear teeth. The dedendum circle is known also as the *root circle.*

Circular Pitch. The distance between the centers of two adjacent teeth measured along the pitch circle.

Tooth Space. The distance between adjacent teeth measured along the pitch circle.

Tooth Thickness. The thickness of a tooth measured along the pitch circle.

Tooth Face. That portion of the curved surface of a tooth which lies *above* the pitch circle.

Tooth Flank. That portion of the curved surface of a tooth which lies *below* the pitch circle.

Clearance. The distance between the top of a tooth from a mating gear to the bottom of the tooth space of a gear being mated.

Diametral Pitch. The number of teeth in a gear to each inch of pitch diameter. Therefore, a 12-pitch gear has 12 teeth for each inch of pitch diameter. On prints, the term "pitch" preceded or followed by a number refers to diametral pitch. Also abbreciated to DP.

DIMENSIONS OF SPUR GEAR PARTS

In Chapter 8, page 131, are tables of rules and formulas for calculating the dimensions of any part of a spur gear. The basic data always indicated on gear drawings are the *number* of teeth and the diametrical pitch. From these specifications, any dimension can be found by substituting the appropriate values in the formulas.

For instance, if the note on the blueprint reads, "2 Pitch—18 Teeth" (meaning a diametral pitch of 2 for a gear with 18 teeth), we can find the pitch diameter by using rule No. 3.

Given: Diametral Pitch (P) = 2

Number of teeth (N) = 18

Wanted: Pitch Diameter (D)

Rule 3: $D = \frac{N}{P}$

Calculation: $D = \frac{18}{2} = 9$

To get the outside diameter from the above data, we use the formula for rule 15.

Given: Diametral Pitch (P) = 2

Number of teeth (N) = 18

Wanted: Outside Diameter (O)

Rule 15: $O = \frac{N + 2}{P}$

Calculation: $O = \frac{18 + 2}{2} = 10$

By using the formulas in the above manner, the dimensions for all the parts of a gear may be found from the data given on the blueprint.

BEVEL GEARS

The principal difference between bevel gears and spur gears is that, whereas the shafts of spur gears are parallel to each other, the shafts of bevel gears are generally at right angles to each other. See Fig. 16.

Thus while the teeth of spur gears may be imagined as being cut on a cylinder, the teeth of bevel gears may be imagined as being cut on a cone, as in Fig. 16.

The various parts of a bevel gear and their terminology are illustrated in Print No. S-35, Chapter 8, page 136.

The tables for computing bevel gear dimensions are shown on pages 136 and 137. As with spur gears, the basic data always given on drawings of bevel gears are the *number of teeth* and the *diametral pitch.* The methods for using bevel gear formulas are the same for spur gear formulas.

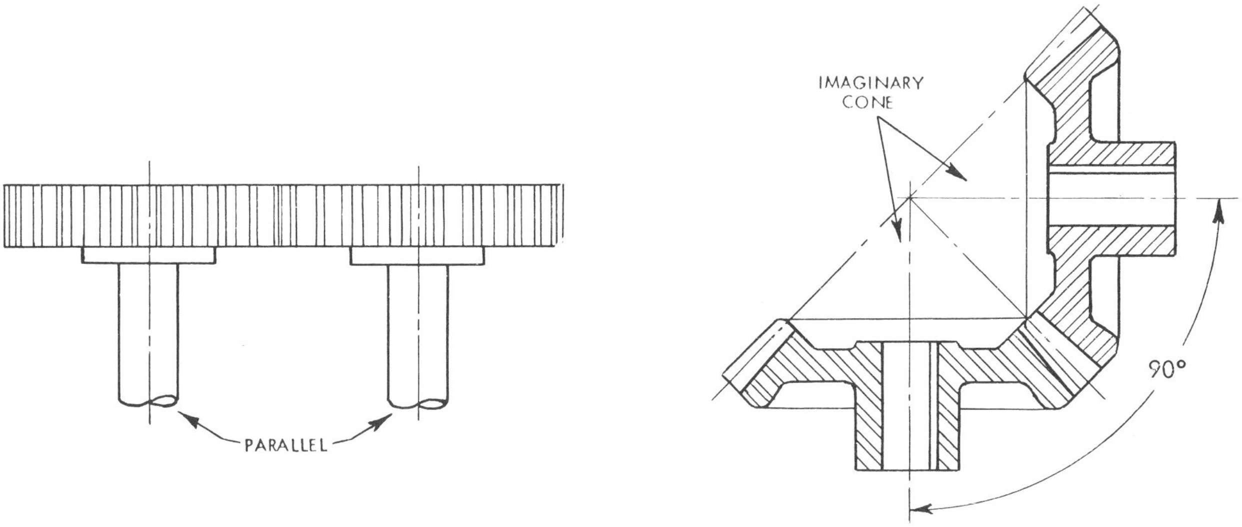
PARALLEL
IMAGINARY CONE
90°

Fig. 16

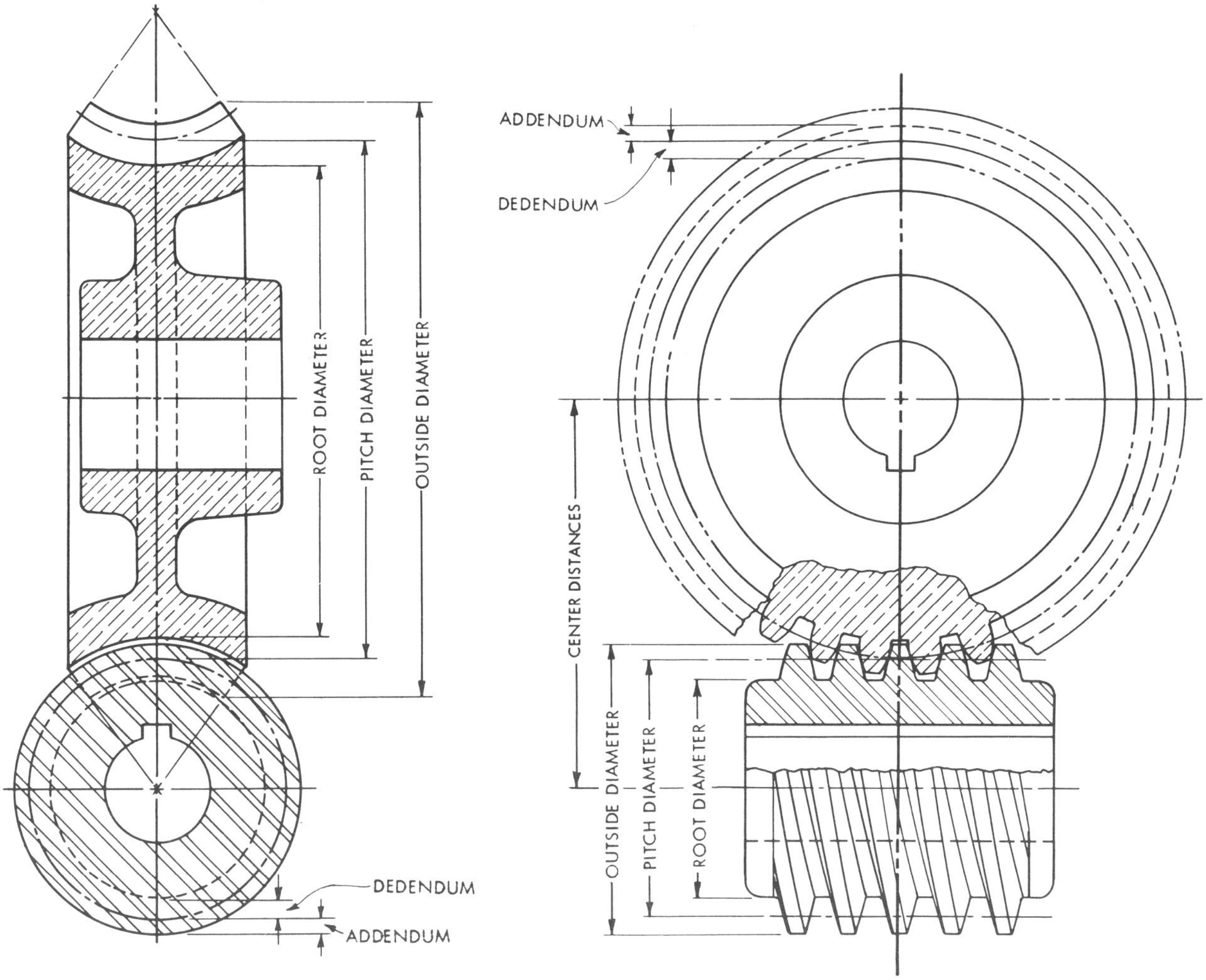
ROOT DIAMETER
PITCH DIAMETER
OUTSIDE DIAMETER
DEDENDUM
ADDENDUM
ADDENDUM
DEDENDUM
CENTER DISTANCES
OUTSIDE DIAMETER
PITCH DIAMETER
ROOT DIAMETER

Fig. 17

WORM GEARING

Worm gearing consists of two parts: the *worm gear,* or *worm wheel,* and the *worm,* both of which are shown in Fig. 17. Note that the axes of the worm gear and the worm do not intersect, as with bevel gears; nor are they parallel, as with spur gears.

The rim of the worm gear is concave to permit the worm to mesh with the gear teeth. Generally the gear is designed to advance one tooth for every revolution of the worm. Worm gearing is therefore especially suited to speed reduction. If, for instance, the worm gear has 12 teeth and it advances one tooth for each revolution of the worm, the velocity is 12 to 1. The velocity ratio therefore depends on the number of teeth in the worm gear and the kind of thread on the worm.

SELF STUDY SUGGESTION NO. 7.

It is suggested that you perform the following study exercises before taking Trade Competency Test for Chapter 6.

Exercise: Figs. 18 and 19 present two views of objects having slanted surfaces. From these views construct the top view, and from the three-view drawing thus obtained, complete the isometric drawings. (Answers are given at the end of the book.)

DO NOT TEAR OUT

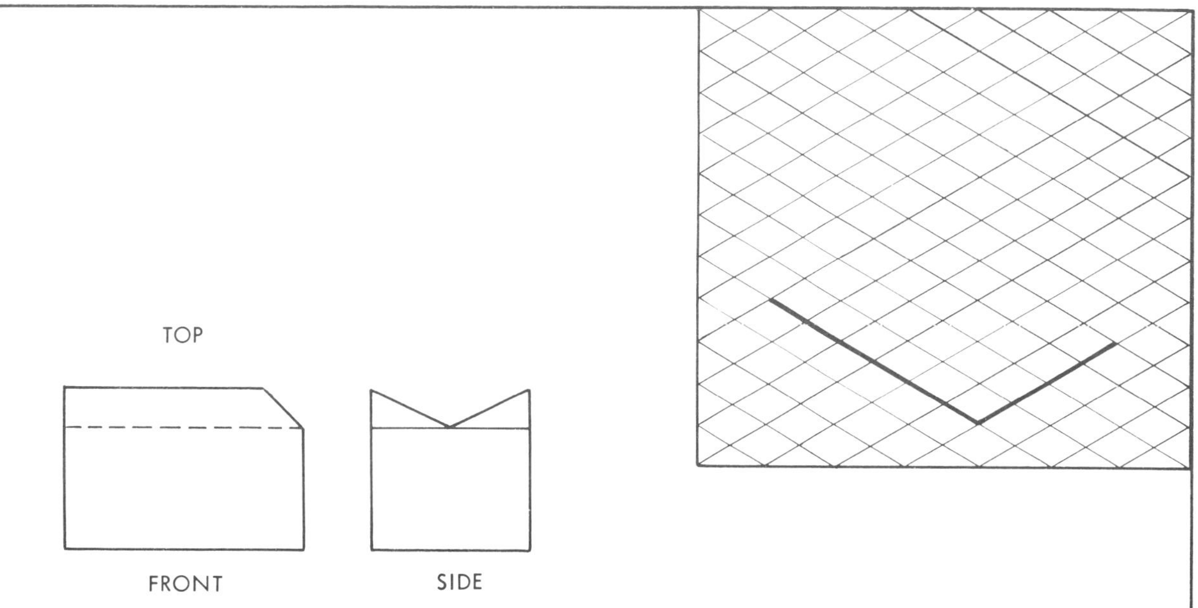

DRAW TOP VIEW AND COMPLETE ISOMETRIC VIEW

Fig. 18

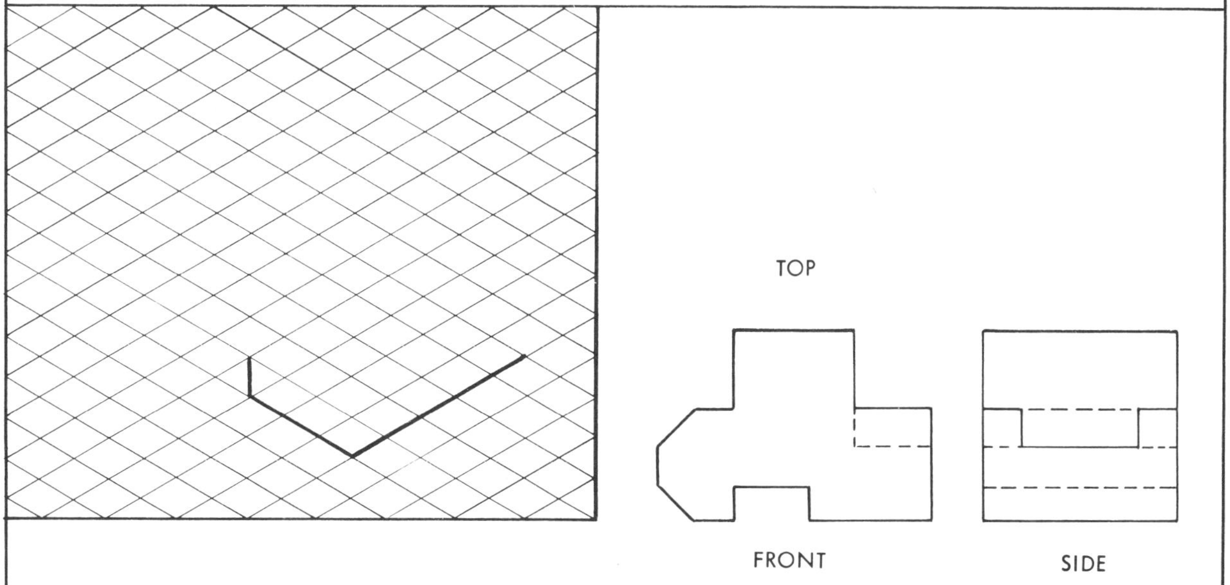

DRAW TOP VIEW AND COMPLETE ISOMETRIC VIEW

Fig. 19

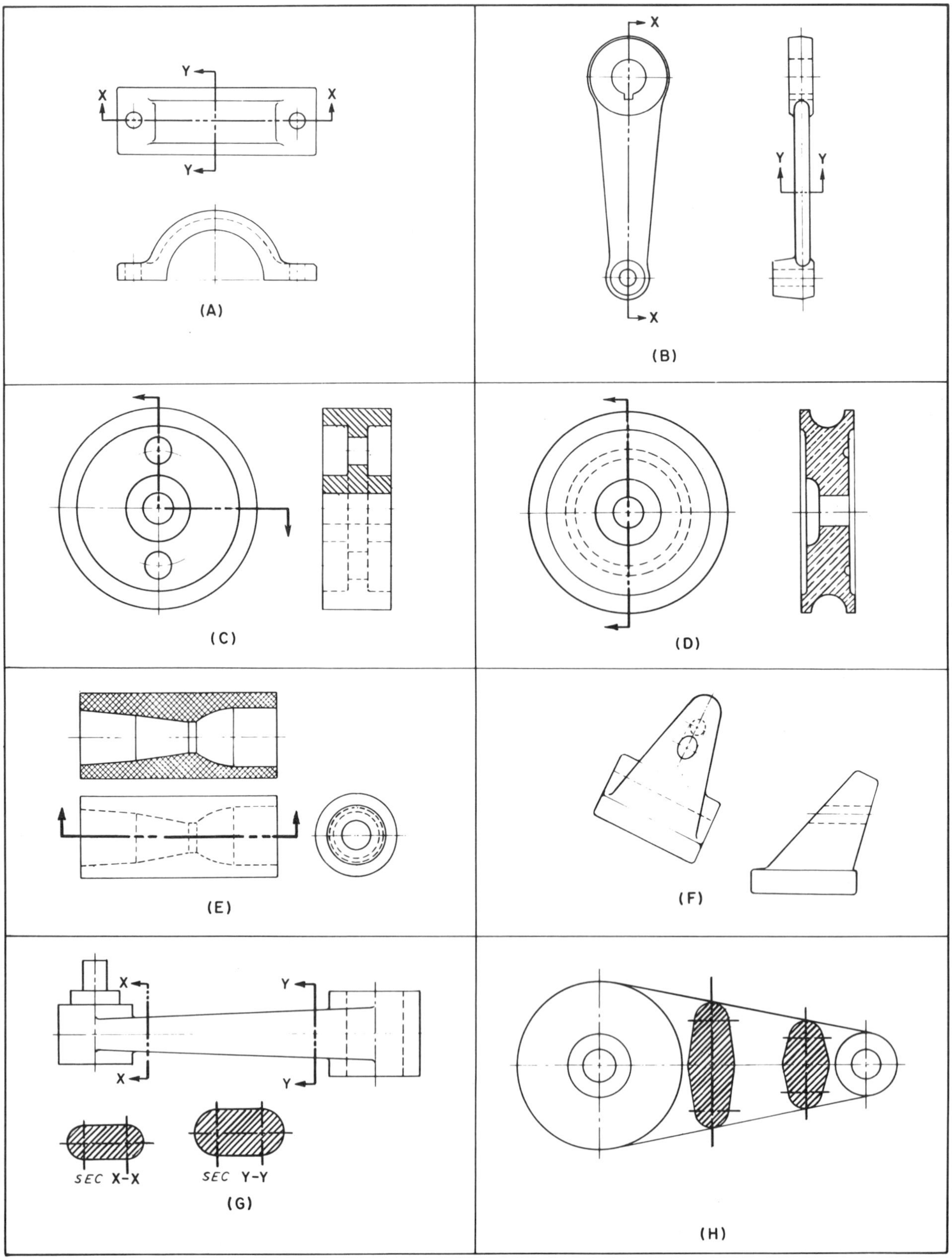

Plate I

TRADE COMPETENCY TEST
(Part I is based on Plate 1, Chapter 6)

Student's Name________________Instructor's Name________________

TEAR OFF HERE

1. **IDENTIFICATION TEST.** Plate 1 consists of alphabetically labeled examples of common types of views. In the space after each letter listed below, you are to place the name of the type of view in Plate No. 1 which corresponds to that letter.

LABEL	NAME OF TYPE OF VIEW	SCORE
A		
B		
C		
D		
E		
F		
G		
H		
	TOTAL	

2. **MULTIPLE CHOICE TEST.** After each of the following incomplete statements and questions, there are four labeled solutions which may be used to complete the statement or answer the question. In the answer column you are to place the letter of the solution which correctly completes the statement or answers the question.

Example:

What kind of material is indicated by the section lines of (D) Plate I?

A. Copper, Brass, Bronze
B. Zinc, Aluminum
C. Steel
D. Cast Iron

1. A detail section is most similar to

A. an offset section.
B. a removed section.
C. a broken-out section.
D. a half section.

	ANSWER	SCORE
	A	
1.		

2. The purpose of an auxiliary view is to show
   - A. a section view of a slanted surface.
   - B. a true outline of a slanted surface.
   - C. an isometric view of a slanted surface.
   - D. an enlarged outline of a slanted surface.

3. Two cutting planes are shown on Fig. (A), Plate I. Which section is essential to complete the drawing?
   - A. Section X-X
   - B. Section Y-Y

4. Two cutting planes are shown on Fig. (B), Plate I. Which section is essential to complete the drawing?
   - A. Section X-X
   - B. Section Y-Y

5. Three views are shown in Fig. (E), Plate I. Which view shows the internal detail of the object most clearly?
   - A. Front View
   - B. Side View
   - C. Section View

6. Which of the following types of dimensional indications is most likely to be found on a unit assembly drawing?
   - A. Detailed dimensions of each part
   - B. Overall dimensions of each part
   - C. Distances between centers
   - D. Tolerances

7. How many teeth does a 50" pitch diameter gear have with 10 diametral pitch teeth?
   - A. 10
   - B. 500
   - C. 50
   - D. 200

8. How are the axes of spur gears aligned with each other?
   - A. Generally they intersect at right angles.
   - B. Generally they intersect at acute angles.
   - C. They are parallel to each other.
   - D. They are neither parallel to each other nor intersecting.

9. What is the outside diameter of a gear that has 22 teeth with a 6 diametral pitch tooth?
   - A. 3.66″
   - B. 3.33″
   - C. 4″
   - D. 13.2″

	ANSWER	SCORE
2.		
3.		
4.		
5.		
6.		
7.		
8.		
9.		
TOTAL		

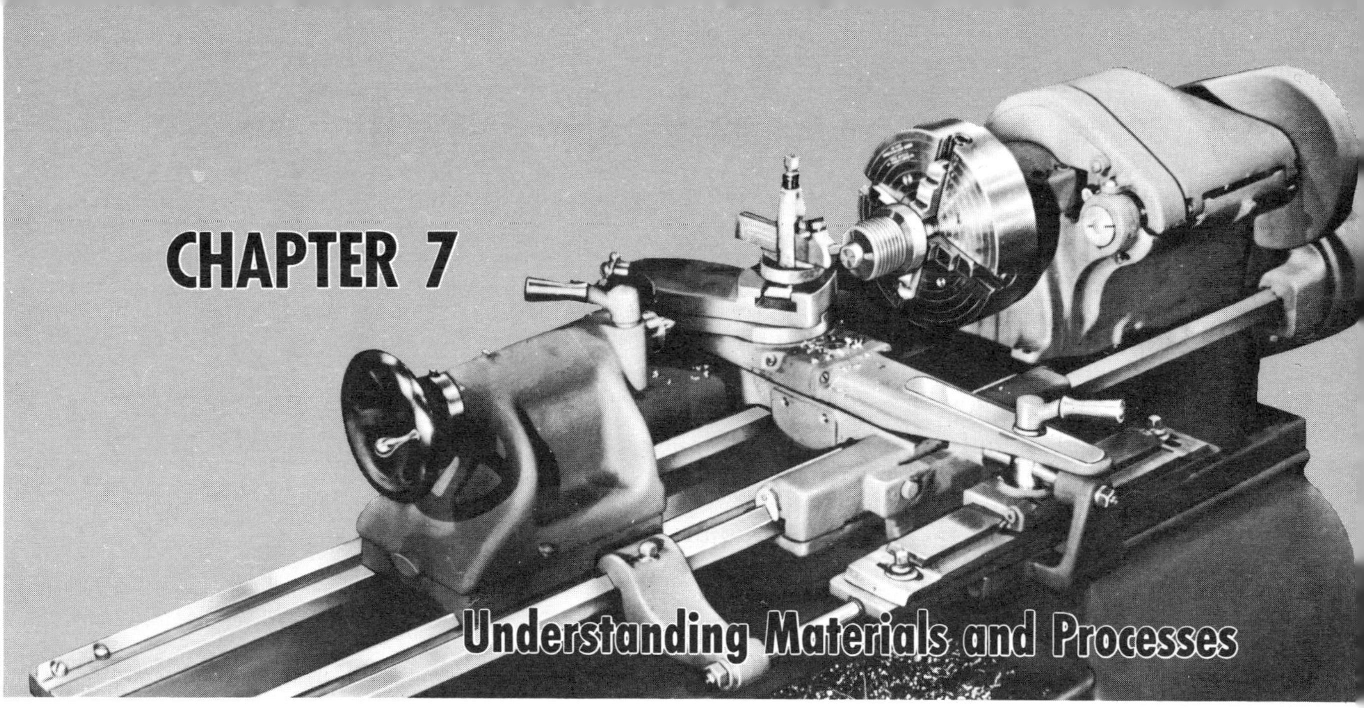

South Bend Lathe, Inc.

THE SELECTION AND USE OF MATERIALS

It isn't absolutely necessary for a Machine Tradesman to have a thorough technical knowledge of the various materials that he finds specified on blueprints or working drawings, but some understanding of why and how a particular material is specified can help him in the long run to do a better and more efficient job.

Basic Materials. The basic materials used in the Machine Trades are *ferrous* and *non-ferrous metals, plastics,* and *rubber.*

With the development of new resins in recent years, there has been a remarkable increase in the use of plastics. These new high-strength materials will often be specified on blueprints in place of metals in a number of applications.

At first rubber was used only as an erasing material for pencil marks—useful to draftsmen in preparing drawings, but in the beginning it seemed an unlikely material to be found specified on a blueprint. Now, many applications have been found for this material, and its resilient and elastic properties have given it its own particular field of usefulness in the Machine Trades.

The most widely used materials, however, are the ferrous metals, which include iron and steel. Nine out of ten tons of all the metal we use today is steel, and this brief consideration of materials will be concerned mainly with steel, as on the bulk of the prints and drawings you work from, steel will be the specified material.

Non-ferrous metals are aluminum, brass, bronze, copper, lead, magnesium, titanium, and zinc. Other important metals, which are used mainly as alloying agents, are chromium, cobalt, manganese, molydenum, nickel, phosphorus, silicon, tin, and vanadium.

Selection of Materials. As the student becomes more experienced in the Machine Trades, he will become familiar with the different characteristics of the materials he handles. If he has a real interest in his job, he will often ask himself just why a certain material was specified for a job—especially if it is a material whose characteristics, he's found from previous experience, make it difficult or time-consuming to work with. Knowing something of the reasons that lie behind the selection of materials specified can help the Machine Tradesman to understand blueprints more completely and to interpret them more intelligently.

It is the design engineer's responsibility to select the materials that are going to be made into a product. This complex assignment requires that he must know what the product is going to be used for, and how it is going to be used; as a designer, he must know the appropriate means of fabricating and treating the materials.

He might decide that a certain material would be just right for the job, and then have to rule it out because of cost factors; either because the basic cost of the material would make the end-product too expensive to sell in competition with other products like it, or because the methods used in fabricating it would be too slow. There isn't much point in designing and manufacturing a product that can't be sold at a competitive price. To take an extreme example, there isn't likely to be much of a market in the automobile industry for solid gold hub caps.

The most important factors the design engineer must keep in mind when selecting materials are: *strength, stability, weight, appearance,* and *cost.* When determining the materials from a cost standpoint, he has even to consider packaging and shipping costs, which are often factors in controlling the design of a product.

Materials on the Blueprint. The designer's final decisions on the choice of materials are stated on the blueprint in the Bill of Materials. In the detail drawings derived from the blueprint, there will be a short note concerning the material required for that particular component. Students should always check the prints in this book for the note that specifies the material.

For example, take a look at the note in Fig. 1 (Detail 9). What does the note mean?

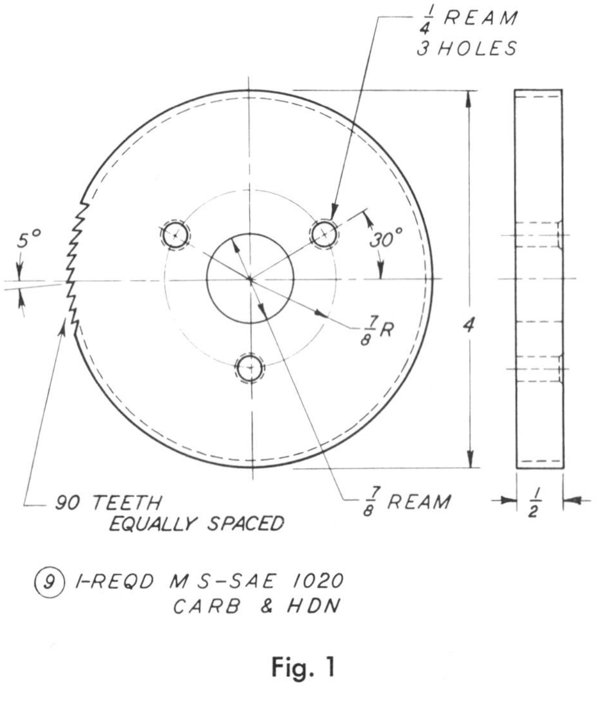

Fig. 1

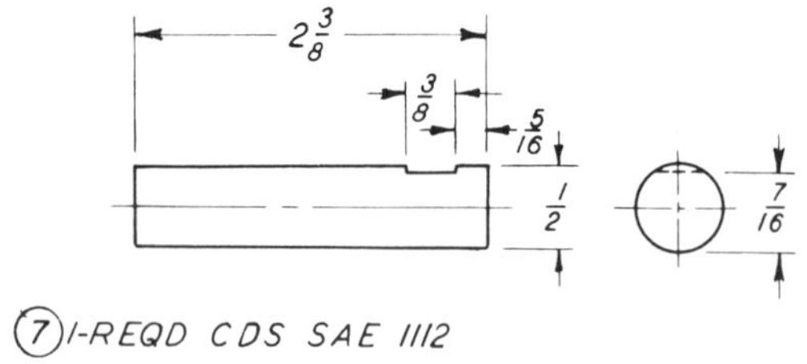

Fig. 2

"1—REQD" means that only one part of this particular shape and function is needed in the completed assemblies. "M.S." means that the part should be made of mild steel. (See Appendix F for "List of Common Abbreviations" where shortened forms like "M.S." are given.) "S.A.E. 1020" means that this particular steel type is classified by the Society of Automotive Engineers in their numbering system for steels as Number 1020. In this system, the first digit (1) indicates that S.A.E. 1020 is a plain carbon steel; the second digit (0) indicates that no other alloying agent is present in it; and the last two digits (20) indicate that it has a carbon content of .15 to .25%. "CARB. & HDN." refers to the heat treatment and means that the part should be carburized and hardened after it is machined.

The note under Detail 7 of the same print designates a different material. See Fig. 2. This means that this component should be made from cold drawn steel ("C. D. S.") of the type classified in the numbering system of the Society of Automotive Engineers as Number 1112.

Here, then, are two component parts of an automatic oiler to be made from two different steel types. The designer's reasons for specifying these materials can be guessed if you examine the Classification of Carbon and Alloy Steels (Table 1).

Under "Special Characteristics" you find that S.A.E. 1020 is very tough but has only fair machining qualities, while S.A.E. 1112 has excellent machining characteristics. It is for such reasons that the designer makes his selection, always seeking the exactly appropriate characteristics for the part to be made. A study of Table 1 will help you under-

TABLE 1. CLASSIFICATION OF CARBON AND ALLOY STEELS*

TYPE OF STEEL	SAE No.	SPECIAL CHARACTERISTICS	COMMON USES
CARBON	1010	Low tensile strength Machines rough	Welding Steel, Sheet Iron, Tacks, Nails, Etc.
	1020	Very tough Machines fair	Fan Blades, Sheet Steel, Pipe, Structural Steel
	1030	Heat treats well Stronger than 1020	Seamless Tubing, Shafting, Gears
	1040	Heat treats average Machines fair	Auto Axles, Bolts, Connecting Rods
	1045	Thin sections must be carefully quenched	Coil Springs, Auger Bits, Screw Drivers
	1055	May be oil tempered	Miscellaneous Coil Springs
	1060	Soft tool steel Does not hold edge	Valve Springs, Lock Washers, Non-edged Tools
	1070	Stands severe shocks Very tough and hard	Wrenches, Anvils, Dies, Knives
	1080	Holds edge well Medium hard	Shovels, Hammers, Chisels, Vise Jaws
	1085	Hard tool steel	Knife Blades, Auto Bumpers, Taps, Saws
	1090	Very hard tool steel Thin edges will be brittle	Coil and Leaf Auto Springs, Taps, Hack-saw Blades, Cutters
FREE CUTTING	1112	Excellent machining characteristics	Screw Machine Stock, Studs, Screws, Bolts
	1115	Stronger and tougher but machines slowly	Above items where more strength is required
	X1314	Machines well, case hardens very well	Used where surface hardness and strength is desired
	X1330	Machines very well	Used where better quality hardness is desired

(continued on next page)

*Adapted from Machine Shop Operations and Setups (American Technical Society)

TABLE 1. (continued)

TYPE OF STEEL	SAE No.	SPECIAL CHARACTERISTICS	COMMON USES
MANGANESE	1330	Withstands hard wear, hammering and shock	Burglar Proof Safes, Railroad Rails (curved)
NICKEL-CHROMIUM	3115	Very hard and strong	Armor Plate
	3130	Very hard and strong	Gears
	3140	Very hard and strong	Springs, Axles
	3150	Very hard and strong	Shafts
MOLYBDENUM	4130	Withstands high heat and hard blows	Very Fine Wire, Ball and Roller Bearings, High Grade Auto and Machinery Parts
	4140		
	4150	Same as above	
	4320	Same as above	
	4615	Same as above	Same as above
	4815	Same as above	Same as above
CHROMIUM	5120	Hard and tough	Burglar Proof Safes, Springs, Cutting Tools, Bolts and Rollers for Bearings
	5140	Resists rust, stains and scratches	
	52100	Hard and very tough	Same as above
STAINLESS-CHROMIUM	51210	Not heat treatable	Stainless Cooking Utensils, Sinks
	51310	Can be heat treated	Same as above
	51335	Can be heat treated, does not corrode	Stainless Steel Cutlery, Tableware, Ball Bearings
	51710	Same as above	Same as above

stand why a steel of a particular type and composition is specified in the blueprints you work from.

The American Iron and Steel Institute ("A.I.S.I.") has a numerical index system which is almost identical to that of S.A.E. Occasionally on a blueprint, steel will be referred to by a trade name, as in Print 115 in this book, where you will find under Detail 1 a note that reads:

KETOS STL HDN & GRD

"Ketos" is the trade name of a particular type of tool steel. Generally, tool steels are manufactured by specialized firms, and are sold largely under trade names.

The two other principal types of ferrous metal you will find specified on blueprints are cast iron and wrought iron. Cast iron is a high-carbon alloy (containing over 1.7% carbon), and wrought iron is a low-carbon alloy (containing 0.02% carbon). Steel lies in between them in carbon content and is classified as follows:

Low-carbon steel—
Under 0.30% carbon
Medium-carbon steel—
0.30 to 0.70% carbon
High-carbon steel—
0.70 to 1.70% carbon

The percentage of carbon in steel is the most important single factor that governs its properties and uses.

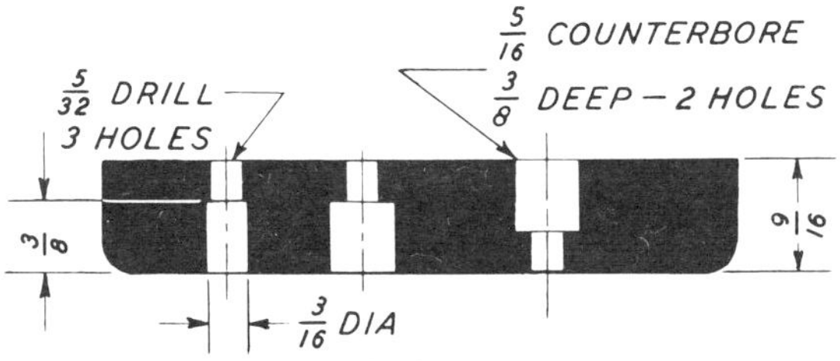

Fig. 3

The non-ferrous metals and the plastics are also listed in the Bill of Materials and in notes on detail drawings by number systems, material designations, or trade names. To take an example, the wrought aluminum alloy frequently used for cooking utensils and sheet metal work has the designation 3003-H18.

Plastics are designated either by the type name of the compound (see Fig. 3) or by a trade name. Because plastics are increasingly used in the Machine Trades, Table 2 gives a summary of the characteristics and typical uses of some common plastics.

Methods of Fabrication. The first operation for a Machine Tradesman on starting on a piece of work is to check it with the blueprint or working drawing to make sure the workpiece is the one specified on the print. For this reason, it is worth while having some information about the methods of fabrication of workpieces, so that at least the terms used will be familiar to you.

Casting. Casting, or founding, consists of giving shape to the piece by pouring the material in its fluid state into a mold. The three principal methods of casting are; sand casting, permanent molding, and die casting.

Cast iron, ductile iron, malleable iron, steel, aluminum, and brass are the materials usually fabricated as sandcastings. The process requires the use of a pattern which is constructed either of wood or metal and is a duplicate of the piece to be made, except that it is slightly larger to allow for the shrinkage of the poured metal as it solidifies. A shrinkage allowance has been determined for each metal. These allowances range from 1/8″ per foot for cast iron to 5/32″ per foot for aluminum. The amount of shrinkage also depends, to a certain extent, upon the shape and size of the casting.

Aluminum and some grey iron castings are made in permanent molds. This provides a more economical way of casting than the usual sand cast method. The metal molds for grey iron are constructed of heat-resistant alloy and coated with an insulating medium after each casting is ejected. This method of casting grey iron is confined to small castings, and even then only to those with high volume requirements.

Permanent molds for aluminum castings are made either of steel or Meehanite metal. Usually they consist of two halves in which cavities have been machined to conform to the shape of the piece to be made. The mold halves are mounted on a hydraulic fixture and clamped together while the molten metal is poured in. When the metal has solidified, the molds are separated and the casting is ejected. The advantages of this method of casting aluminum are lower costs resulting from increased production rate, closer dimensional control, improved appearance, and an increase in physical properties.

Zinc and many aluminum alloys are easily die cast. The dies are made of an alloy steel and hardened to prevent wear. In the casting operation the dies are locked together, and the molten metal is forced into the die cavity, making it possible to cast thinner sections.

Forging. Forging is the process of forming material into shape by pressing or hammering it. The metal is usually heated to a plastic state, but some metals are soft enough to be forged without heating. Steel, aluminum, brass, and titanium are the metals most frequently fabricated by this method.

Closed die forging is by far the most widely used method of forging, and requires the use

TABLE 2. CHARACTERISTICS OF COMMON PLASTICS

TYPE	CHARACTERISTICS	TYPICAL USES
PHENOL FORMALDEHYDE Wood flour filled Rag filled	Wide range of Properties Inexpensive	Housings Knobs Laminated Products Cast Products
CELLULOSE NITRATE	Fast Burning Difficult to mold	Fountain Pens Brush Handles Spectacle Frames
CELLULOSE ACETATE	Slow Burning Good Machining Qualities Tough	Wide range of applications
UREAS	Durable Good Strength Water Resistant	Plywood Lighting Fixtures Machine Housing
METHACRYLATES	High Priced High Transparency	"Lucite," "Plexiglass" Transparent Enclosures Lenses
POLYSTYRENE	Exceptional Clarity Good Molding	Refrigerator Parts Closures Housings Toys
POLYAMIDES	High Priced Good Abrasion Resistance	"Nylon" Bushings Gears
VINYLIDENE CHLORIDE	Outstanding Resistance to Chemicals	"Saran" Tubing Chemical Equipment
EPOXY	Excellent Adhesion To Metals	Tooling (including Jigs & Fixtures) Cast Parts

of steel dies that are operated with a drop hammer or with a mechanical or hydraulic press. The dies are made of tough heat-resistant steel and impressions are machined into them which conform to the shape of the part to be forged. All forging operations tend to improve the properties of the metal, and when strength is a factor this justifies the increased cost of this method of fabrication.

Stamping. Stamping is the technique of shaping parts from sheets, plates, strips, or bars of metal in dies operated either in hydraulic or mechanical presses. Steel, aluminum, brass, and zinc are the materials used to make metal stampings. Since metal of uniform thickness is used in the manufacture of stampings, the parts themselves can have little change in section thickness.

Machining. The shape of a part will often lend itself to machining directly from standard stock and is the most economical means of fabrication. It can be done by the many machine tools available. Bar stock, either ferrous or nonferrous, produced by rolling mills, is normally used. The most popular tool for high production requirements is the automatic screw machine. This machine will produce parts requiring turning, drilling, boring, and threading, and can perform all these operations simultaneously.

Injection Molding. The basic method of fabricating thermoplastic materials is by injection molding. This method requires the use of hardened steel molds in which cavities have been machined and highly polished. These cavities are duplications of the parts to be made except for the allowance of shrinkage of the material to be molded. This allowance varies from .004″ to .025″ per inch depending on the type of plastic used. Provisions are made in the molds for circulating water to maintain the proper temperature. The molds are operated in either a vertical or horizontal press, and they are held firmly together during the molding operation. The plastic, in a granular state, is fed into a heating chamber. The temperature of this chamber is regulated to a level that causes the granular material to plasticize. In this state it is forcibly injected into the mold cavity and allowed to harden before being removed.

Extrusion. Certain non-ferrous metals that include aluminum, brass, bronze, and copper are readily fabricated by the extrusion process. This requires a metal ingot which is usually cast to a specific shape to fit the chamber of the extrusion press. The ingot is first heated to its plastic state and then forced through an aperture of the required shape. Copper water tubing and structural shapes of aluminum are two of the principal products produced by this means. The exterior surface of the extruded part has a very fine finish and can be made to close tolerances. The pressure required to perform the extruding operation also causes a substantial improvement in the physical properties of the material.

Heat Treatment. This is a general term which covers the processing of metal by heat and chemicals to change the physical properties of the material. Heat treatment processes include annealing, carburizing, casehardening, hardening, normalizing, and tempering. Appendix E, "Glossary of Common Machine Trade Terms," gives definitions of the most common heat treatment processes.

REVIEW OF BASIC PRINCIPLES

This is an appropriate time to review some of the most important principles of blueprint reading that you have learned so far. A constant awareness of these principles and what they mean will go a long way in improving your ability to read and interpret blueprints. We will briefly review each of these principles which have been dealt with more thoroughly in the preceding chapters.

Blueprints in Production. A production blueprint is a drawing of an object which provides complete and precise information necessary for the production of that object. In effect, then, blueprints are *instructions to be followed* by people who will be concerned with the production of the object. Basically blueprints convey these instructions in two ways: (1) by providing views of the object, showing its exact shape and size; and (2) by providing written dimensions, notes, and symbols, which give information not shown in the views.

Visualizing Blueprints. One of the basic skills of blueprint reading is the ability to *visualize* an object from the views of it given on the blueprint. Visualization means nothing more than the ability *to form a mental picture of an object*. In blueprint reading, a mental picture of an object is formed by studying the different views of an object and relating them in your mind in such a manner that, when properly related, they give you an accurate image of what the object would actually look like. Visualization is, therefore, of great importance in blueprint reading as a means for understanding the general shape and structure of an object.

Understanding Views. The ability to understand the different views on a blueprint goes hand in hand with the ability to visualize the object. In other words, the better you can visualize a blueprint, the better you can understand the various views, and vice versa. We learned from Chapter 3 that, in effect, the different views on a blueprint allow you to look at the object from many sides at once. This allows you to study and understand the relationships between the different dimensions and parts of the object. This advantage is particularly important in the production or assembling of a complicated object.

Working Drawings. You learned in Chapter 4 that a working drawing provides the complete information necessary for the production of an object. You learned also that a working drawing is made up basically of three parts: (1) lines which make up the views of the object, (2) dimensions which indicate the size of the object, and (3) notes and symbols which provide additional information necessary for the production of the object. To understand working drawings, therefore, you should understand the "alphabet of lines" and what each type of line indicates; you should understand the different types of dimensional markings, such as decimal dimensions and angular dimensions; finally, you should understand the meaning of the various types of notes and symbols that are commonly found on blueprints, such as thread, finishing, and material symbols.

Supplementary Information. Supplementary information includes special notes and symbols other than those mentioned above. These include many of the "finer points" of blueprint reading which are, nevertheless, necessary to the understanding of how a particular object should be produced. This aspect of blueprint reading includes such details as special machining operations, revision notes, bills of material, scaling and reference to standard specification sheets. These details are so important to the machine tradesmen that the failure to take them into account on a production job may well mean scrapping the work.

Special Views and Assembly Drawings. This aspect of blueprint reading concerns special methods of representing objects to convey information not adequately conveyed by the ordinary methods of drafting. Mostly, these include section views, detail views, auxiliary views, and assembly drawings. Gear drawings are also included in this heading since much of the information necessary to the production of gears is not shown directly on the drawing. This information, however, can be found by studying the notes and gear specifications on gear drawings. On the basis of these notes and specifications, you can refer to the tables of gear formulas and then compute the precise dimensions that are required. The table of spur gear formulas is found on page 131 while the tables of bevel gear formulas are found on pages 136-137.

CHAPTER 8

Actual Production Blueprints

In this chapter we have included an assortment of thirty-seven prints, of which the majority are actual production blueprints used in the Machine Trades. These prints are typical in that they include assembly drawings, gear drawings, and many other drawings of precise machine parts which require a number of machining operations. They are provided to give you further practice in blueprint reading and to illustrate the various elements of blueprints which we have dealt with.

It is suggested that the material in this chapter be assigned by the instructor to adapt to specific teaching schedules and conditions. The Trade Competency Tests on the prints may therefore be used either as self-test exercises or as materials for examination. The large number of prints and the variety of machine parts represented allow for considerable flexibility in adapting to course schedules and subject matter.

It is further suggested that the assembly drawings and gear drawings be assigned when the instructor feels that the student is adequately prepared for them.

A Final Note. You will occasionally find variations in the manner in which blueprints are drawn. Despite the very considerable standardization of drafting procedures, some industries use different methods of conveying information on blueprints. *So be prepared, in this book and on the job, to find prints and drawings that do not conform to the general principles outlined in this text.* Always check carefully to make sure you understand the blueprint before proceeding with the job.

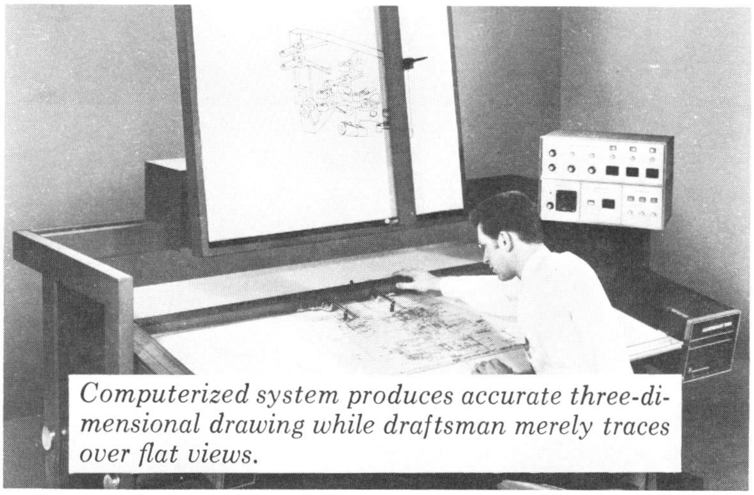

Computerized system produces accurate three-dimensional drawing while draftsman merely traces over flat views.

Perspective Systems, Inc.

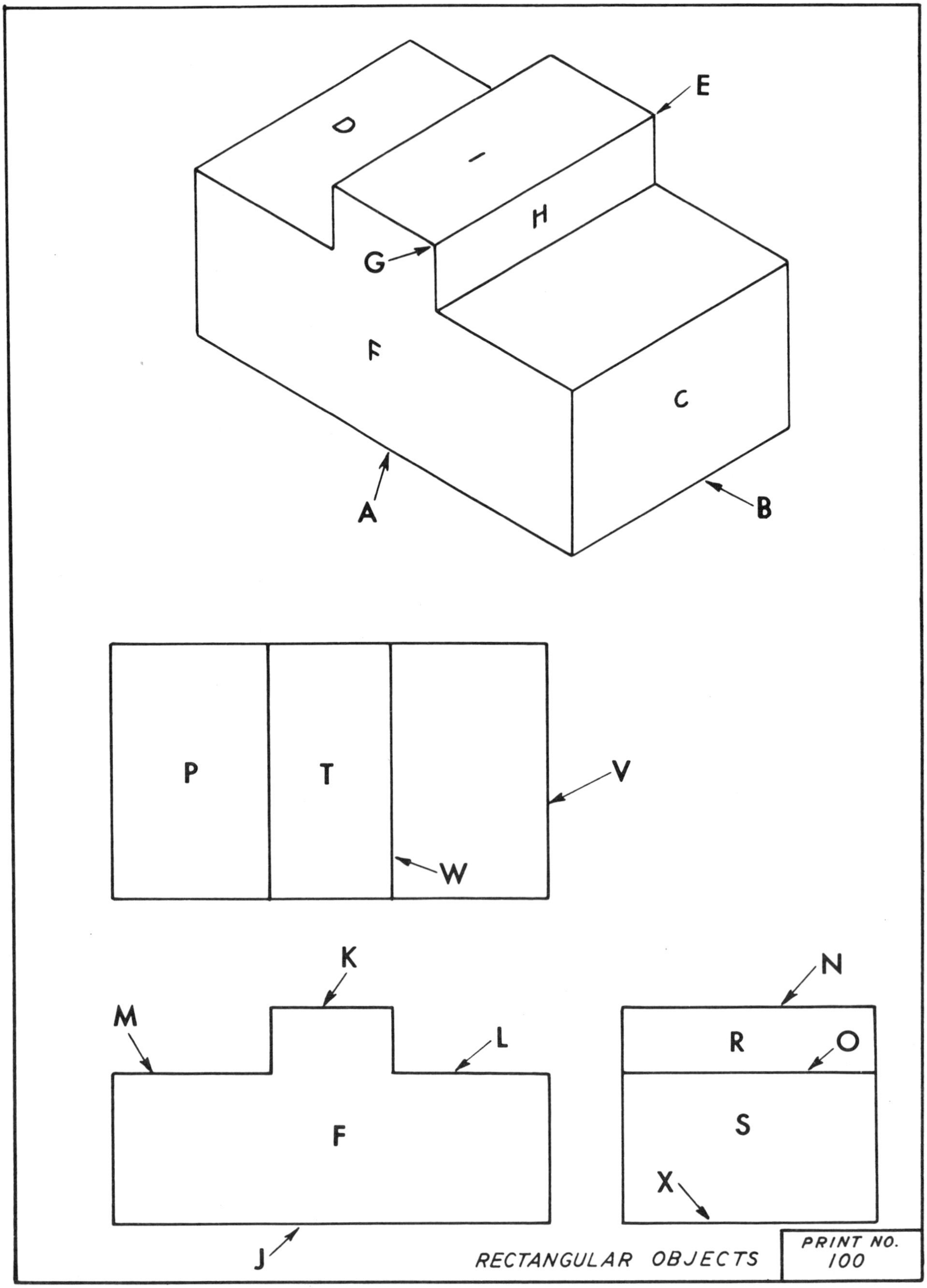
D
I
E
H
G
F
C
A
B
P
T
V
W
K
M
L
N
R
O
F
S
X
J
RECTANGULAR OBJECTS
PRINT NO.
100

TRADE COMPETENCY TEST

For Print No. 100

T E A R O F F H E R E

Student's Name____________________ Instructor's Name____________________

Questions	Answers
1. Surface C is represented by what surface in the right side view?	1. ________
2. Surface D is represented by what surface in the top view?	2. ________
3. Surface I is represented by what surface in the top view?	3. ________
4. Surface H is represented by what surface in the right side view?	4. ________
5. Surface C is represented by what line in the top view?	5. ________
6. Surface D is represented by what line in the front view?	6. ________
7. Surface I is represented by what line in the front view?	7. ________
8. Surface H is represented by what line in the top view?	8. ________
9. Surface I is represented by what line in the right side view?	9. ________
10. Surface D is represented by what line in the right side view?	10. ________
11. Edge A is represented by what line in the front view?	11. ________
12. Edge B is represented by what line in the right side view?	12. ________

TRADE COMPETENCY TEST

For Print No. 101

Student's Name____________________ Instructor's Name____________________

Questions	Answers
No. 1	
1. Give the length, height, and width dimensions of the object.	1. __________
2. Give the dimensions A, B, and C.	2. A_____ B_____ C_____
3. Surface D is represented by what line in the right side view?	3. __________
4. Surface E is represented by what line in the right side view?	4. __________
No. 2	
5. Give the length, height, and width dimensions of the object.	5. __________
6. Give dimensions for A, B, C, D, and E.	6. A_____ B_____ C_____ D_____ E_____
7. Surface F is represented by what line in the right side view?	7. __________
8. Surface G is represented by what line in the right side view?	8. __________
No. 3	
9. Give dimensions for A, B, C, D, and E.	9. A_____ B_____ C_____ D_____ E_____ F_____
10. Surface F is represented by what line in the top view?	10. __________
11. How is surface H represented in the right side view?	11. __________
12. Give the width of the ¼″ slot or groove.	12. __________

No. 4

13. From the isometric drawing make a working drawing on one of the pieces of graph paper in the back of the book. Insert all dimensions.

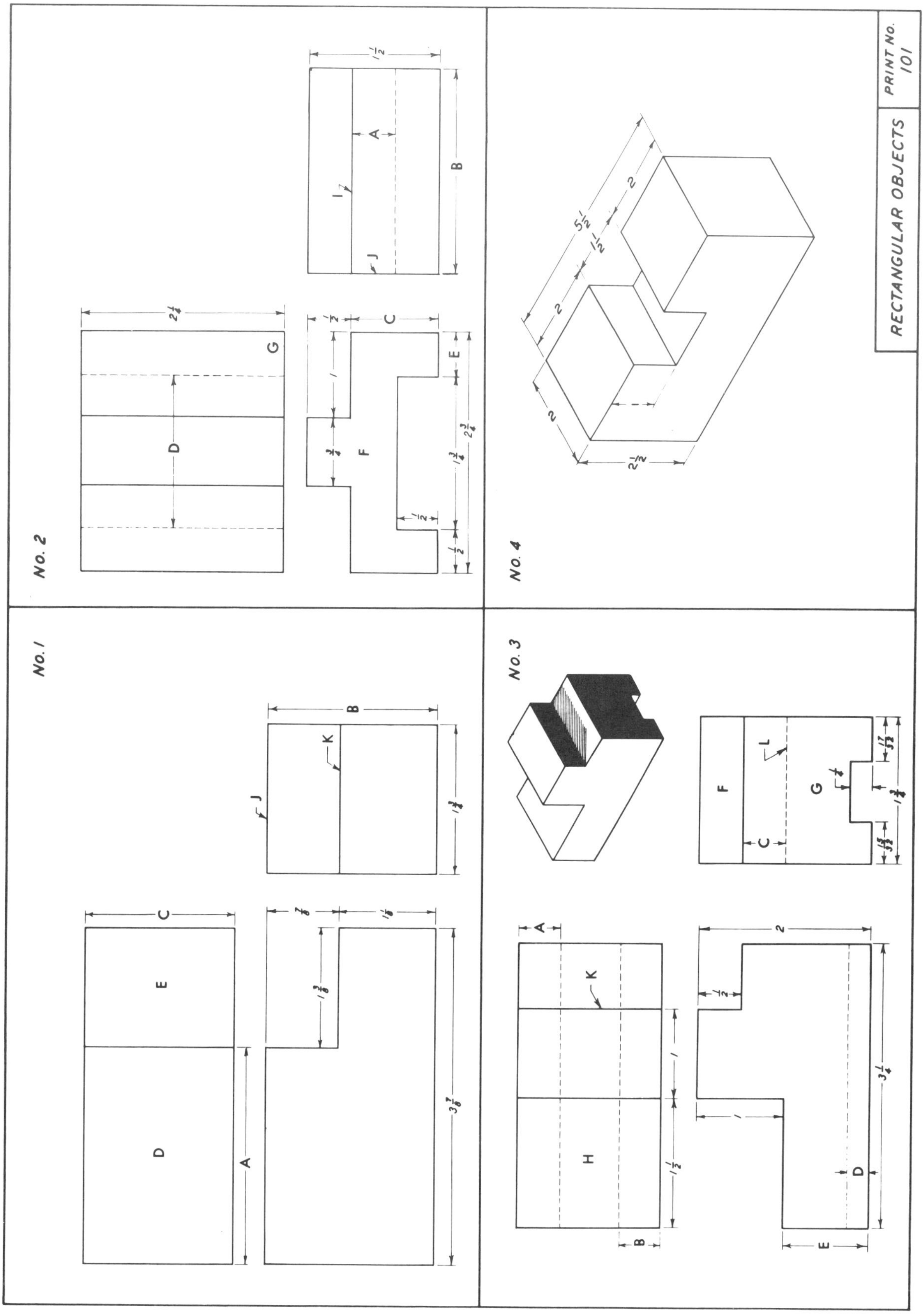
RECTANGULAR OBJECTS
PRINT NO. 101
No. 1
No. 2
No. 3
No. 4

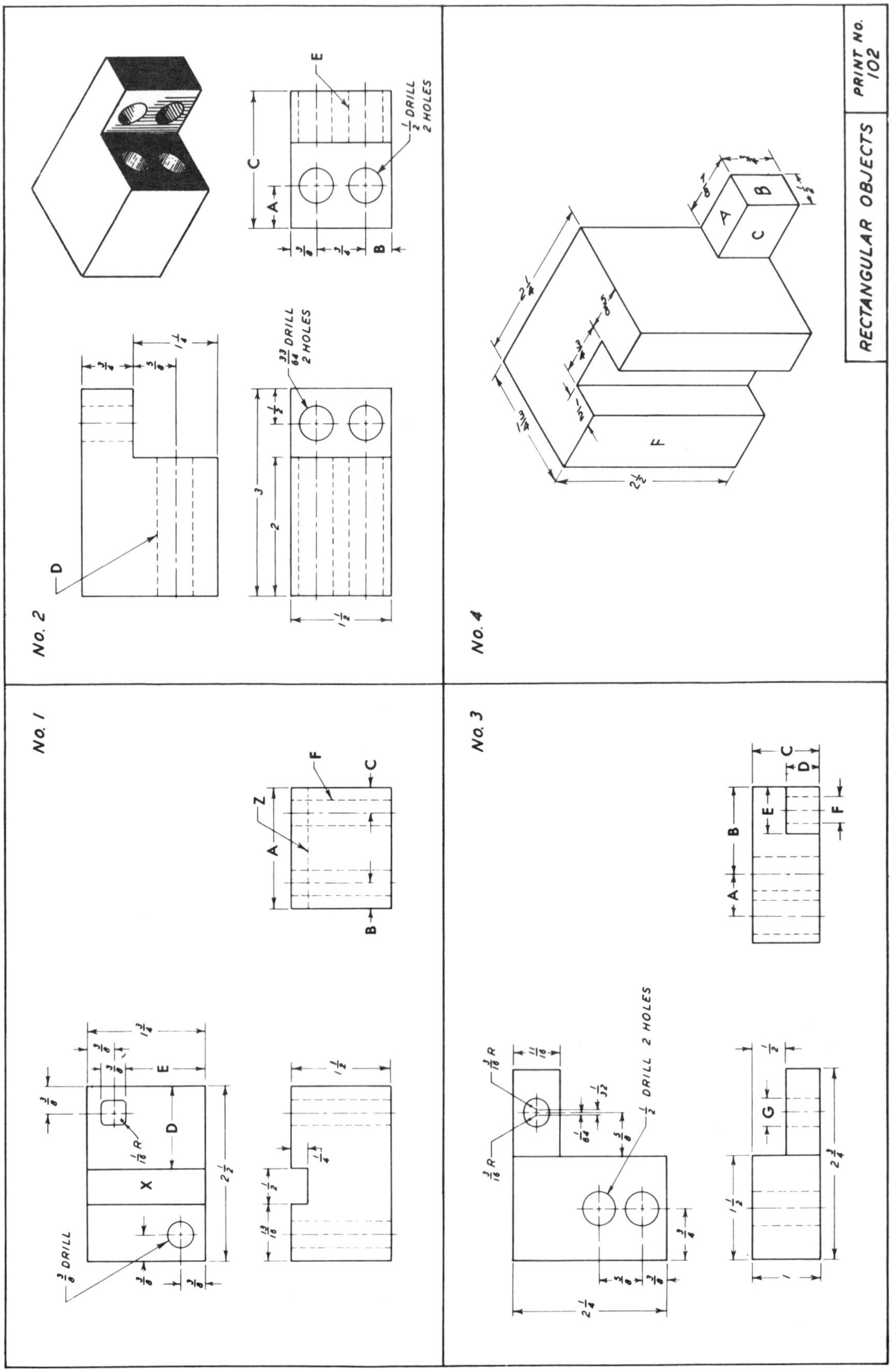
PRINT NO. 102
RECTANGULAR OBJECTS
NO. 1
3/8 DRILL
NO. 2
33/64 DRILL 2 HOLES
1/2 DRILL 2 HOLES
NO. 3
1/2 DRILL 2 HOLES
NO. 4

TRADE COMPETENCY TEST

For Print No. 102

Student's Name________________ Instructor's Name________________

Questions	Answers
No. 1	
1. Give dimensions for A, B, C, D, and E.	1. A______ B______ C______ D______ E______
2. Does dashed line indicated by F represent a round or square hole?	2. ______
3. Surface X is represented by what line in right side view?	3. ______
No. 2	
4. Give dimensions A, B, and C.	4. A______ B______ C______
5. Give diameters of holes indicated by D and E.	5. D______ E______
6. Give length of holes indicated by D and E.	6. D______ E______
7. What is the distance between the centerlines of the ½″ holes?	7. ______
No. 3	
8. Give dimensions A, B, C, D, E, F, and G.	8. A______ B______ C______ D______ E______ F______ G______
9. Give the length, height, and width dimensions of object.	9. ______
10. Give dimensions for locating the ½″ holes.	10. ______

No. 4

11. Using a piece of the graph paper in the back of the book and the isometric drawing, make a three-view working drawing of the object. Show all necessary dimensions. Place the letters shown in the isometric on the proper surfaces of the working drawing.

TEAR OFF HERE

TRADE COMPETENCY TEST

For Print No. 103

Student's Name________________________ Instructor's Name________________________

Questions	Answers
No. 1	
1. How is corner indicated by A represented in top view?	1. ________________
2. How is corner indicated by B represented in top view?	2. ________________
3. How many surfaces has the object?	3. ________________
4. How is corner indicated by B represented in the front view?	4. ________________
No. 2	
5. Give dimensions A and B.	5. A________ B________
6. Surface C is represented by what line in the right side view?	6. ________________
7. What is the smallest diameter hole drilled in the object?	7. ________________
No. 3	
8. Give dimensions A, B, and C.	8. A________ B________ C________
9. List drills necessary to machine this detail. For reamed holes choose drill 1/64″ smaller in diameter than reamer.	9. ________________

No. 4

10. Using a piece of the graph paper in the back of the book and the isometric sketch, make a three-view working drawing of the object. Show all dimensions. Place the letters shown in the isometric on the proper surfaces of the working drawing.

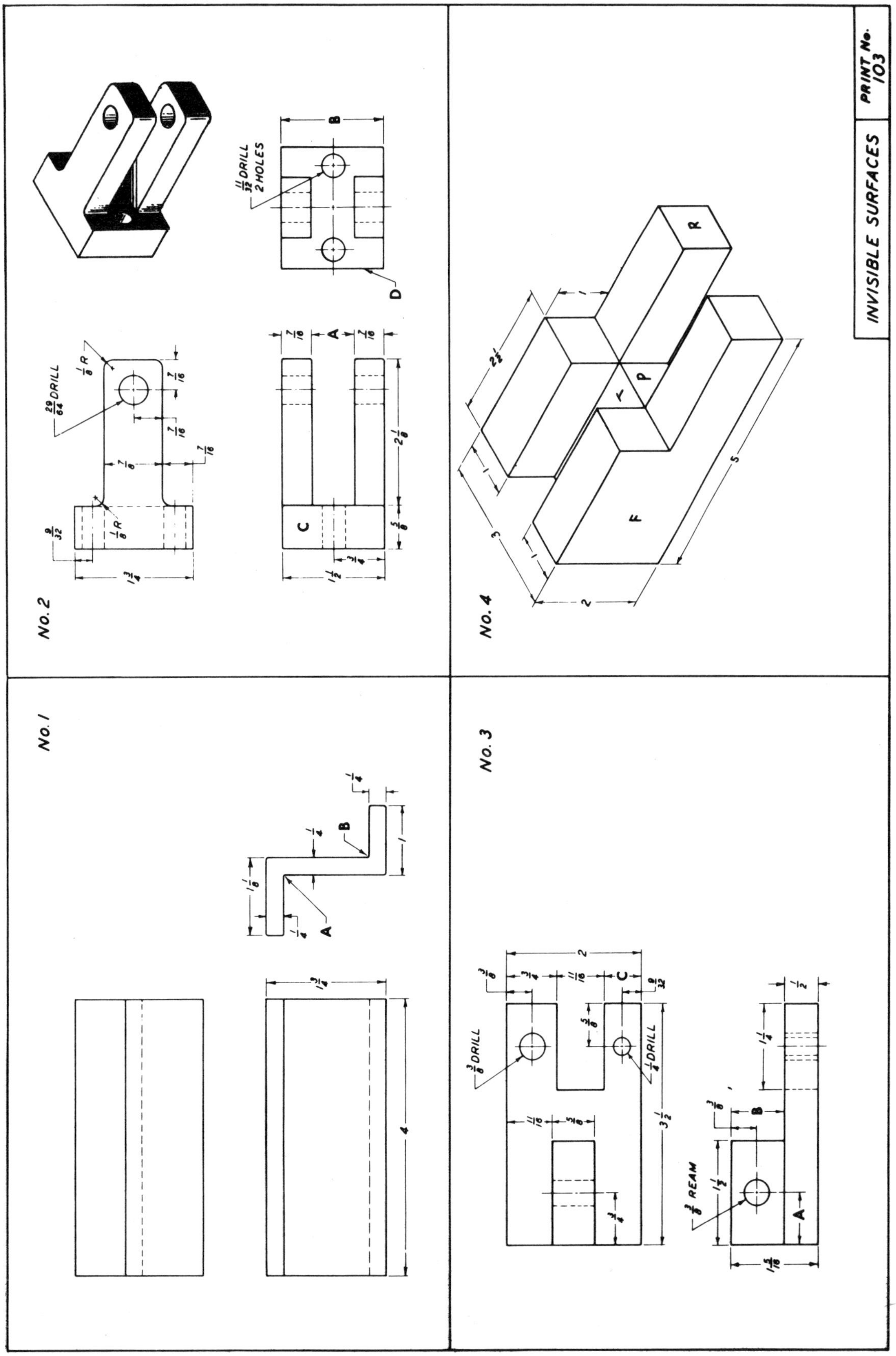
PRINT No. 103
INVISIBLE SURFACES
No. 1
No. 2
No. 3
No. 4
29/64 DRILL
11/32 DRILL 2 HOLES
3/8 DRILL
1/4 DRILL
3/8 REAM

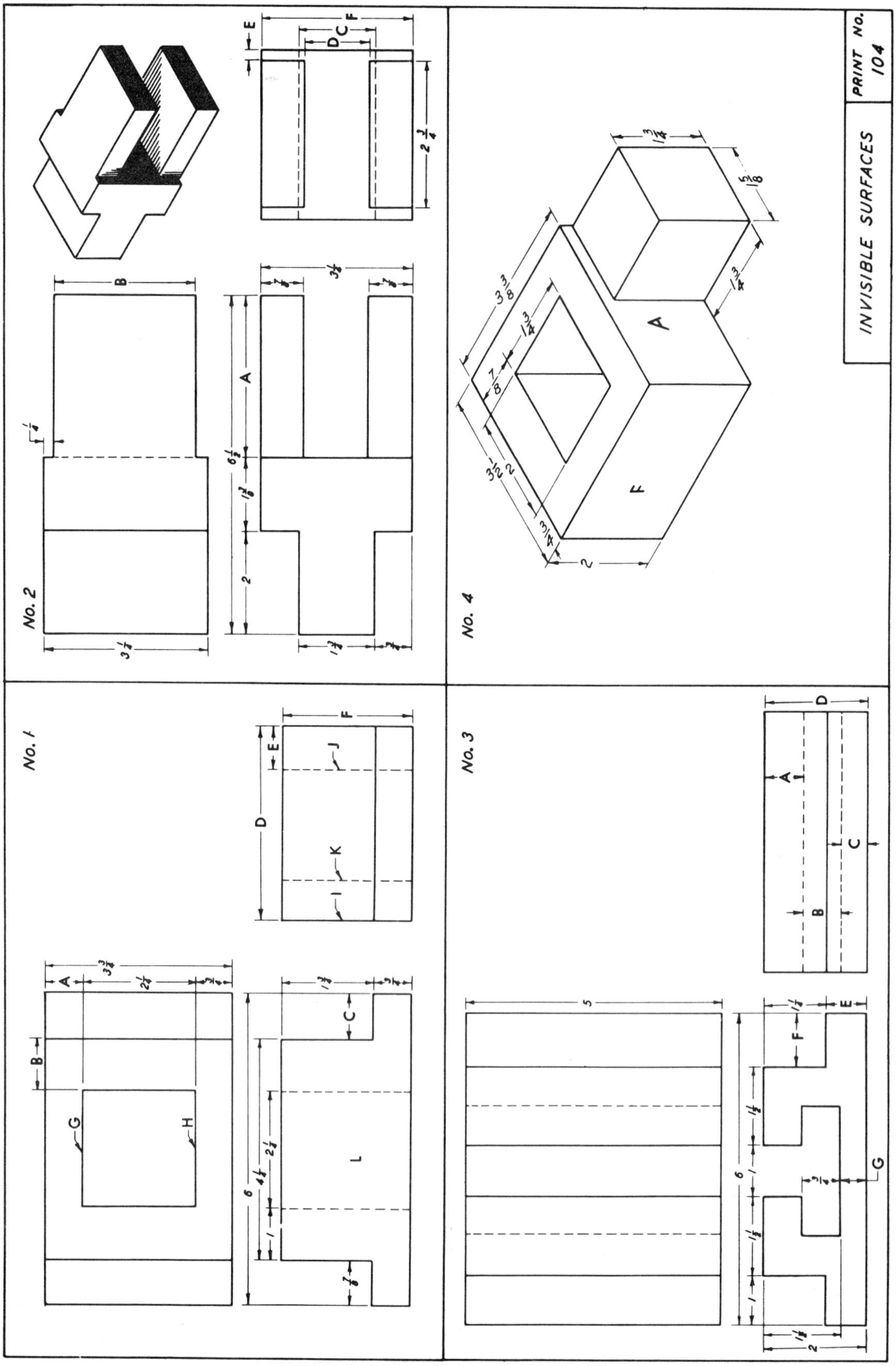
PRINT NO.
104
INVISIBLE SURFACES
No. 1
No. 2
No. 3
No. 4

TRADE COMPETENCY TEST

For Print No. 104

Student's Name____________________ Instructor's Name____________________

T E A R O F F H E R E

Questions	Answers
No. 1	
1. Give dimensions A, B, C, D, E, and F.	1. A____ B____ C____ D____ E____ F____
2. Line G is represented by what line in the right side view?	2. ________
3. Line H is represented by what line in the right side view?	3. ________
4. Surface L is represented by what line in the right side view?	4. ________
5. Give the width and depth of the square hole.	5. ________
No. 2	
6. Give dimensions A, B, C, D, E, and F.	6. A____ B____ C____ D____ E____ F____
7. Give the length, width, and height dimensions of the object.	7. ________
No. 3	
8. Give dimensions A, B, C, D, E, F, and G.	8. A____ B____ C____ D____ E____ F____ G____
9. Give the length, width, and height dimensions of the object.	9. ________
10. How many views are necessary to describe this object?	10. ________
No. 4	
11. Give the size of the rectangular hole.	11. ________

12. Using a piece of the graph paper in the back of the book and the isometric drawing, make a three-view working drawing of the object. Show all dimensions. Place the letters shown in the isometric on the proper surfaces of the working drawing.

TRADE COMPETENCY TEST

For Print No. 105

Student's Name____________________ Instructor's Name____________________

This print has four objects that have slant surfaces. Object #2 illustrates that slant surfaces can be measured by degrees and inches, or by inches alone. The projection indicated by "H" in the right side view in #2 is called a shank. The $1\frac{1}{4} \times 1 \times \frac{1}{4}$ projection in #4 is referred to as a tongue. These projections provide the means of holding tools in place on a machine or fixture and permit easy removal.

Questions	Answers
No. 1	
1. Give dimensions A, B, and C.	1. A_____ B_____ C_____
2. Does dimension C measure true length of slant surface?	2. __________
3. Explain why.	3. __________
4. Surface X is represented by what surface in right side view?	4. __________
5. Surface G is represented by what line in right side view?	5. __________
No. 2	
6. Give dimensions A, B, C, and D.	6. A_____ B_____ C_____ D_____
7. Surface E is represented by what surface in the right side view?	7. __________
8. Surface G is represented by what line in the right side view?	8. __________
9. Surface H is represented by what line in the top view?	9. __________
10. How many slant surfaces has the object?	10. __________
No. 3	
11. Draw right side view. Use graph paper in back of book.	11. __________
No. 4	
12. Give dimensions A, B, C, and D.	12. A_____ B_____ C_____ D_____
13. Draw right side view. Use graph paper in back of book or ordinary drafting paper.	13. __________
14. What is the length dimension?	14. __________
15. From what view did you get the length dimension?	15. __________
16. How thick is the left end of the object?	16. __________

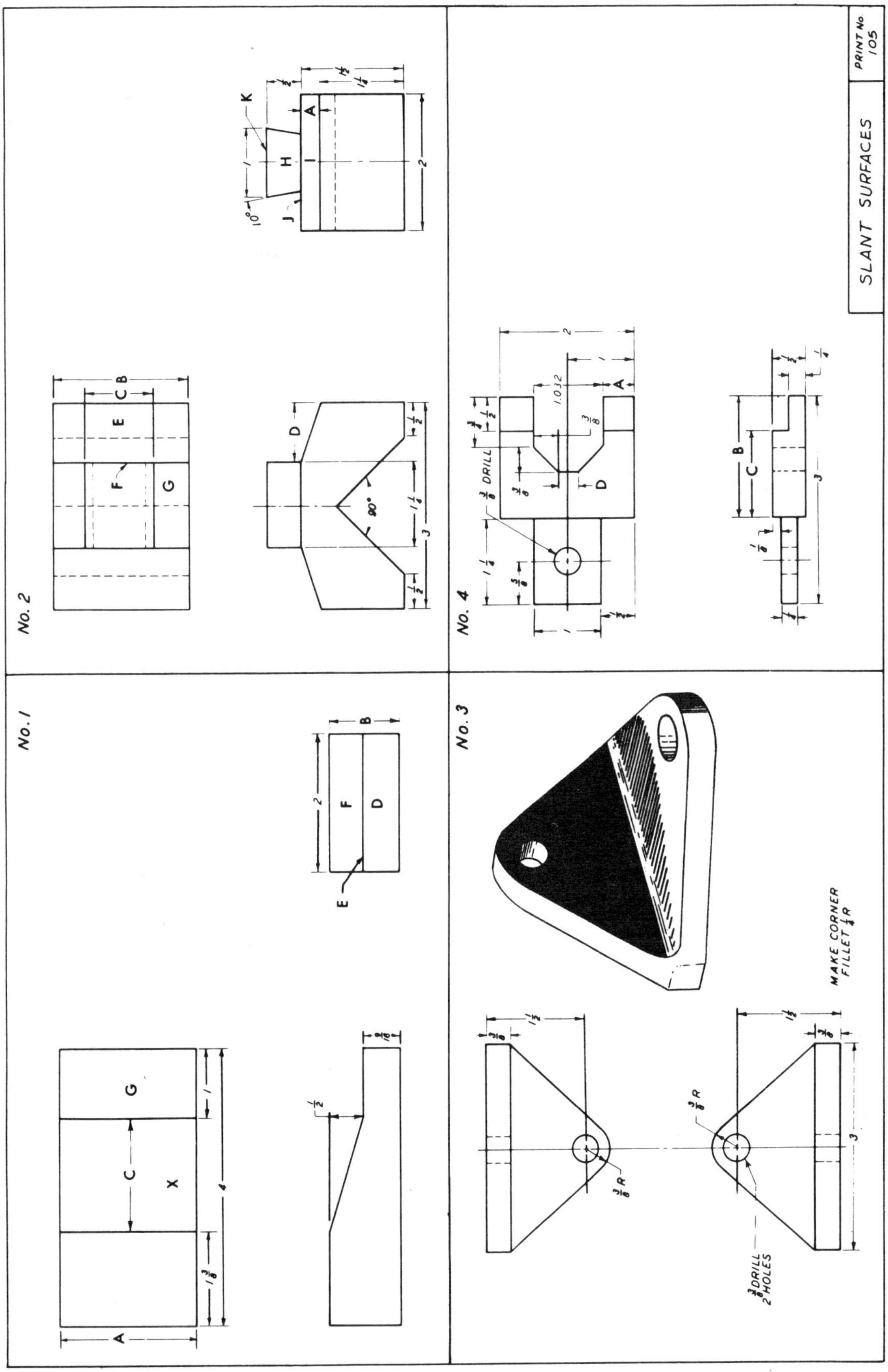
PRINT No 105
SLANT SURFACES
No. 1
No. 2
No. 3
No. 4
MAKE CORNER FILLET 1/4 R
3/8 DRILL 2 HOLES
3/8 DRILL

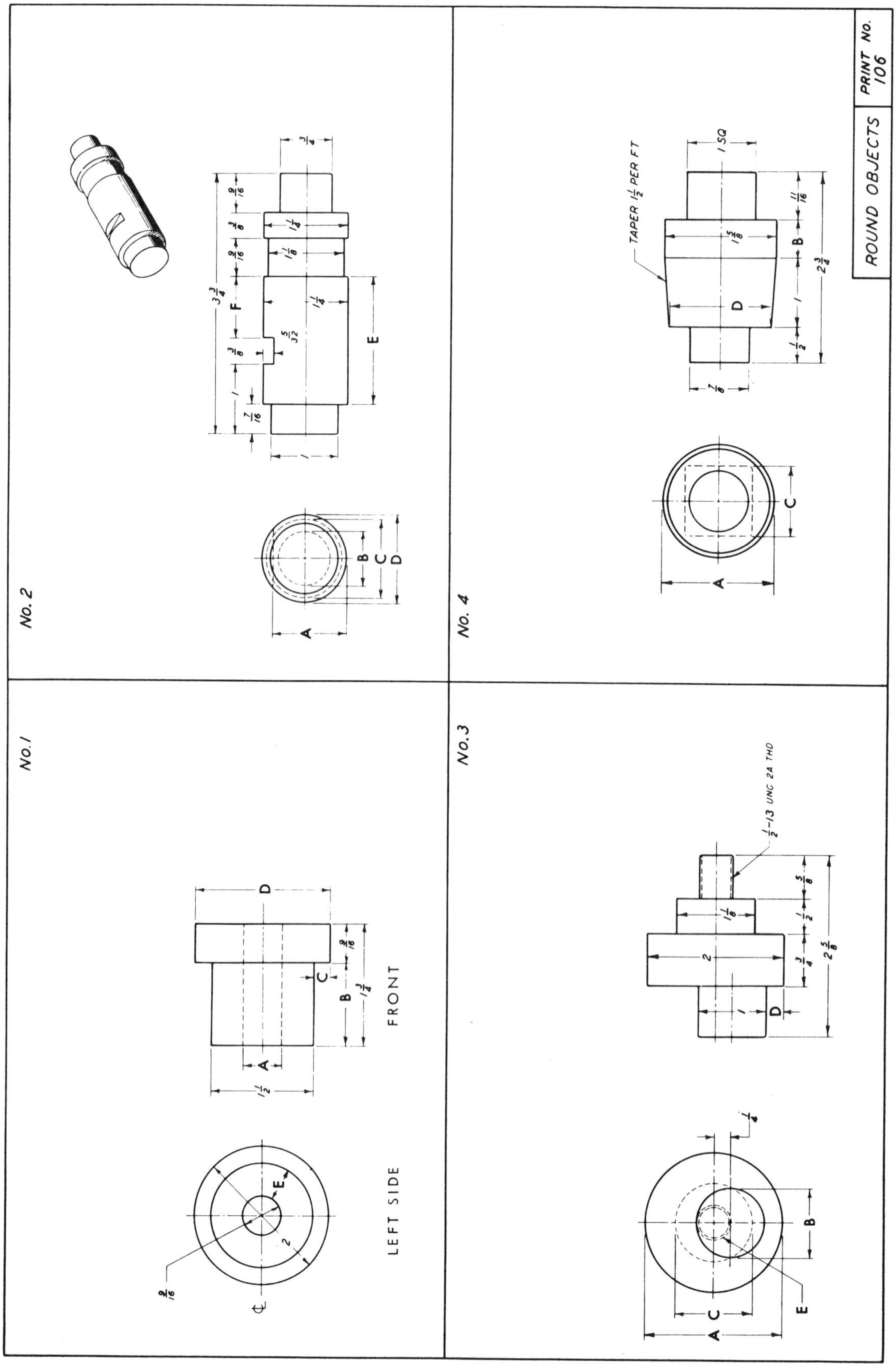
PRINT NO. 106
ROUND OBJECTS
No. 1
FRONT
LEFT SIDE
No. 2
No. 3
1/2-13 UNC 2A THD
No. 4
TAPER 1/2 PER FT
1 SQ

TRADE COMPETENCY TEST

For Print No. 106

Student's Name____________________ Instructor's Name____________________

TEAR OFF HERE

Questions	Answers
No. 1	
1. Give dimensions A, B, C, D, and E.	1. A_____ B_____ C_____ D_____ E_____
2. Can you tell from the front view whether the object is round or rectangular?	2. _____
No. 2	
3. Give dimensions A, B, C, D, E, and F.	3. A_____ B_____ C_____ D_____ E_____ F_____
No. 3	
4. Give dimensions A, B, C, and D.	4. A_____ B_____ C_____ D_____
5. What do dashed lines indicated by E represent?	5. _____
6. How many eccentric circles are shown?	6. _____
7. How many concentric circles are shown?	7. _____
No. 4	
8. Give dimensions A, B, and C.	8. _____
9. Give the over-all size of the object.	9. _____
10. From the specified taper, calculate dimension D.	10. _____

The objects shown in Print 106 are machine parts on which the lathe and milling work has been completed. The operation sheet for these parts lists the following material and sequence of operation:

1. Material: 2″ diameter cold drawn steel bar 2″ long. Machine: 10″ lathe. Center with #8 drill. Drill 9/16″ hole; turn 1½″ diameter; face end; reverse in chuck and face large end.
2. Material: 1¼″ cold drawn bar, 4″ long. Machine: 10″ lathe. Turn ¾″ and 1⅛″ diameters and face end. Reverse in chuck, turn 1″ diameter and face end. Vertical mill: mill slot with ⅜″ diameter end mill.
3. Material: 2¼″ round SAE 1045 Steel, 2¾″ long. Machine: 10″ lathe. Face ends to length. Drill press as machine to drill centers, #12 center drill for concentric and eccentric diameters. Lathe: Turn 1″ eccentric diameter. Reverse, place on concentric centers, finish turn. Turn threads.
4. Material: 1¾″ diameter x 3″ die steel. Machine: 10″ lathe. Turn 1⅝″ diameter 1½″ long. Face end. Reverse in chuck, turn taper. Turn ⅞″ diameter. Face end. Machine: horizontal shaper. Machine 1″ square.

TRADE COMPETENCY TEST

For Print No. 107

Student's Name____________________ Instructor's Name____________________

The use of cores and the method of molding impose few limitations in the design of castings and, therefore, they are frequently irregular in shape. The use of cores reduces machining costs and weight of the finished part. The hole in the Bearing Bracket in #2 is cored 1″ in diameter and, therefore, eliminates the drilling operation prior to boring. The base shown in #3 illustrates the weight saving possibilities by the use of castings.

Questions	Answers
No. 1	
1. Give dimensions for A, B, C, and D.	1. A______ B______ C______ D______
2. Give dimensions for E, F, and G.	2. E______ F______ G______
3. What is the length of the bore?	3. ______
No. 2	
4. Give dimensions for A, B, C, D, E, and F.	4. A______ B______ C______ D______ E______ F______
5. What is radius J?	5. ______
6. Surface G is represented by what line in the right side view?	6. ______
7. Surface H is represented by what line in the right side view?	7. ______
8. What view shows the shape of the base of the object?	8. ______
No. 3	
9. Give the distance between the centerlines of the holes in the top of the object.	9. ______
10. Give dimensions for A, B, C, D, and E.	10. A______ B______ C______ D______ E______

No. 4

11. Using the graph paper in the back of the book or ordinary drafting paper, make a full size top view of the object.

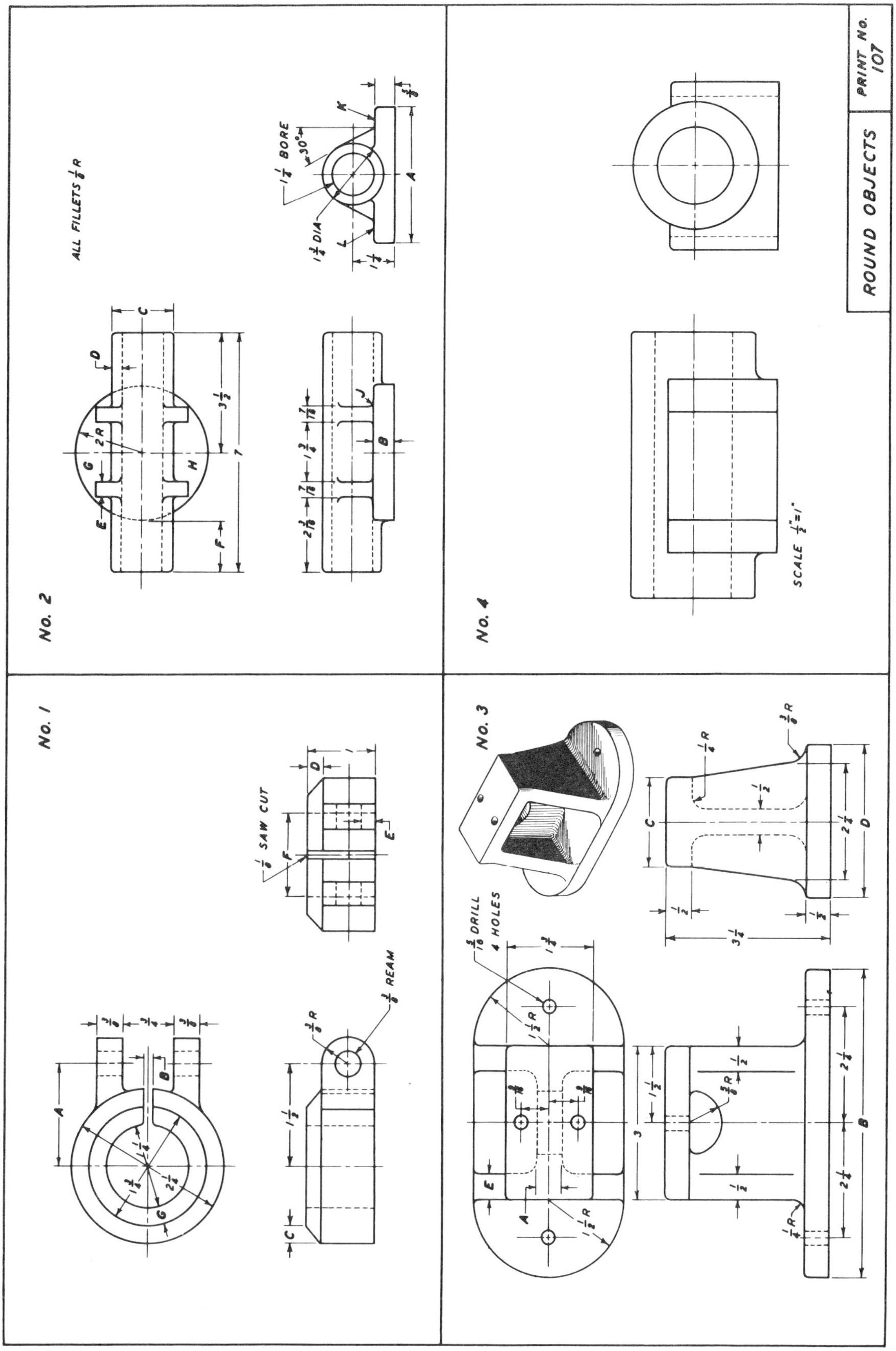
PRINT NO. 107
ROUND OBJECTS
NO. 1
NO. 2
NO. 3
NO. 4
ALL FILLETS 1/8 R
SCALE 1/2"=1"
1/8 SAW CUT
1/8 REAM
3/16 DRILL 4 HOLES
1 1/4 BORE
1 1/2 DIA

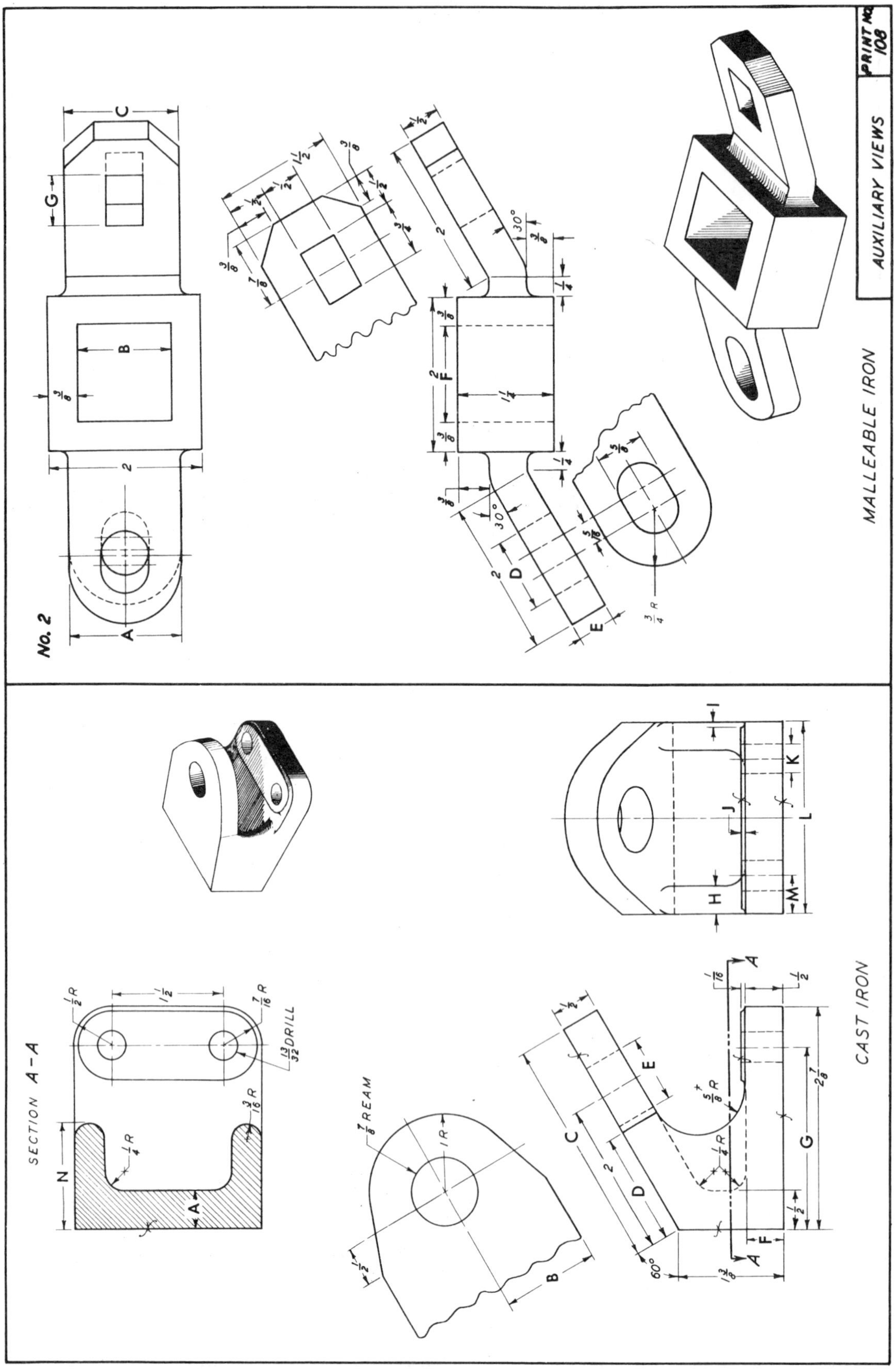
PRINT NO. 108
AUXILIARY VIEWS
No. 2
MALLEABLE IRON
CAST IRON
SECTION A-A
$\frac{13}{32}$ DRILL
$\frac{7}{8}$ REAM

TRADE COMPETENCY TEST

For Print No. 108

Student's Name________________ Instructor's Name________________

Auxiliary views are aids in visualizing an object because they carry out the principle of projecting a view of an object in its true shape. You will note the top flange of the casting in #1 could not be shown or dimensioned accurately in any view other than the auxiliary view. The "f" denoting finish indicates to the pattern maker that material must be added to these surfaces of the casting to allow for machining. For iron or aluminum castings of this size, the amount would be approximately ⅛″. Steel castings would require slightly more finish because of a typically rough surface and more difficulty in maintaining dimensional control.

TEAR OFF HERE

Questions	Answers
No. 1	
1. Give dimensions for A, B, C, D, E, and F.	1. A______ B______ C______ D______ E______ F______
2. Give dimensions for G, H, I, J, K, L, and M.	2. G______ H______ I______ J______ K______ L______ M______
3. Is dimension N given on the drawing?	3. ______
4. What symbol indicates which surfaces of this object are to be finished?	4. ______
5. Name the five views on this drawing.	5. 1.______ 2.______ 3.______ 4.______ 5.______
6. What material do the cross hatched sections represent?	6. ______
No. 2	
7. Give dimensions for A, B, C, D, E, and F.	7. A______ B______ C______ D______ E______ F______
8. Does the dimension G measure true length of slot?	8. ______
9. What is the size of the rectangular hole in the right-hand portion of the object?	9. ______
10. What is the radius dimension of the rounded end of the object?	10. ______
11. Is the center section round or square?	11. ______
12. Which view shows the answer to Question 11?	12. ______

TRADE COMPETENCY TEST

For Print No. 109

Student's Name____________________ Instructor's Name____________________

Questions

Detail No. 1

1. What does dashed line A represent? 1. __________
2. Give dimension B. 2. __________
3. To what depth is the thread to be cut? 3. __________
4. Give the angle specified for the ground surface. 4. __________
5. What symbol is used to specify the surface to be ground? 5. __________

Detail No. 2

6. Give dimension D. 6. __________
7. What diameter of saw is to be used in cutting slot? 7. __________
8. How many milling cutter sizes will this detail accommodate? 8. __________

Detail No. 3

9. Give the largest diameter and the length of the detail. 9. __________
10. What does dashed line C indicate? 10. __________
11. Give correct tap drill size to be used for tapped hole. 11. __________

Assembly

12. Give the dimensions E, F, G, H, I, J, K, L, and M. 12. E_____ F_____ G_____ H_____ I_____ J_____ K_____ L_____ M_____
13. Give the approximate dimension of N. 13. __________
14. In what order should detail numbers be put in the circles on the assembly drawing? 14. __________
15. Give fractional sizes of cutters this chuck will accommodate. 15. __________

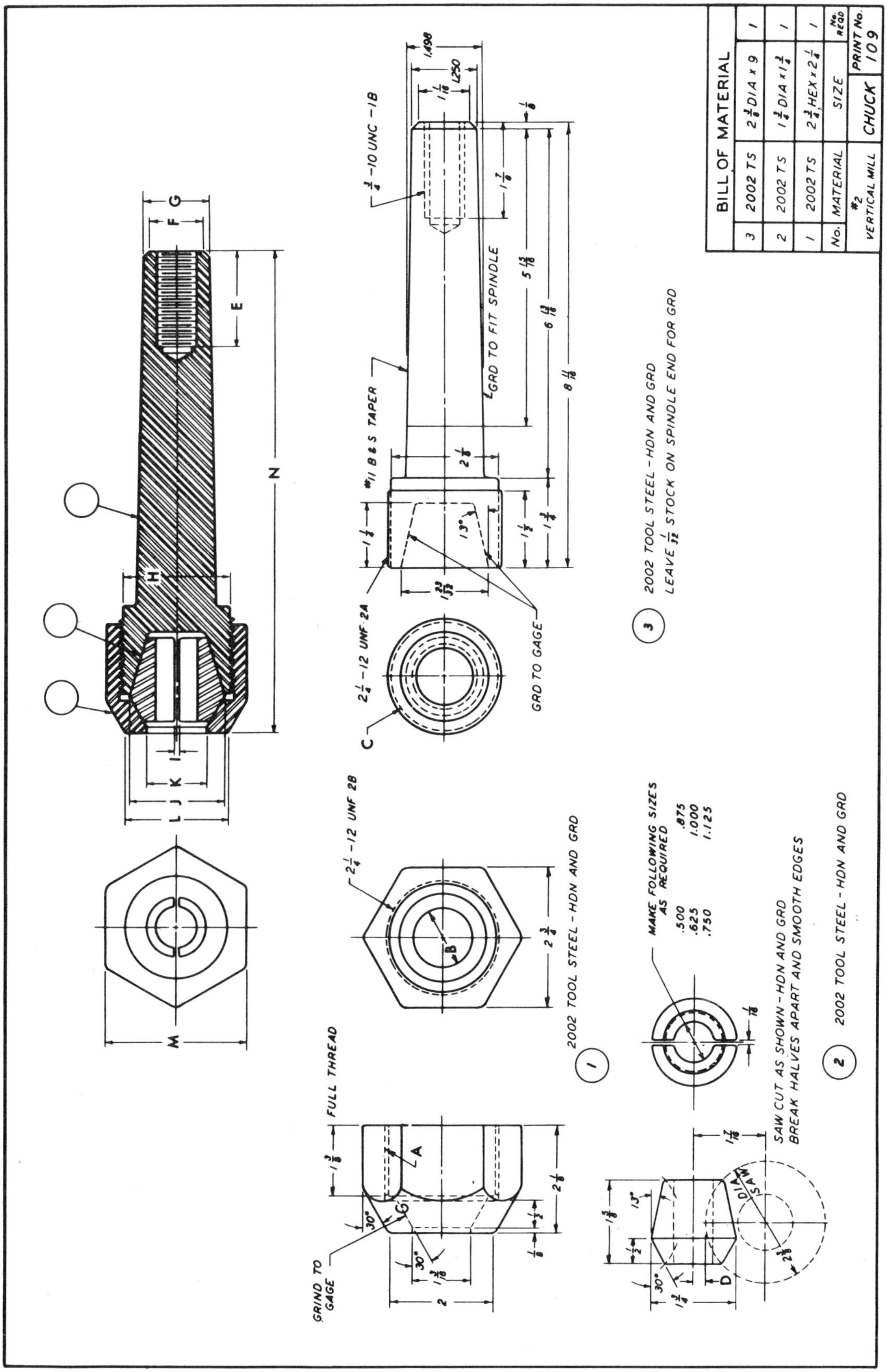
BILL OF MATERIAL
3 2002 TS 2 3/8 DIA x 9 1
2 2002 TS 1 3/4 DIA x 1 3/4 1
1 2002 TS 2 3/4 HEX x 2 1/4 1
No. MATERIAL SIZE No. REQD
#2 VERTICAL MILL
CHUCK
PRINT No. 109
3/4-10 UNC-1B
GRD TO FIT SPINDLE
#11 B & S TAPER
2002 TOOL STEEL - HDN AND GRD
LEAVE 1/32 STOCK ON SPINDLE END FOR GRD
2 1/4-12 UNF-2A
GRD TO GAGE
2 1/4-12 UNF-2B
2002 TOOL STEEL - HDN AND GRD
MAKE FOLLOWING SIZES AS REQUIRED
.500 .625 .750 .875 1.000 1.125
SAW CUT AS SHOWN - HDN AND GRD
BREAK HALVES APART AND SMOOTH EDGES
2002 TOOL STEEL - HDN AND GRD
FULL THREAD
GRIND TO GAGE
SAW DIA

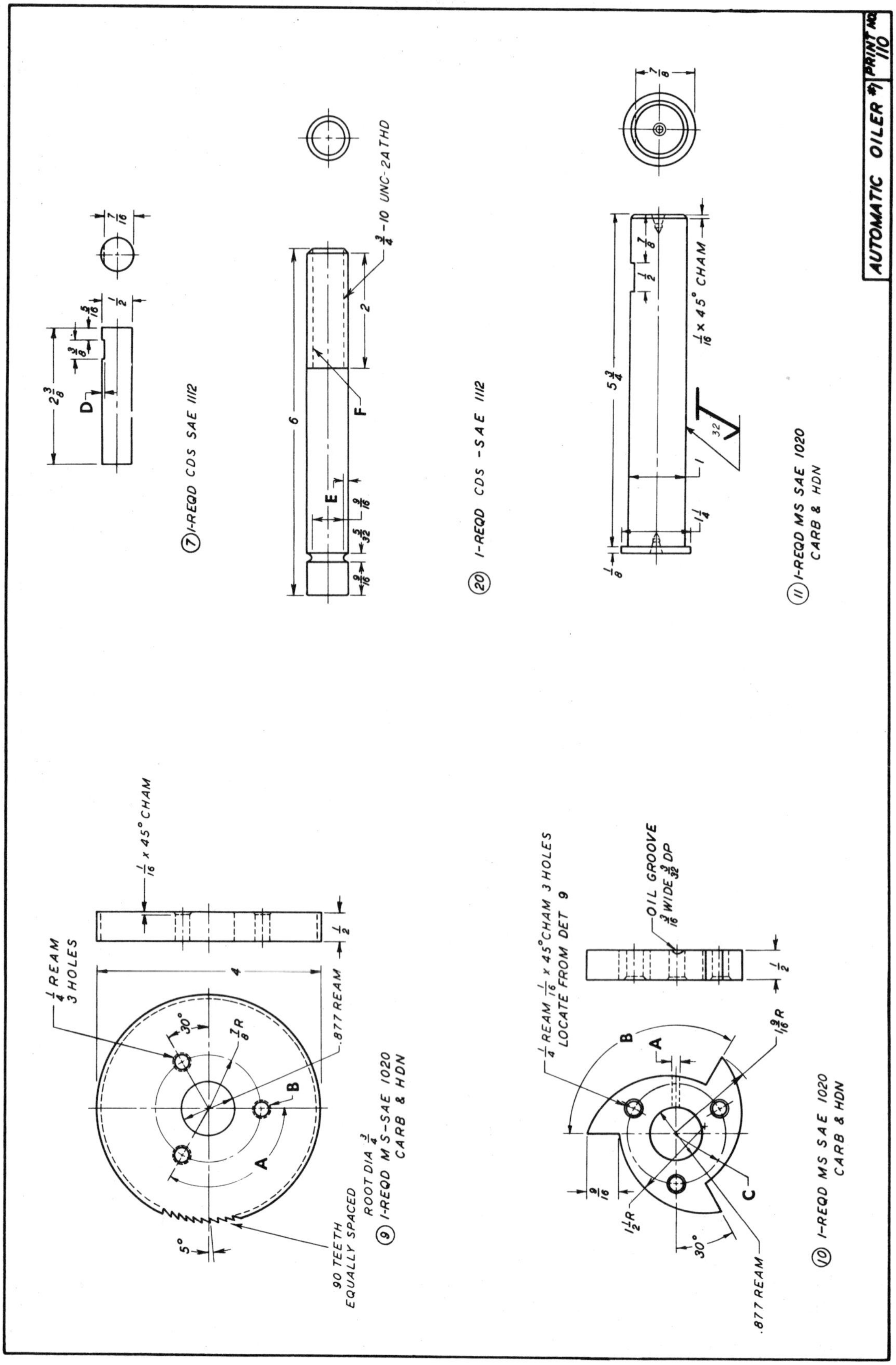
AUTOMATIC OILER #1
PRINT NO. 110
⑦ 1-REQD CDS SAE 1112
⑳ 1-REQD CDS - SAE 1112
3/4-10 UNC-2A THD
⑪ 1-REQD MS SAE 1020 CARB & HDN
1/16 x 45° CHAM
1/4 REAM 3 HOLES
.877 REAM
90 TEETH EQUALLY SPACED
ROOT DIA 3/4
⑨ 1-REQD MS-SAE 1020 CARB & HDN
1/4 REAM 1/16 x 45° CHAM 3 HOLES LOCATE FROM DET 9
OIL GROOVE 3/16 WIDE 3/32 DP
⑩ 1-REQD MS SAE 1020 CARB & HDN

TRADE COMPETENCY TEST

For Print No. 110

Student's Name____________________________ Instructor's Name____________________________

You will note that the three ¼″ drilled holes in Detail 10 are located from Detail 9. This is done to insure the proper alignment of holes in each piece. The two details after finishing are placed in proper alignment by slipping them over a ⅞″ diameter bar, or mandrel, and clamping them together. The ¼″ diameter holes are then drilled through each piece at the same time.

Note the dashed lines at each end of Detail 11. These indicate a center drill. This piece is finish ground after heat treating. In order to perform this operation it will be suspended on the centers of a lathe or external grinder. Centers are not needed for Detail 7 or Detail 20 because these are made from cold drawn steel with surface finish and a dimensional accuracy that, in this case, does not require finishing.

TEAR OFF HERE

Questions	Answers
Detail No. 9	
1. How many degrees in angle A?	1. __________
2. How many degrees between the teeth?	2. __________
3. What do dashed lines B indicate?	3. __________
4. How many holes are to be reamed?	4. __________
5. Give the overall dimensions of the wheel.	5. __________
Detail No. 10	
6. Give dimension A.	6. __________
7. How many degrees in angle B?	7. __________
8. What kind of material is required for this detail?	8. __________
9. Why are the three small holes shown by two solid lines in the front view?	9. __________
Detail No. 7	
10. Give overall dimensions of this detail.	10. __________
11. What is dimension D?	11. __________
12. What kind of material is used?	12. __________
Detail No. 20	
13. What does dashed line F indicate?	13. __________
14. What is the length of dimension E?	14. __________
Detail No. 11	
15. Give the total length of this detail.	15. __________
16. How is this piece to be finished?	16. __________

TRADE COMPETENCY TEST

For Print No. 111

Student's Name____________________ Instructor's Name____________________

Read all notes and study the methods of showing sections. You will note that, except in the case of section A-A and section B-B, the sections are drawn on an extension line from the cutting plane. This is a practical way of showing sections of this kind.

Examine the sectional view indicated by L. You will note that the 13/16″ dim. measures the distance I on the front view and therefore the 11/16″ dim. measures the distance A on the top view.

Questions	Answers
1. Give the length of the following dimensions: A, B, C, D, E, F, G, H, I, J, and K.	1. A____ B____ C____ D____ E____ F____ G____ H____ I____ J____ K____
2. Is surface X in front of or behind surface Y, as seen in the front view?	2. ________
3. Is surface U in front of or behind surface V in the front view?	3. ________
4. How many sectional views are shown?	4. ________
5. Which view shows that surface Y is above the centerline of bosses U and V?	5. ________
6. Can you tell from section A-A what the true shape of surface V is?	6. ________
7. What is the distance between the centerlines of bosses U and V?	7. ________
8. What is the parallel distance between surface U and surface V?	8. ________

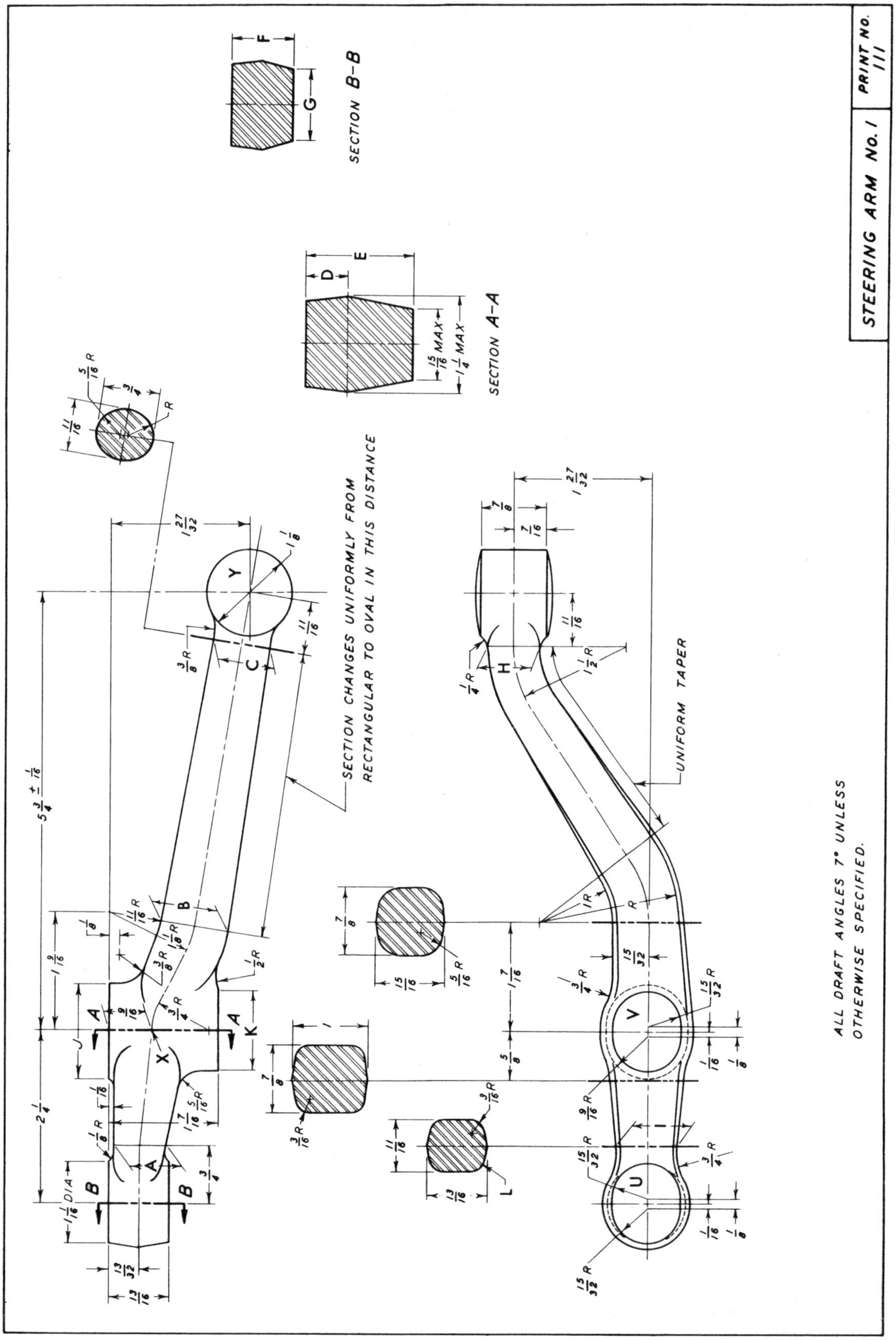
STEERING ARM NO. 1
PRINT NO. 111
SECTION A-A
SECTION B-B
SECTION CHANGES UNIFORMLY FROM RECTANGULAR TO OVAL IN THIS DISTANCE
UNIFORM TAPER
ALL DRAFT ANGLES 7° UNLESS OTHERWISE SPECIFIED.

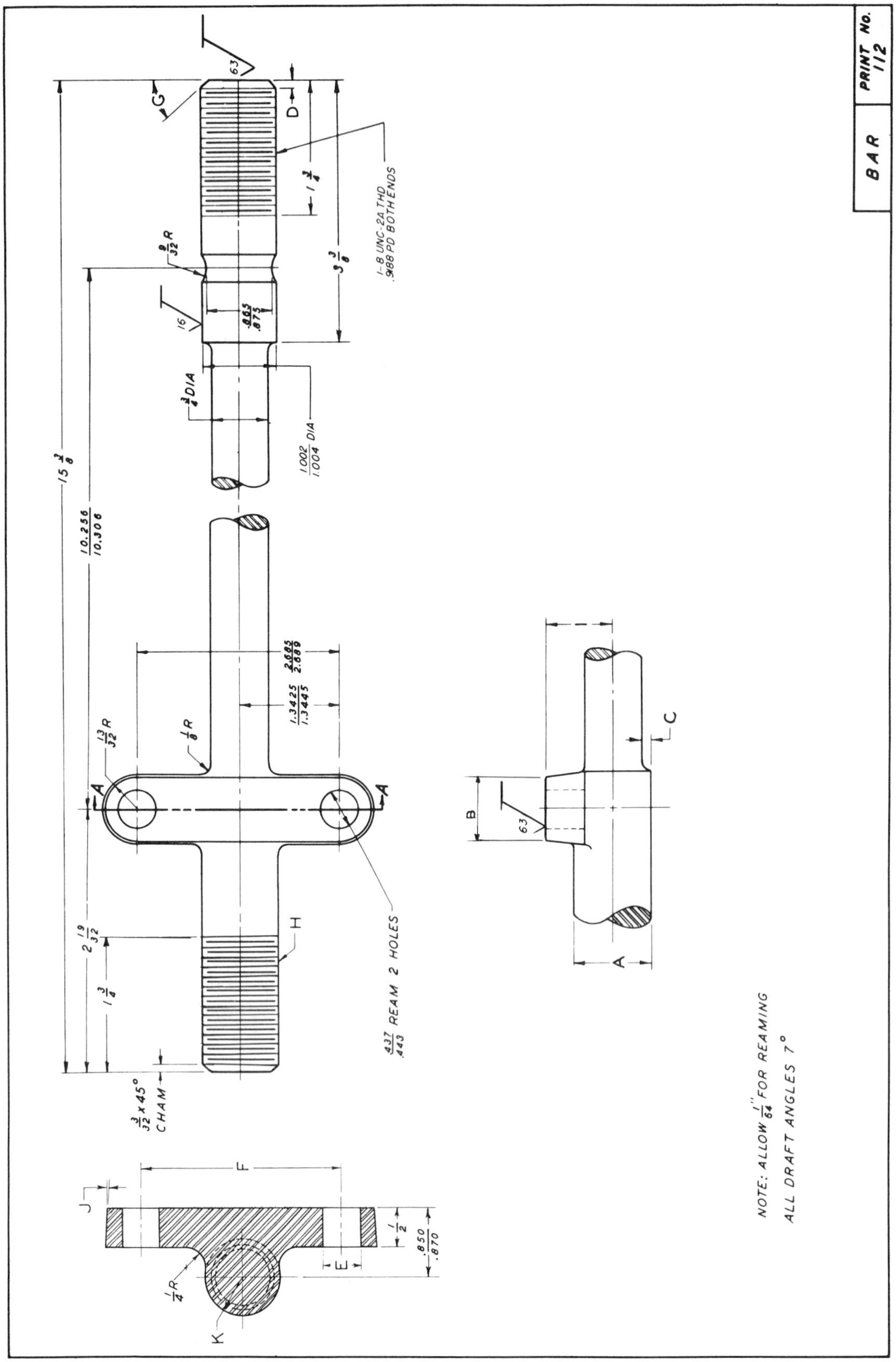
PRINT No. 112
BAR
1-8 UNC-2A THD
.988 PD BOTH ENDS
15 1/8
10.256
10.306
3 3/8
1 1/4
16
63
.865
.875
3/4 DIA
1.002
1.004 DIA
2.685
2.689
1.3425
1.3445
13/32 R
1/8 R
A
.437
.443 REAM 2 HOLES
H
2 19/32
1 3/4
3/32 x 45°
CHAM
D
G
B
C
E
F
J
K
1/4 R
1/2
.850
.870
NOTE: ALLOW 1/64" FOR REAMING
ALL DRAFT ANGLES 7°

TRADE COMPETENCY TEST

For Print No. 112

Student's Name________________ Instructor's Name________________

TEAR OFF HERE

Questions	Answers
1. Give dimensions A, B, C, D, and E.	1. A______ B______ C______ D______ E______
2. Give the MAX dimension of F.	2. ______
3. How many degrees in angle G?	3. ______
4. Give the number of threads per inch and the pitch diameter of the threads indicated by H.	4. ______
5. What is the maximum dimension of I?	5. ______
6. Give the number of degrees in angle J.	6. ______
7. Give the dimension of radius K.	7. ______
8. What do the arrows on the ends of section line A-A indicate?	8. ______
9. What is the tolerance for the spacing of the ℄ of the holes?	9. ______

The forged blank used to make the automotive bar requires two types of forging operations. The first increases the size of the forging bar and the second reduces, shapes, and sizes it. A bar of alloy steel ⅞″ in diameter and of proper length, heated to forging temperature (2100°-2250°) is placed between the dies of an upset forging machine similar to that shown in Fig. 1. The upset tool striking the end of the bar of hot steel, upsets it into the shape and diameter of the die impression. This operation is performed on each end of the forging stock to increase the bar size to the cross section necessary to make the finished forging. The upset forged blank is again heated to forging temperature and finish forged by either a drop hammer or a mechanical forging press.

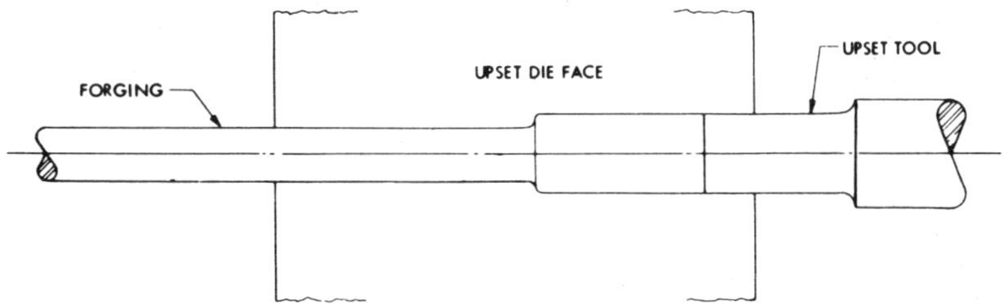

Fig. 1

TRADE COMPETENCY TEST

For Print No. 113

Student's Name______________________ Instructor's Name______________________

Note that Print 113 has a complete view of one detail and a simplified drawing of the opposite detail. This is general practice when right and left hand parts are required.

Questions	Answers
1. Give dimensions A, B, C, D, E, F, G, and H.	1. A_____ B_____ C_____ D_____ E_____ F_____ G_____ H_____
2. Give radius J.	2. __________
3. How many degrees in angle K?	3. __________
4. Give drill and tap size for hole indicated by I.	4. __________
5. Allowing 11/32 on the width of each piece for cutting off and machining the four pieces for this fixture, what length bar is required.	5. __________
6. Which view of detail 110 shows the depth of the slots and how deep are they?	6. __________
7. Is the contour formed by the 3½″ spherical radius concave or convex?	7. __________
8. Are the threads to be carburized and hardened or left soft?	8. __________
9. What is the angular spacing between the ⅜ × ⅛ slots?	9. __________

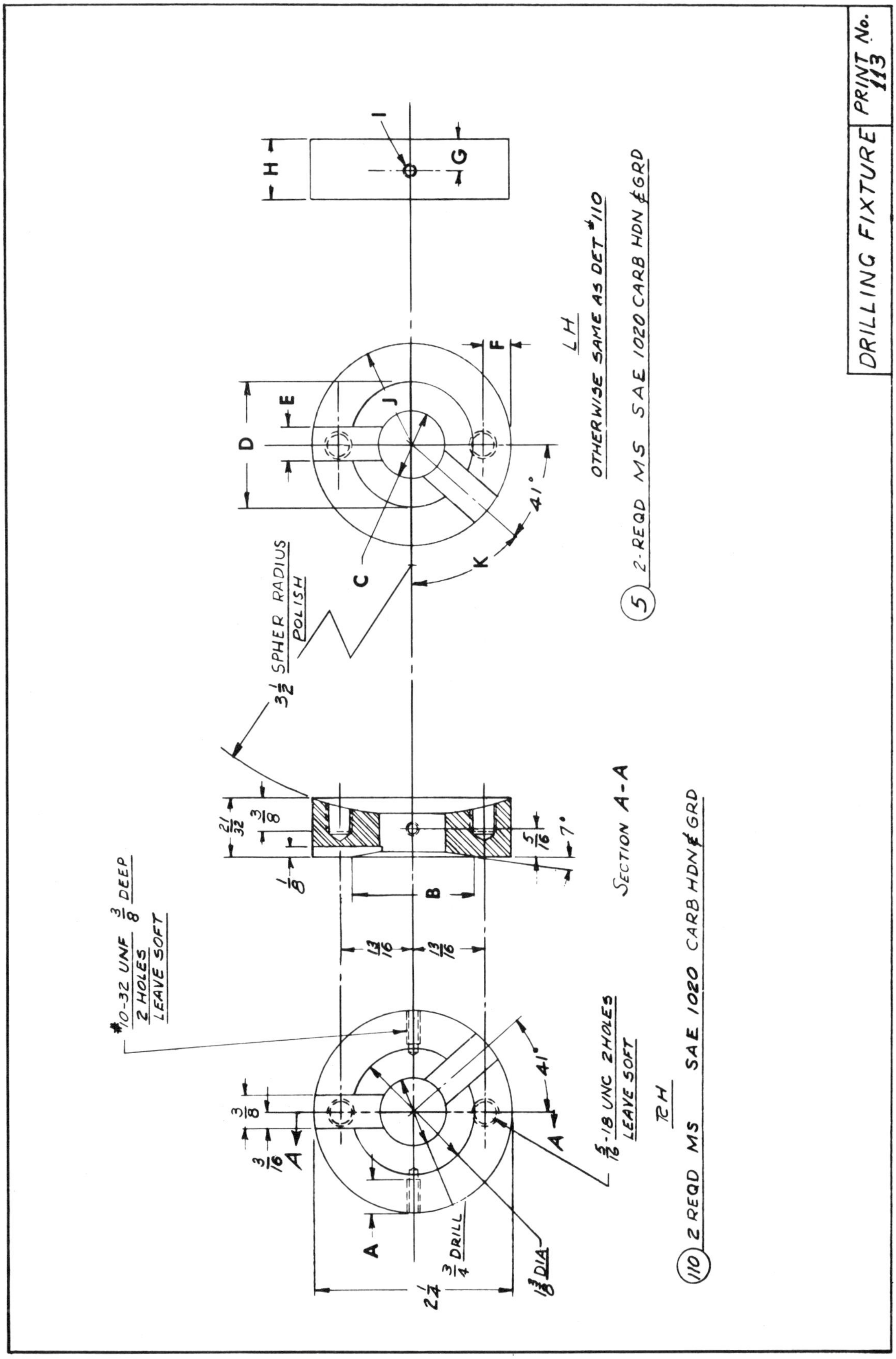
DRILLING FIXTURE
PRINT No. 113
LH
OTHERWISE SAME AS DET #110
5 2-REQD MS SAE 1020 CARB HDN & GRD
3 1/2 SPHER RADIUS POLISH
SECTION A-A
#10-32 UNF 3/8 DEEP
2 HOLES
LEAVE SOFT
5/16-18 UNC 2HOLES
LEAVE SOFT
RH
110 2 REQD MS SAE 1020 CARB HDN & GRD
3/4 DRILL
1 3/8 DIA

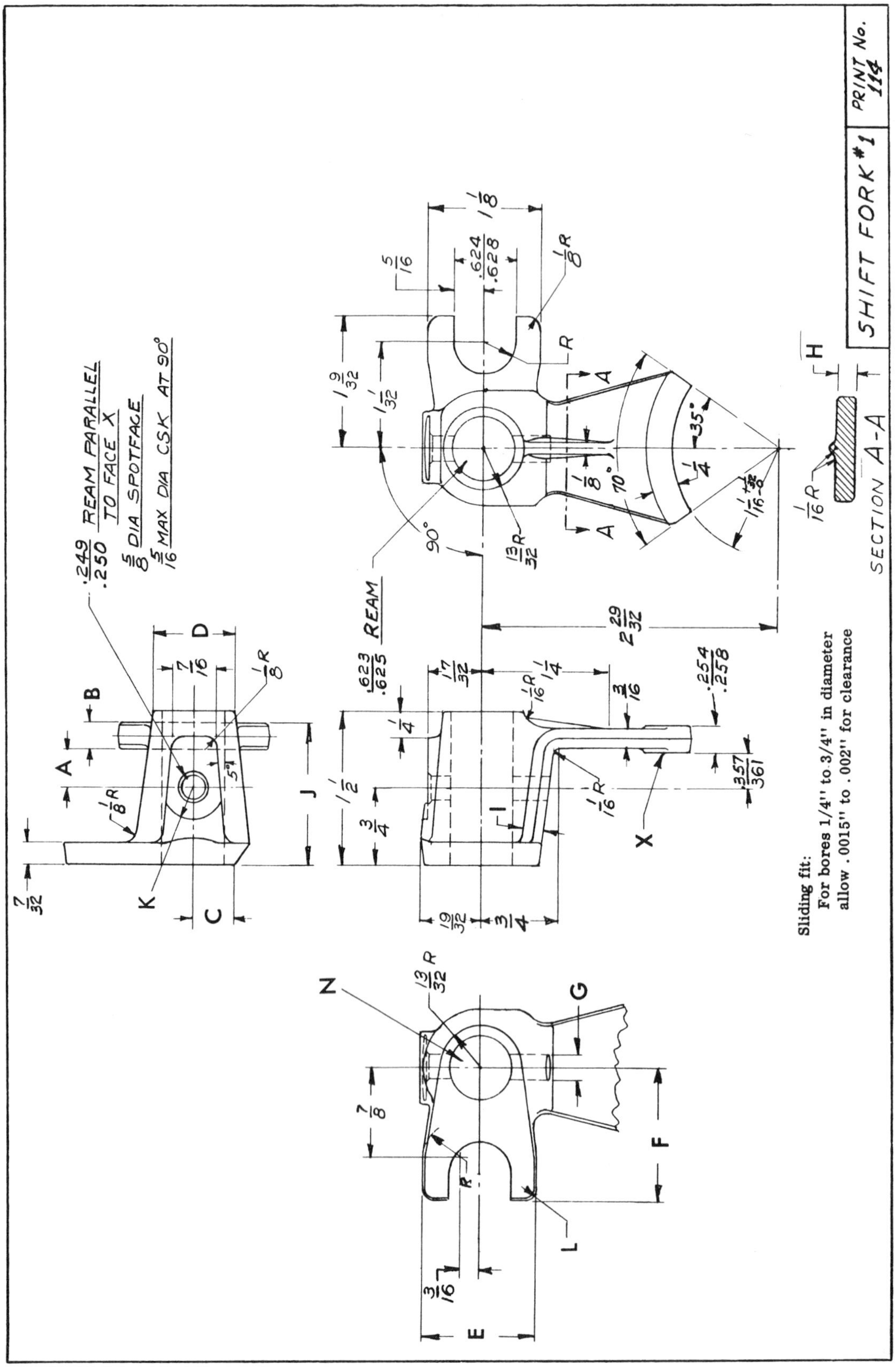
PRINT No. 114
SHIFT FORK #1
.249/.250 REAM PARALLEL TO FACE X
5/8 DIA SPOTFACE
5/16 MAX DIA CSK AT 90°
REAM
SECTION A-A
Sliding fit:
For bores 1/4" to 3/4" in diameter
allow .0015" to .002" for clearance

TRADE COMPETENCY TEST

For Print No. 114

Student's Name____________________ Instructor's Name____________________

Questions	Answers
1. Give dimensions A, B, C, D, E, F, G, H, I, and J.	1. A______ B______ C______ D______ E______ F______ G______ H______ I______ J______
2. Give radius dimensions K and L.	2. K______ L______
3. Give diameter of hole indicated by N.	3. ______
4. What is the tolerance for this hole?	4. ______
5. Which views show that this hole goes all the way through the piece?	5. ______
6. The centerline and locaton of the .249/-.250 hole must be held in close relation to what surface?	6. ______
7. There is a 1-1/32″ dimension given for the location of a radius. Which view shows it?	7. ______
8. What is the minimum radius located by this 1-1/32″ dimension?	8. ______
9. What is the maximum radius located by this 1-1/32″ dimension?	9. ______
10. Is this radius shown in the top view?	10. ______
11. For free movement of the fork on the shaft what size should the shaft be?	11. ______
12. The function of this part is to shift the second speed gear into a meshing position in an automotive transmission. It is subjected to high stresses and shock. Would you recommend that it be made as a steel forging, malleable casting, ductile iron casting or steel casting.	12. ______

TEAR OFF HERE

TRADE COMPETENCY TEST

For Print No. 115

Student's Name____________________ Instructor's Name____________________

Questions	Answers
Detail No. 1	
1. Give dimensions A, B, C, D, E, F, G, H, I, J, K, L, M, N, O, and P.	1. A____ B____ C____ D____ E____ F____ G____ H____ I____ J____ K____ L____ M____ N____ O____ P____
2. How many degrees in angle Q?	2. ____
3. Give the height, length, and width dimensions of each die.	3. ____
4. Is the right side of recess B-B in front view vertical or slanting?	4. ____
5. How is the projection on the bottom of the object, with dimensions ¾″ × 2⅝″, shown in top view? Give approximate location in top view.	5. ____
6. In the section A-A note that portion with a 2-1/16″ R. dimension. What is the width of this as shown in the top view?	6. ____
Detail No. 2	
7. Give the dimensions R, S, T, and U.	7. R____ S____ T____ U____
8. How is surface Y represented in bottom view?	8. ____
9. How is surface V represented in left side view?	9. ____
10. What is the distance from surface W to surface X?	10. ____
11. Is surface X in front of or behind surface W as seen in the top view?	11. ____
12. Is surface X in the same plane as surface Z?	12. ____

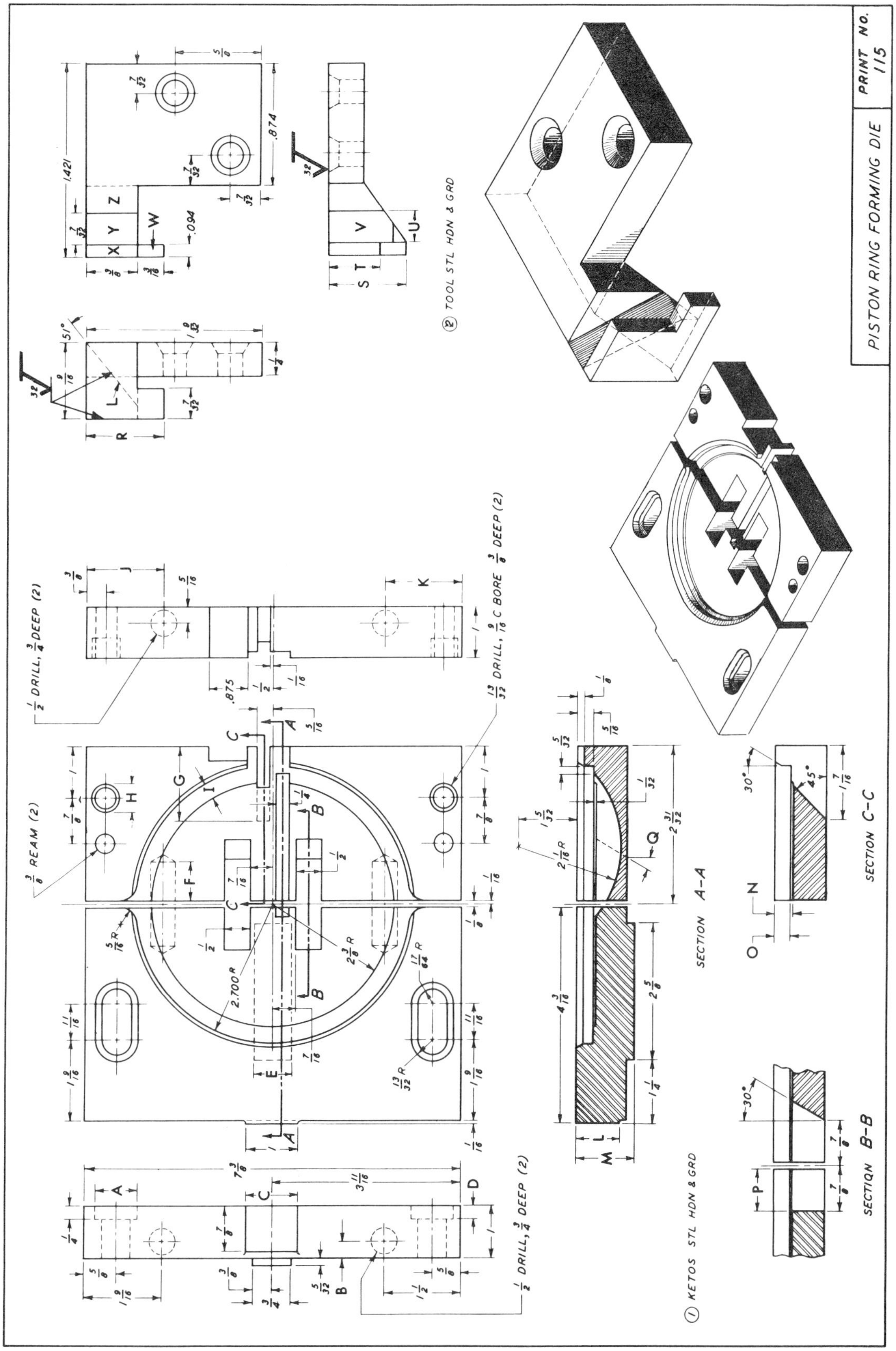
PISTON RING FORMING DIE
PRINT NO. 115
① KETOS STL HDN & GRD
② TOOL STL HDN & GRD
SECTION A–A
SECTION B–B
SECTION C–C
3/8 REAM (2)
1/2 DRILL, 3/4 DEEP (2)
13/32 DRILL, 9/16 C BORE 3/8 DEEP (2)

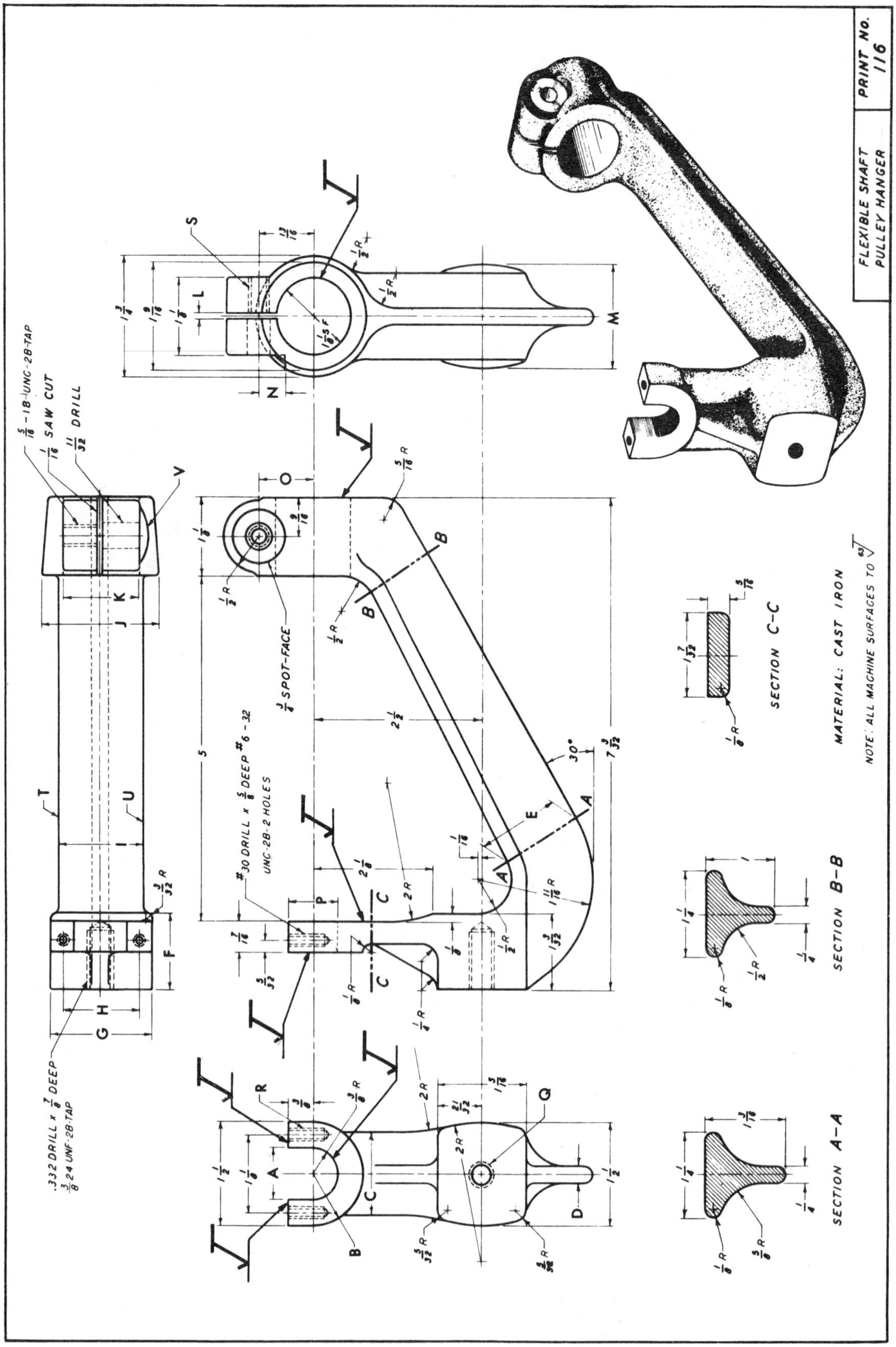
PRINT NO. 116
FLEXIBLE SHAFT PULLEY HANGER
MATERIAL: CAST IRON
NOTE: ALL MACHINE SURFACES TO 63
5/16 - 18 UNC-2B TAP
1/16 SAW CUT
11/32 DRILL
.332 DRILL x 7/8 DEEP
3/8 24 UNF-2B TAP
#30 DRILL x 5/8 DEEP #6-32
UNC-2B-2 HOLES
3/4 SPOT-FACE
SECTION A-A
SECTION B-B
SECTION C-C

TRADE COMPETENCY TEST

For Print No. 116

Student's Name____________________ Instructor's Name____________________

In product design, appearance is always given important consideration by the engineer. The Flexible Shaft Pulley Hanger, though only a component part of a shop tool, illustrates the care and skill that designers exercise to make a product attractive. The curved section of the arm, the 2″ radius on the exposed area of the boss at left and the taper on the clamp boss at right in the top view all contribute to the strong and attractive appearance of the part. Though these irregular contours might add to the initial cost of the pattern equipment for the part, they have a negligible effect on the cost of the finished piece.

TEAR OFF HERE

Questions	Answers
1. Give dimensions A, B, C, D, E, F, G, H, I, J, K, L, M, N, O, and P.	1. A____ B____ C____ D____ E____ F____ G____ H____ I____ J____ K____ L____ M____ N____ O____ P____
2. Give specifications for tapped holes indicated by Q, R, and S.	2. Q____ R____ S____
3. Are lines indicated by T and U parallel?	3. ____
4. How is this shown?	4. ____
5. What causes curved line indicated by V?	5. ____
6. How is the 1⅛ hole shown in the front view?	6. ____
7. Find S on the right side view. Why are double horizontal dashed lines shown only in one half of the section?	7. ____
8. How many holes is it necessary to tap in the object?	8. ____

TRADE COMPETENCY TEST

For Print No. 117

Student's Name________________ Instructor's Name________________

A knowledge of the functional requirements and the methods of fabrication is essential to analyze the design of the 6,000 lb. Steam Forging Hammer Ram. The height of the ram must be great enough to give adequate bearing length on the column guides. Width and length dimensions of the ram must be of a size that will accommodate a die to make the maximum size forging within the capacity of the hammer. It is necessary that the ram be in balance. In other words, if each corner were rested on a separate scale they should register approximately the same weight. The ram is made from a forged steel blank and must be machined overall. Planers are used on large work of this kind; therefore the ram is designed with flat surfaces so that most of the finishing can be performed with this type of tool.

Questions	Answers
1. Give dimensions A, B, C, D, E, F, G, H, I, J, K, L, and M.	1. A______ B______ C______ D______ E______ F______ G______ H______ I______ J______ K______ L______ M______
2. How is surface S represented in the right side view?	2. ______
3. How is surface S represented in the top view?	3. ______
4. How is surface R represented in the top view?	4. ______
5. How is surface Y represented in the top view?	5. ______
6. How is surface U represented in the right side view?	6. ______
7. How is surface Q seen in the front view?	7. ______
8. At a price of 30 cents per lb., what is the approximate cost of this Ram?	8. ______
9. How deep is the 8″ diameter hole?	9. ______
10. What is the angle of the sides of the recessed portion at the top of the ram?	10. ______
11. What is the included angle between the sides of the grooves on the sides of the ram?	11. ______

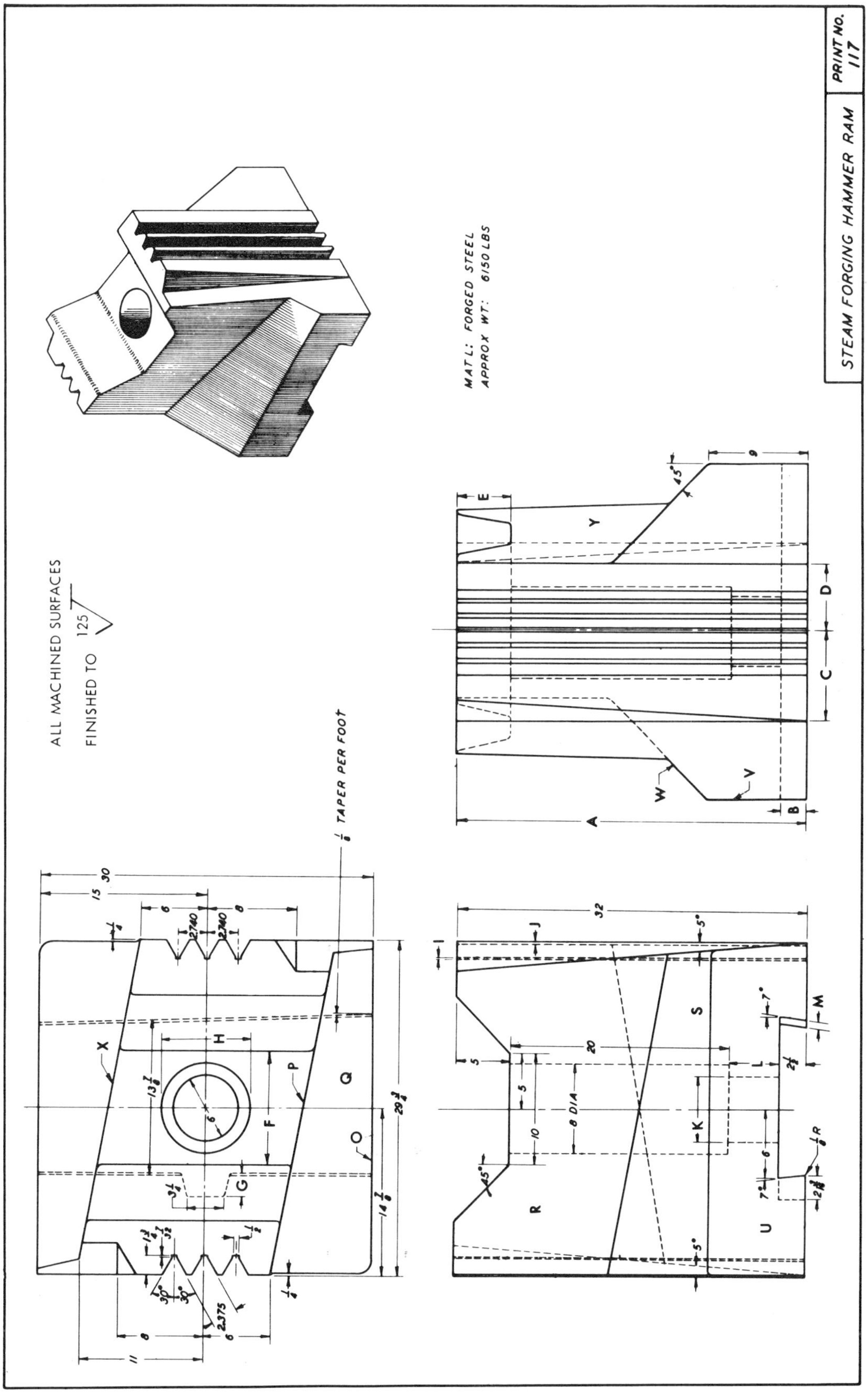
PRINT NO. 117
STEAM FORGING HAMMER RAM
MAT'L: FORGED STEEL
APPROX WT: 6150 LBS
ALL MACHINED SURFACES
FINISHED TO 125
1/8 TAPER PER FOOT

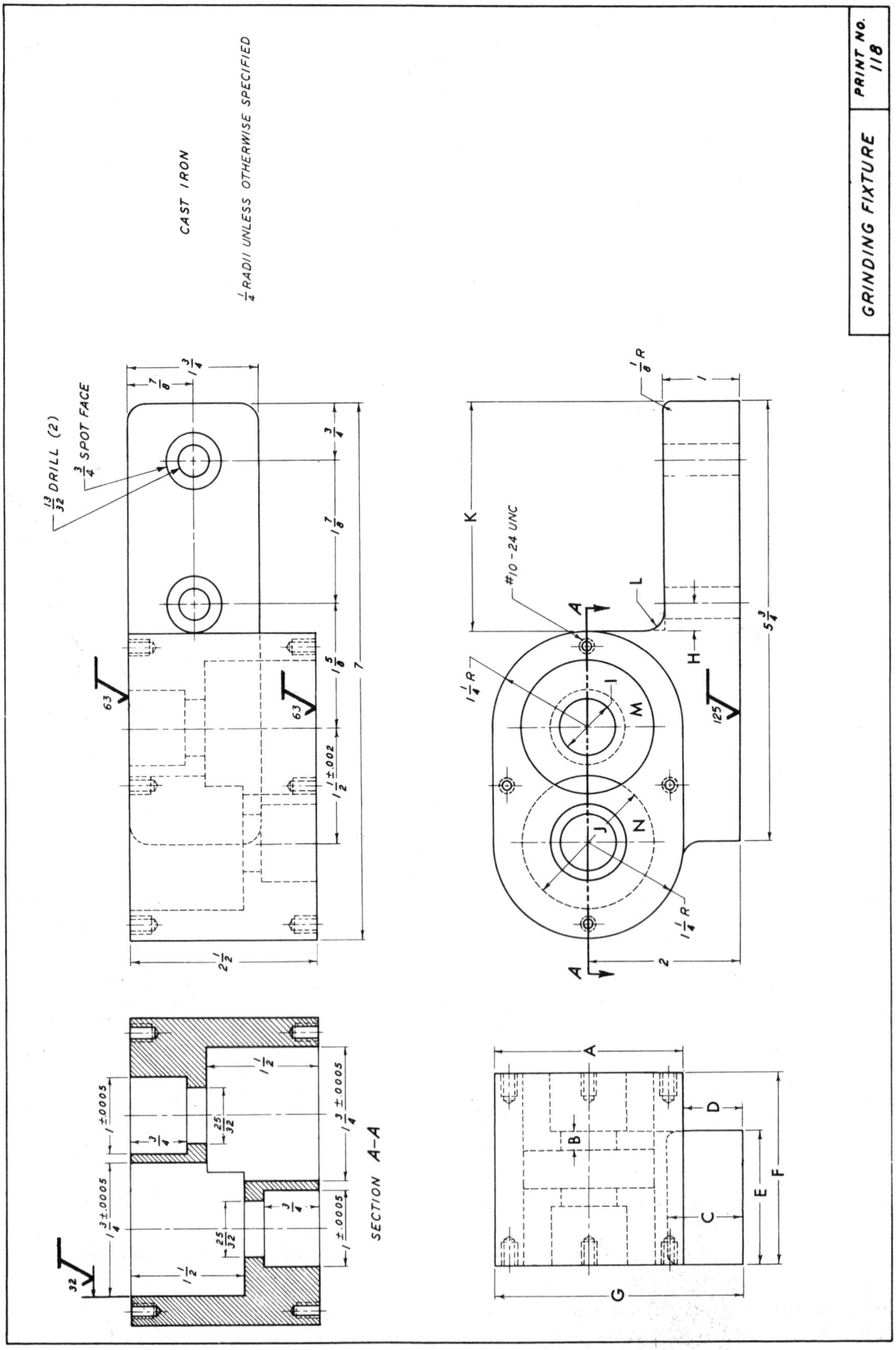
PRINT NO. 118
GRINDING FIXTURE
CAST IRON
1/4 RADII UNLESS OTHERWISE SPECIFIED
13/32 DRILL (2)
3/4 SPOT FACE
#10-24 UNC
SECTION A-A

TRADE COMPETENCY TEST

For Print No. 118

Student's Name________________ Instructor's Name________________

The dictionary's definition for a fixture is "anything fixed in one place." This definition is applicable to the mechanical industry except that it is seldom used in reference to basic manufacturing equipment. Attachments, such as jigs, clamping devices or anything not a standard part of a machine (but an adaptation to the machine to perform certain work) are considered as fixtures.

TEAR OFF HERE

Questions	Answers
1. Give dimensions A, B, C, D, E, F, G, H, I, J, K.	1. A______ B______ C______ D______ E______ F______ G______ H______ I______ J______ K______
2. What is the radius L?	2. ______
3. Is surface M in front of, or back of surface N as you see the object in the front view?	3. ______
4. Disregarding bored and drilled holes, how many surfaces are to be machined?	4. ______
5. How many holes are to be tapped?	5. ______
6. What do the large circles in the top view represent?	6. ______
7. What is the thickness of the base of the fixture and from what view is it taken?	7. ______
8. Give the dimensions of the lower surface of the base of the object.	8. ______

TRADE COMPETENCY TEST

For Print No. 119

Student's Name____________________ Instructor's Name____________________

When close tolerances must be maintained between angular surfaces, as illustrated in the guide slot in the right-end view of Detail 1, the same procedure is followed as that for measuring screw threads. This is known as the wire method. A ½″ diameter bar (see dashed circle) is used to measure the angular surface. It can be readily seen that a small radius on the corner of the slot at the reference dimension of 2.028 would cause an inaccurate measurement of the slant surface.

Questions	Answers
Detail No. 1	
1. Give dimensions A, B, C, D, E, F, G, H, I, J, K, L, M, and N.	1. A____ B____ C____ D____ E____ F____ G____ H____ I____ J____ K____ L____ M____ N____
2. Is surface O above or below surface P as you look at the top view?	2. ________
3. What material is used for this detail?	3. ________
4. What is the depth of the dovetail machined in this fixture?	4. ________
5. How is surface P represented in the right side view?	5. ________
Detail No. 2	
6. Give dimensions Q, R, S, T, and U.	6. Q____ R____ S____ T____ U____
7. Give the largest diameter and length dimensions of this detail.	7. ________
8. What is the common name for this type of shaft?	8. ________
9. What is the amount of eccentricity of this shaft?	9. ________
10. Give the length dimensions of those parts of the shaft that are concentric.	10. ________
11. How long is the eccentric section of the shaft?	11. ________

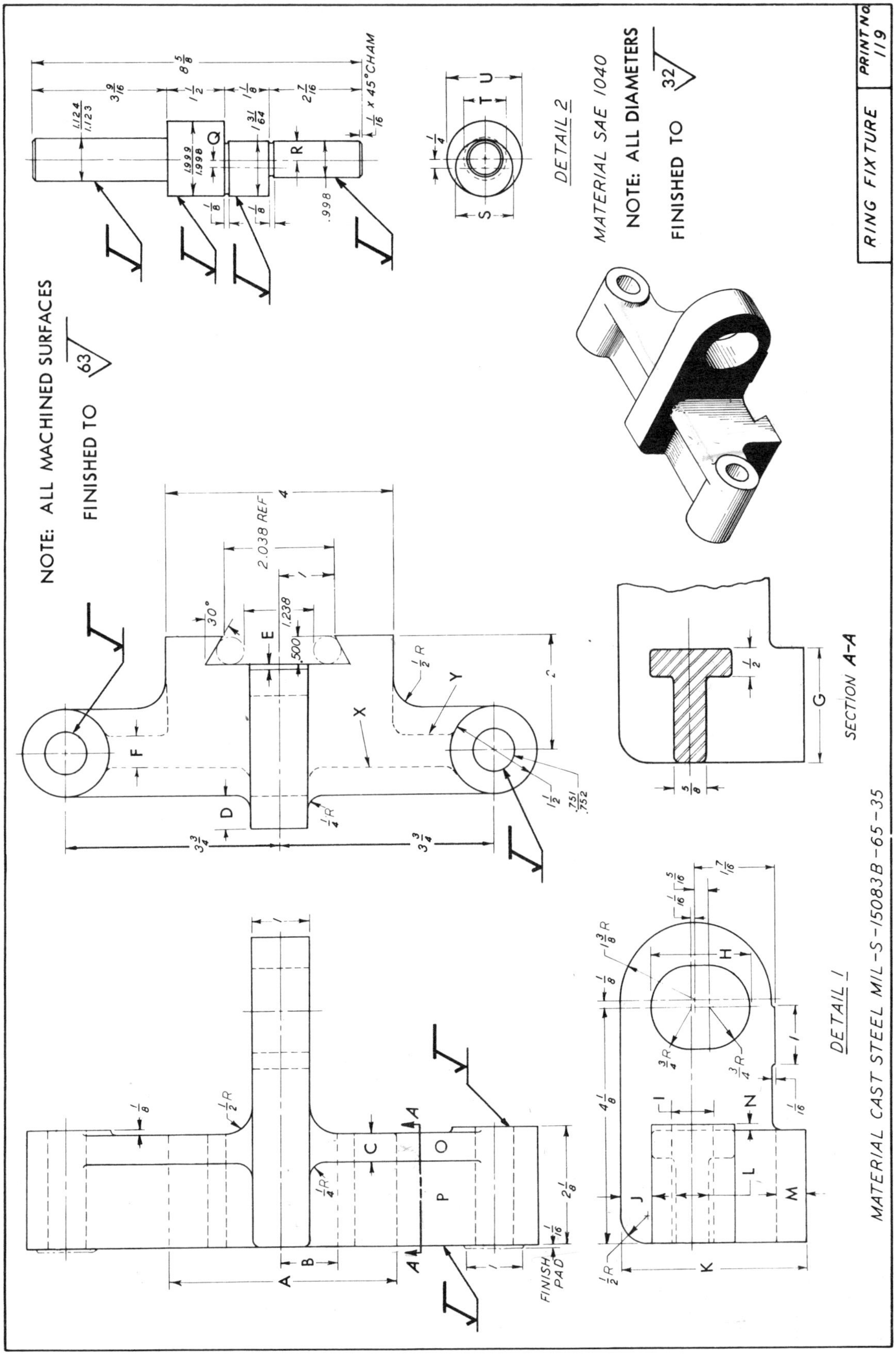
NOTE: ALL MACHINED SURFACES
FINISHED TO 63
DETAIL 2
MATERIAL SAE 1040
NOTE: ALL DIAMETERS
FINISHED TO 32
SECTION A-A
DETAIL 1
MATERIAL CAST STEEL MIL-S-15083B-65-35
FINISH PAD
RING FIXTURE
PRINT NO. 119

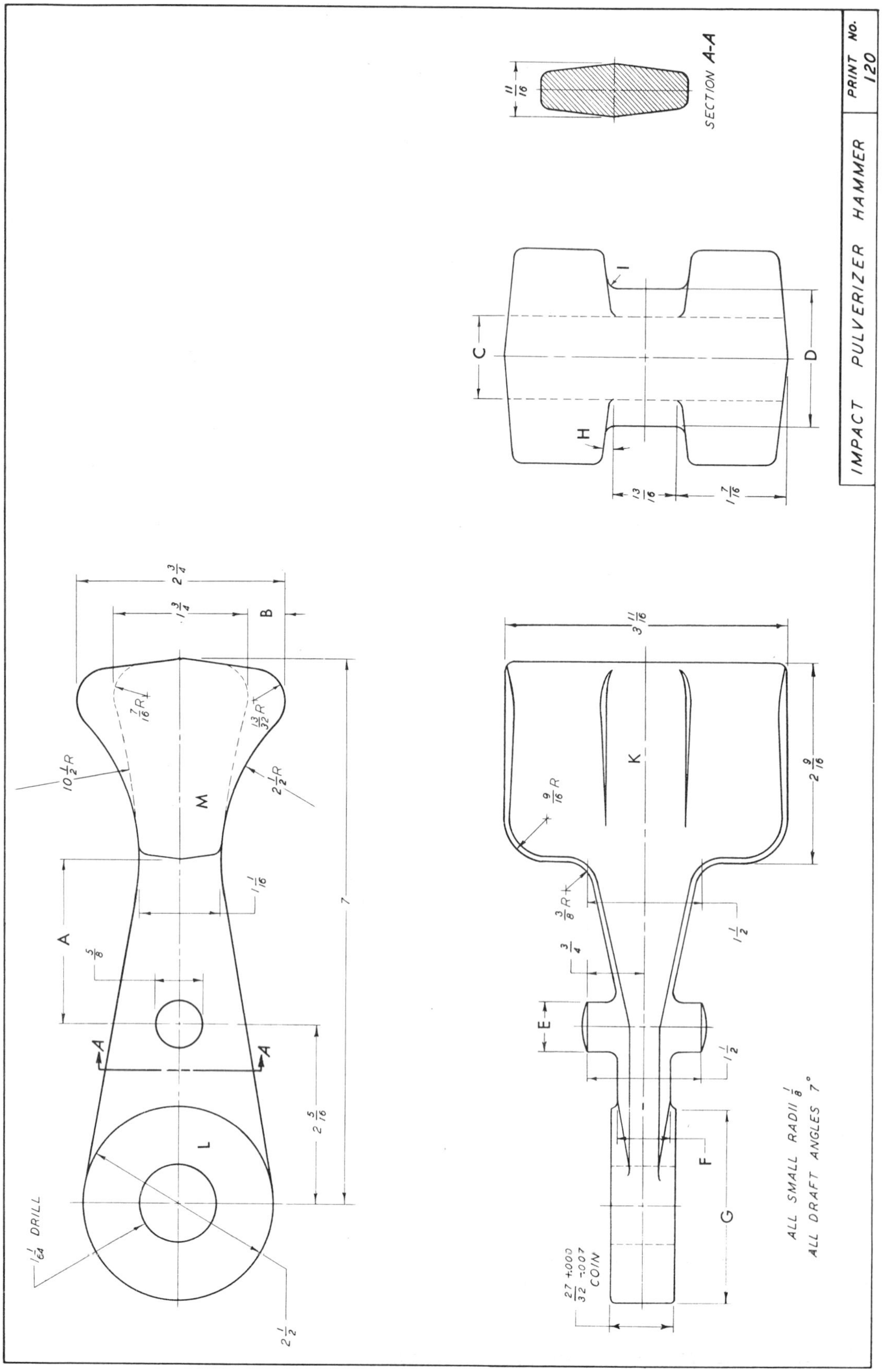
IMPACT PULVERIZER HAMMER
PRINT NO. 120
SECTION A-A
ALL SMALL RADII 1/8
ALL DRAFT ANGLES 7°
1 1/64 DRILL
27/32 +.000 -.007 COIN
2 3/4
1 3/4
10 1/2 R
2 1/2 R
7/16 R
13/32 R
1 1/16
5/8
7
2 5/16
2 1/2
11/16
13/16
1 7/16
3 11/16
2 9/16
9/16 R
3/8 R
3/4
1 1/2
1 1/2
A
B
C
D
E
F
G
H
I
K
L
M

TRADE COMPETENCY TEST

For Print No. 120

Student's Name____________________ Instructor's Name____________________

Coining to remove draft, to control thickness, and to provide surface smoothness is widely used for forgings. Because of the ductility of both aluminum and steel forgings, this operation is usually performed without pre-heating.

TEAR OFF HERE

Questions	Answers
1. Give dimensions A, B, C, D, E, F, and G.	1. A____ B____ C____ D____ E____ F____ G____
2. How many degrees in angle H?	2. ____
3. Give radius I.	3. ____
4. Is surface L in front of or behind surface M as you look at top view?	4. ____
5. Which one of these surfaces is a slant surface?	5. ____
6. Is surface K a recess or projection?	6. ____
7. How many holes are to be drilled in this hammer?	7. ____
8. Is the side view a complete projection?	8. ____
9. How is surface L finished?	9. ____
10. How is the recessed portion in the hammer head shown in top view?	10. ____
11. Give the over-all length of the object.	11. ____

TRADE COMPETENCY TEST

For Print No. 121

Student's Name____________________ Instructor's Name____________________

Questions	Answers
Detail No. 35	
1. Give dimensions A, B, C, D, E, and F.	1. A______ B______ C______ D______ E______ F______
2. What does curved line G indicate?	2. ____________
3. Give the length, height, and width dimensions of object.	3. ____________
4. What do the two solid horizontal lines at the center of the front view represent?	4. ____________
Detail No. 32	
5. Give dimension H, I, J, K, L, M, N, and O.	5. H______ I______ J______ K______ L______ M______ N______ O______
6. Give over-all dimensions of the piece necessary to make this detail, allowing 1/16″ for necessary finish.	6. ____________
7. State the number and diameter of reamed holes in this detail.	7. ____________
8. Give the height dimensions on front view.	8. ____________
9. Give the locating dimensions for the 3/16″ pin.	9. ____________
10. Is the 3/16″ pin on the front or back of the front view?	10. ____________
11. Is the 3/16″ pin shown in the left side view?	11. ____________
12. The yoke portion of the Tie Rod End shown in Detail 35 is assembled in this detail. How much clearance is specified?	12. ____________

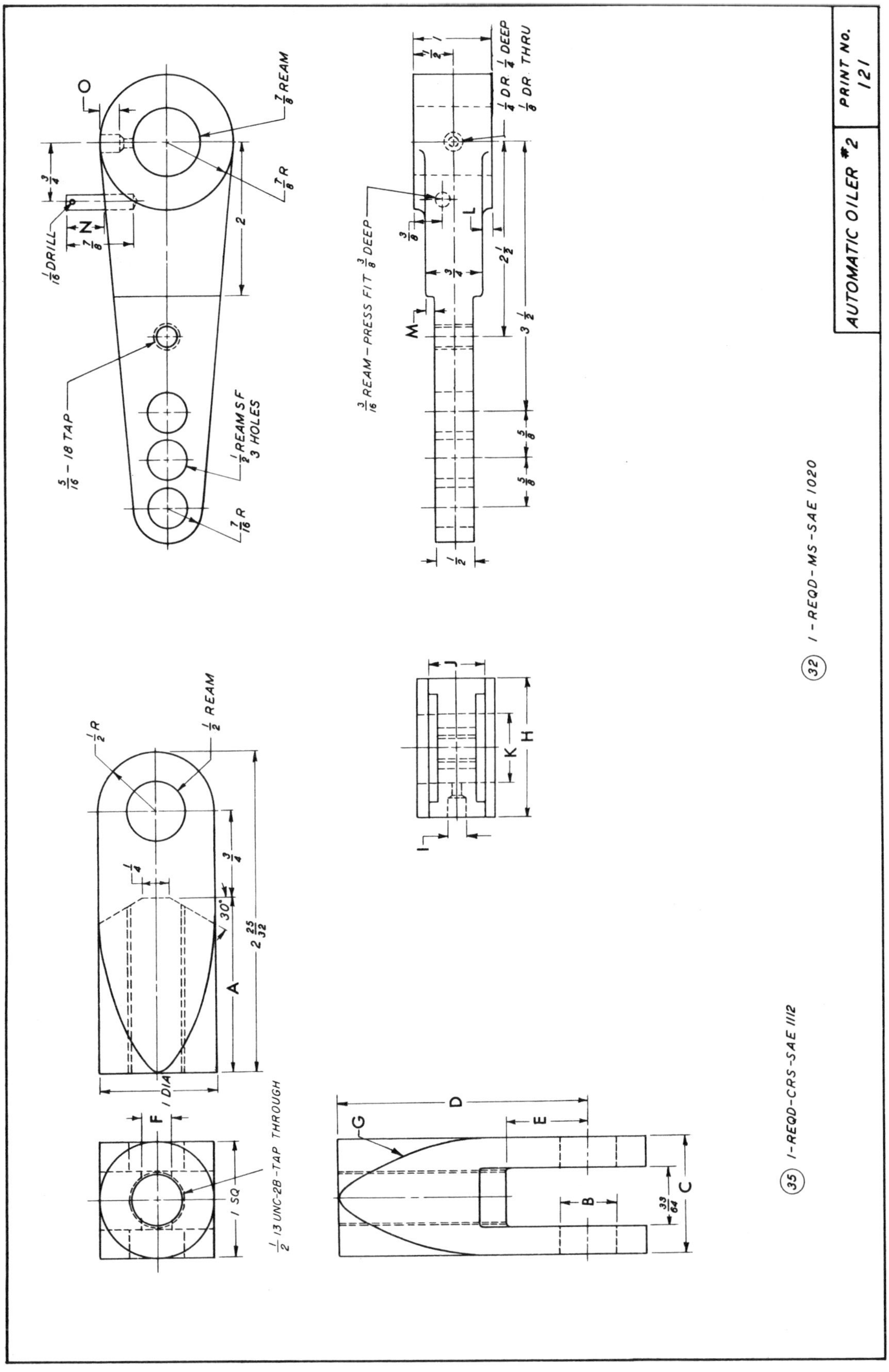
PRINT No. 121
AUTOMATIC OILER #2
7/8 REAM
7/8 R
1/16 DRILL
5/16 - 18 TAP
1/2 REAM SF 3 HOLES
7/16 R
1/4 DR. 1/4 DEEP
1/8 DR. THRU
3/16 REAM - PRESS FIT 3/8 DEEP
32 1-REQD-MS-SAE 1020
1/2 R
1/2 REAM
1 DIA
1 SQ
1/2 13 UNC-2B-TAP THROUGH
35 1-REQD-CRS-SAE 1112

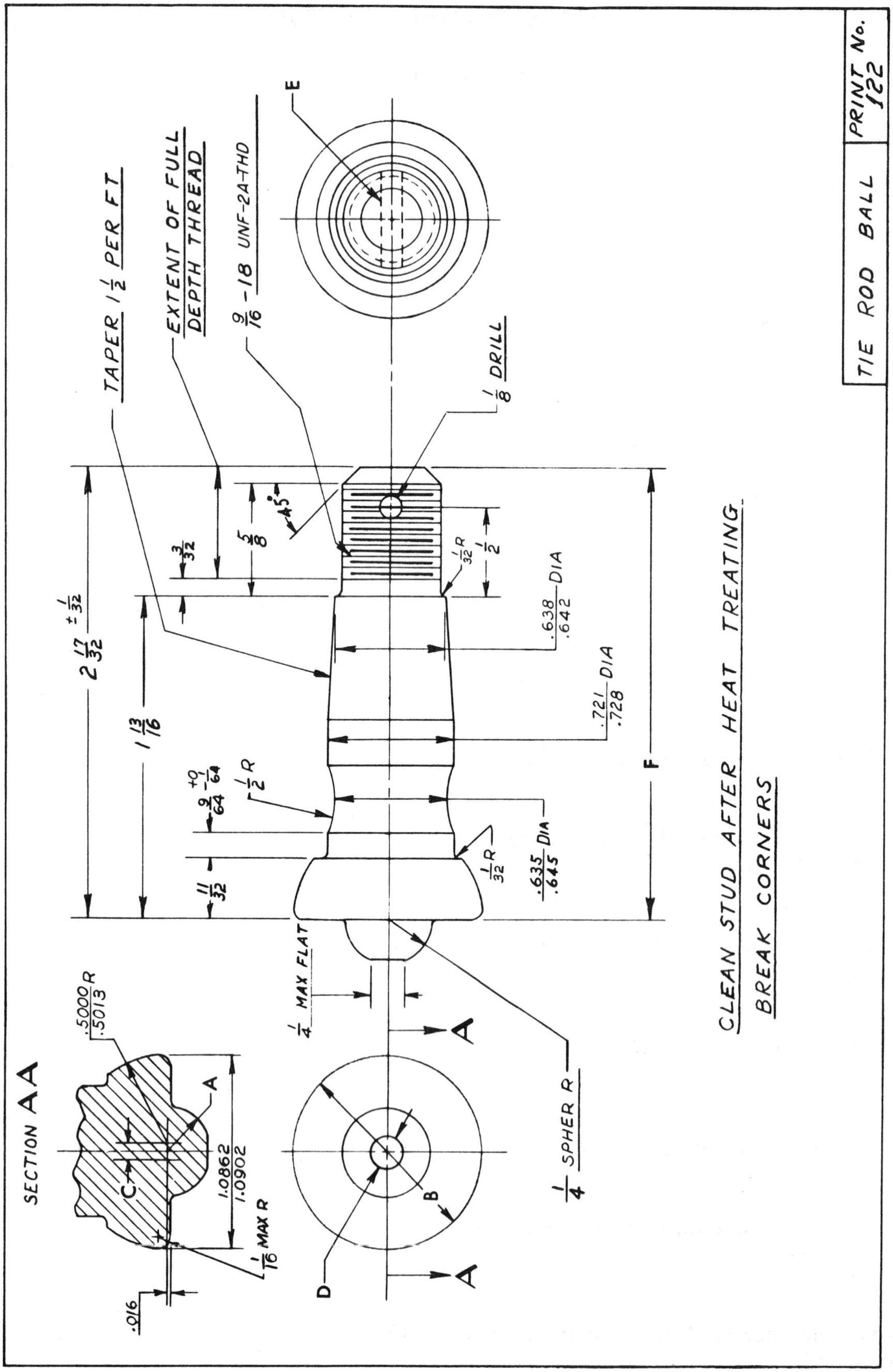
PRINT No. 122
TIE ROD BALL
TAPER 1 1/2 PER FT.
EXTENT OF FULL DEPTH THREAD
9/16 -18 UNF-2A-THD
1/8 DRILL
45°
5/8
3/32
1/32 R
1/2
.638 / .642 DIA
.721 / .728 DIA
2 17/32 ±1/32
1 13/16
9/64 +0 −1/64
1/2 R
11/32
1/32 R
.635 / .645 DIA
F
E
CLEAN STUD AFTER HEAT TREATING
BREAK CORNERS
1/4 MAX FLAT
1/4 SPHER R
A
A
B
D
SECTION AA
.5000 / .5013 R
1.0862 / 1.0902
C
1/16 MAX R
.016

TRADE COMPETENCY TEST

For Print No. 122

Student's Name________________ Instructor's Name________________

T E A R O F F H E R E

Questions	Answers
1. Give the dimension of radius A.	1. ________
2. Give the maximum length of dimensions B, D, and F.	2. B______ D______ F______
3. Give maximum dimension of C.	3. ________
4. What do dashed lines indicated by E represent?	4. ________
5. How many tapered surfaces are shown?	5. ________
6. How many spherical surfaces are shown?	6. ________
7. Can the over-all length of the object be given readily from the drawing?	7. ________
8. Could other dashed circles be shown in the right side view?	8. ________
9. What is the outside diameter of the threaded part?	9. ________
10. What is the angle of the chamfer?	10. ________
11. What is the minimum diameter of the largest part?	11. ________
12. What tolerance is allowed on the largest diameter?	12. ________

TRADE COMPETENCY TEST

For Print No. 123

Student's Name____________________ Instructor's Name____________________

A cold bar of steel 1″ long heated to forging temperature will measure approximately 1-1/64″ long. When cooled it will return to its original length. For this reason tools or dies used for hot working metal must have an allowance added to offset the shrinkage that occurs in the piece after forming. To demonstrate this Print 123 shows a hot punch used to make a depression in a shaft.

Questions	Answers
1. Give dimensions A, B, C, D, E, F, and G.	1. A____ B____ C____ D____ E____ F____ G____
2. Allowing 1/8″ material for machining, what would be the smallest sized rectangular piece of steel required to make the object?	2. ____
3. What do dashed lines indicated by X represent?	3. ____
4. What does dashed line Y represent?	4. ____
5. Give the length dimension of the thread.	5. ____
6. Give diameters of cylindrical parts.	6. ____
7. With shrinkage added what are the actual measurements?	7. 27/32R____ 3/16____ 1/2 R____ 9/16____

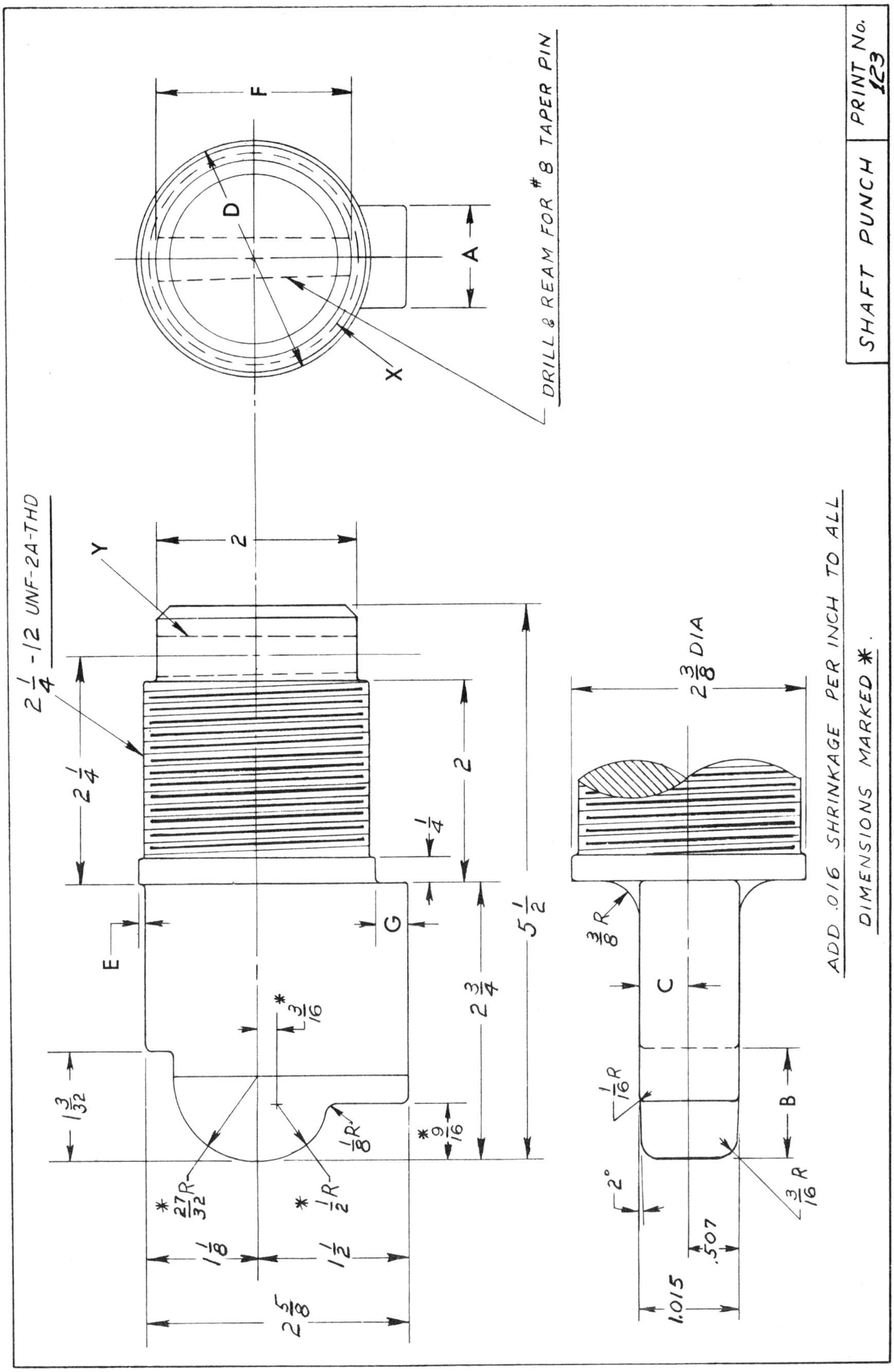
SHAFT PUNCH
PRINT No. 123
DRILL & REAM FOR #8 TAPER PIN
2 1/4 -12 UNF-2A-THD
2 3/8 DIA
ADD .016 SHRINKAGE PER INCH TO ALL DIMENSIONS MARKED *

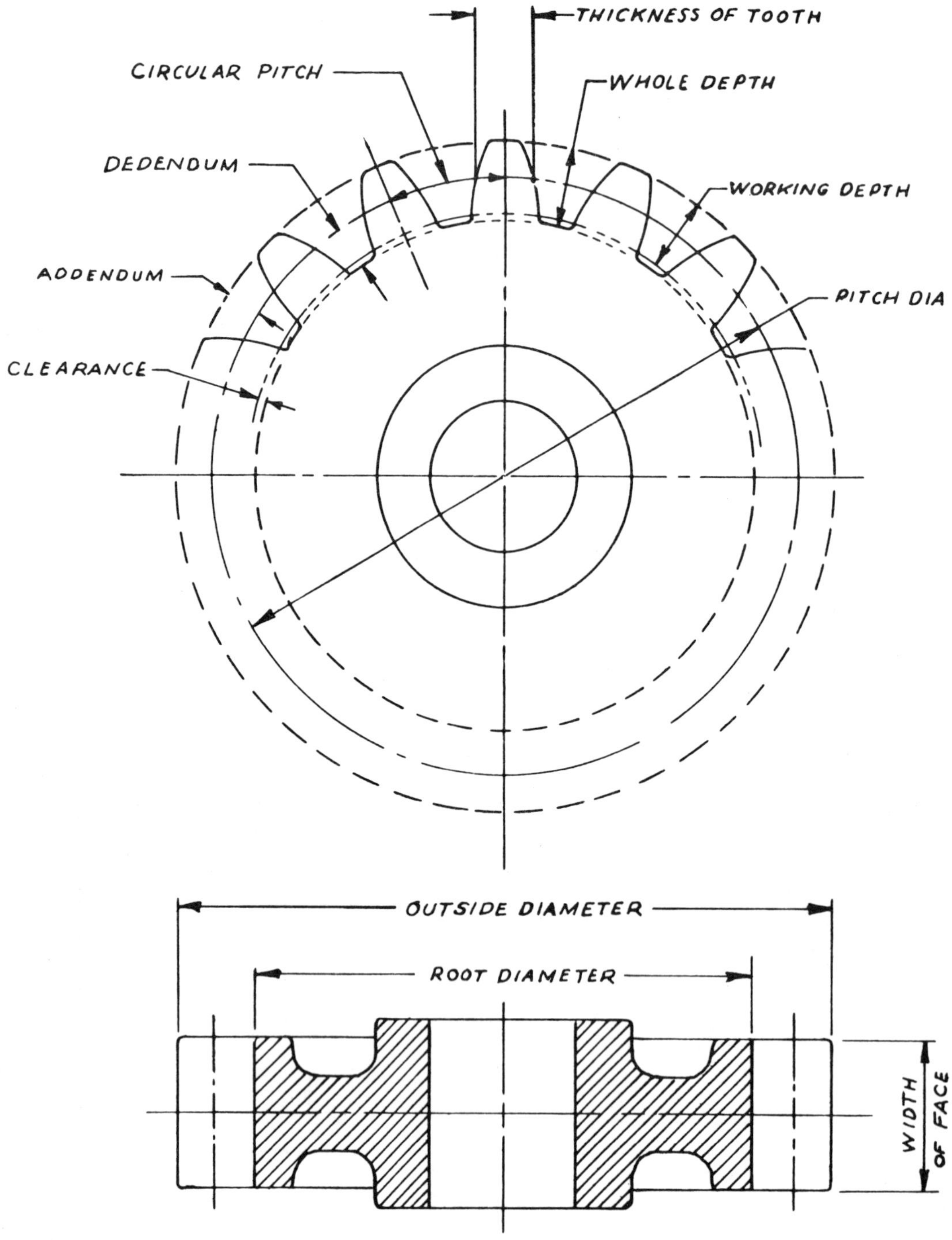

SPUR GEAR TERMS

PRINT No.
S-30

Rules and Formulas for Dimensions of Spur Gears*

No. of Rule	To Find	Rule	Formula
1	Diametral Pitch	Divide 3.1416 by circular pitch.	$P = \frac{3.1416}{P'}$
2	Circular Pitch	Divide 3.1416 by diametral pitch.	$P' = \frac{3.1416}{P}$
3	Pitch Diameter	Divide number of teeth by diametral pitch.	$D = \frac{N}{P}$
4	Pitch Diameter	Multiply number of teeth by circular pitch and divide the product by 3.1416.	$D = \frac{NP'}{3.1416}$
5	Center Distance	Add the number of teeth in both gears and divide the sum by two times the diametral pitch.	$C = \frac{N_g + N_p}{2P}$
6	Center Distance	Multiply the sum of the number of teeth in both gears by circular pitch and divide the product by 6.2832.	$C = \frac{(N_g + N_p)P'}{6.2832}$
7	Addendum	Divide 1 by diametral pitch.	$S = \frac{1}{P}$
8	Addendum	Divide circular pitch by 3.1416.	$S = \frac{P'}{3.1416}$
9	Clearance	Divide 0.157 by diametral pitch.	$F = \frac{0.157}{P}$
10	Clearance	Divide circular pitch by 20.	$F = \frac{P'}{20}$
11	Whole Depth of Tooth	Divide 2.157 by diametral pitch.	$W = \frac{2.157}{P}$
12	Whole Depth of Tooth	Multiply 0.6866 by circular pitch.	$W = 0.6866 P'$
13	Thickness of Tooth	Divide 1.5708 by diametral pitch.	$T = \frac{1.5708}{P}$
14	Thickness of Tooth	Divide circular pitch by 2.	$T = \frac{P'}{2}$
15	Outside Diameter	Add 2 to the number of teeth and divide the sum by diametral pitch.	$O = \frac{N+2}{P}$
16	Outside Diameter	Multiply the sum of the number of teeth plus 2 by circular pitch and divide the product by 3.1416.	$O = \frac{(N+2)P'}{3.1416}$
17	Diametral Pitch	Divide number of teeth by pitch diameter	$P = \frac{N}{D}$
18	Circular Pitch	Multiply pitch diameter by 3.1416 and divide by number of teeth.	$P = \frac{3.1416\,D}{N}$
19	Pitch Diameter	Subtract two times the addendum from outside diameter.	$D = O - 2S$
20	Number of Teeth	Multiply pitch diameter by diametral pitch.	$N = P \times D$
21	Number of Teeth	Multiply pitch diameter by 3.1416 and divide the product by circular pitch.	$N = \frac{3.1416\,D}{P'}$
22	Outside Diameter	Add two times the addendum to the pitch diameter.	$O = D + 2S$
23	Length of Rack	Multiply number of teeth in rack by 3.1416 and divide by diametral pitch.	$L = \frac{3.1416\,N}{P}$
24	Length of Rack	Multiply the number of teeth in the rack by circular pitch.	$L = NP'$

*Reprinted from MACHINERY'S Handbook.
Copyrighted by The Industrial Press, New York.

See Print No. S-30 for Terms

Symbols Used in Spur Gear Calculations

P = Diametral pitch
C = Center distance
N = Number of teeth (if the number of teeth in both gear and pinion are referred to, Ng = number of teeth in gear, and Np = number of teeth in pinion)
F = Clearance
T = Thickness of tooth
D = Pitch diameter
P[1] = Circular pitch
S = Addendum
W = Whole depth of tooth
O = Outside diameter of gear

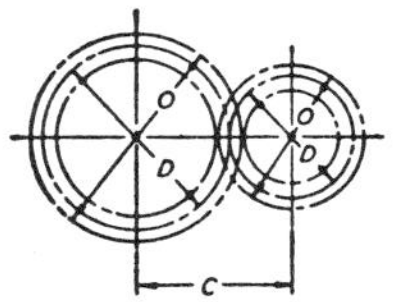

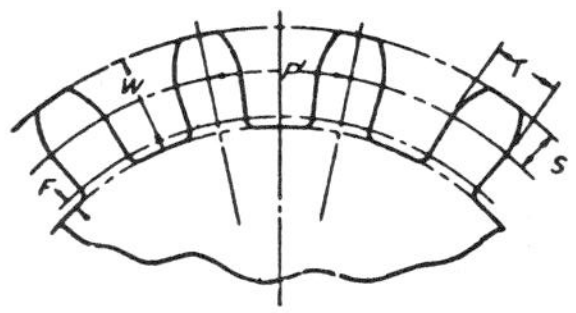

Rules and Formulas for Internal Spur Gears*

(Where rules and formulas are not given, they are the same as for external gears)

No. of Rule	To Find	Rule	Formula
5	Center Distance	Subtract the number of teeth in the pinion from the number of teeth in the gear and divide the remainder by 2 times the diametral pitch.	$C = \frac{N_g - N_p}{2P}$
6	Center Distance	Multiply the difference of the numbers of teeth in the gear and pinion by the circular pitch and divide the product by 6.2832.	$C = \frac{(N_g - N_p)P'}{6.2832}$
15	Inside Diameter	Subtract 2 from the number of teeth and divide the remainder by the diametral pitch.	$I = \frac{N-2}{P}$
16	Inside Diameter	Subtract 2 from the number of teeth, multiply the remainder by the circular pitch, and divide the product by 3.1416.	$I = \frac{(N-2)P'}{3.1416}$
19	Pitch Diameter	Add twice the addendum to the inside diameter.	$D = I + 2S$
22	Inside Diameter	Subtract twice the addendum from the pitch diameter.	$I = D - 2S$

*Reprinted from MACHINERY'S Handbook.
Copyrighted by The Industrial Press, New York.

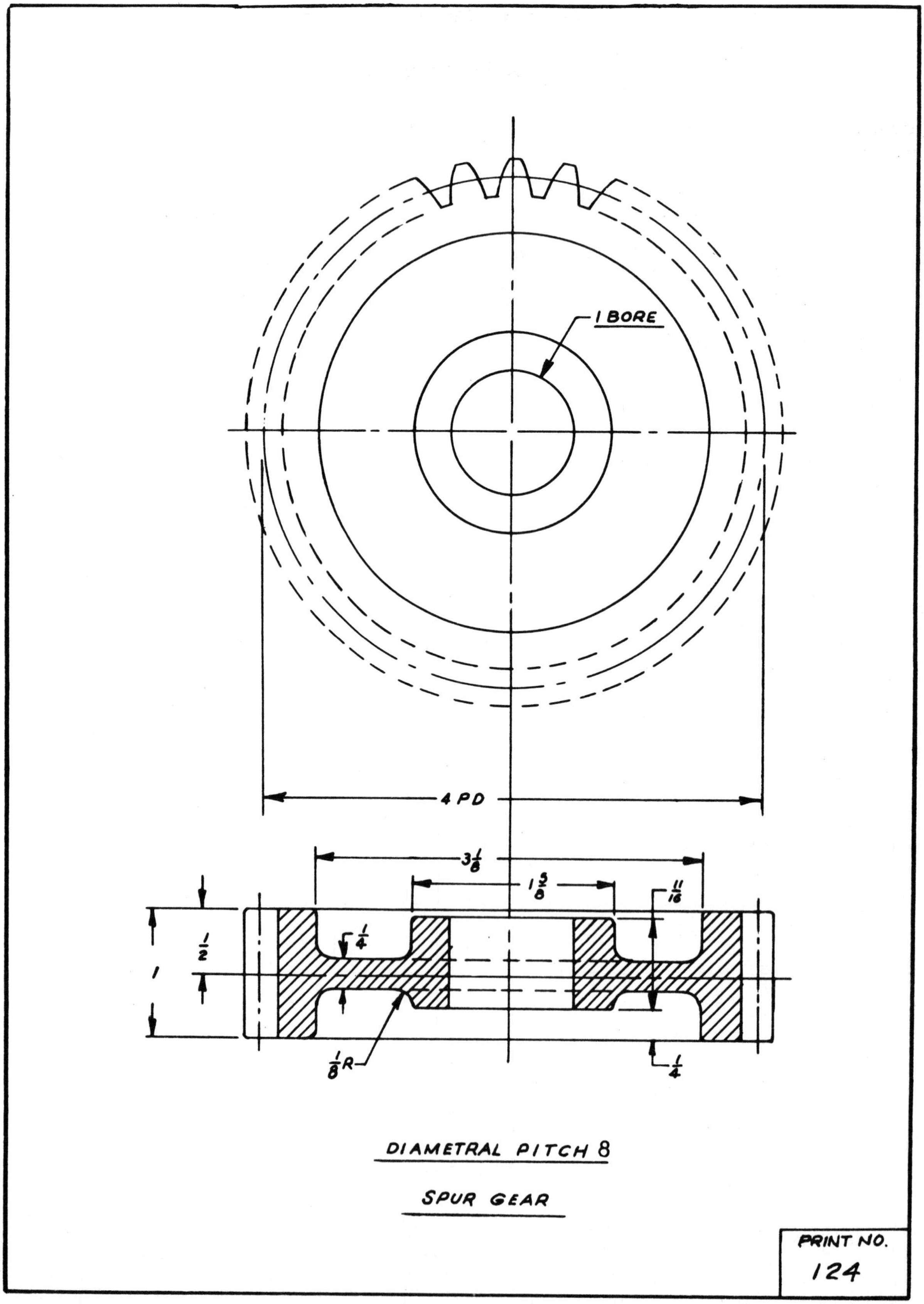
1 BORE
4 PD
3 1/8
1 5/8
11/16
1/4
1/2
1
1/8 R
1/4
DIAMETRAL PITCH 8
SPUR GEAR
PRINT NO.
124

TRADE COMPETENCY TEST

For Print No. 124

Student's Name____________________ Instructor's Name____________________

T E A R O F F H E R E

Questions	Answers
1. Give the number of teeth in gear.	1. __________
2. Give the circular pitch of tooth.	2. __________
3. Give the addendum of tooth.	3. __________
4. Give the dedendum of tooth.	4. __________
5. Give the clearance of tooth.	5. __________
6. Give the thickness of tooth.	6. __________
7. Give the pitch diameter of gear.	7. __________
8. Give the outside diameter of gear.	8. __________
9. Give the root diameter of gear.	9. __________
10. Give the width of the face.	10. __________
11. What type gear is shown?	11. __________
12. a. If this gear had 40 teeth, what would the pitch diameter be?	12. a__________
b. The outside diameter?	b__________
13. Is the hub length symmetrical about the gear ℄ ?	13. __________
14. What is the length dimension of the hub?	14. __________
15. Does the hub have a keyway?	15. __________
16. Does the gear have spokes or arms?	16. __________

TRADE COMPETENCY TEST

For Print No. 125

Student's Name____________________ Instructor's Name____________________

Questions	Answers
1. What is the length of dimension A?	1. __________
2. What are the over-all dimensions of the gear?	2. __________
3. What kind of gear is shown?	3. __________
4. Do the section lines represent the material specified for the gear?	4. __________
5. How many teeth are to be milled in the gear?	5. __________
6. Give the addendum of the tooth.	6. __________
7. Give the dedendum of the tooth.	7. __________
8. Give the whole depth of the tooth.	8. __________
9. Give the clearance of the tooth.	9. __________
10. Give the thickness of the tooth at the pitch line.	10. __________
11. How is the surface of the teeth finished?	11. __________
12. What is the pressure angle of the tooth?	12. __________
13. What is the diameter of the hole?	13. __________
14. How is the gear hardened?	14. __________

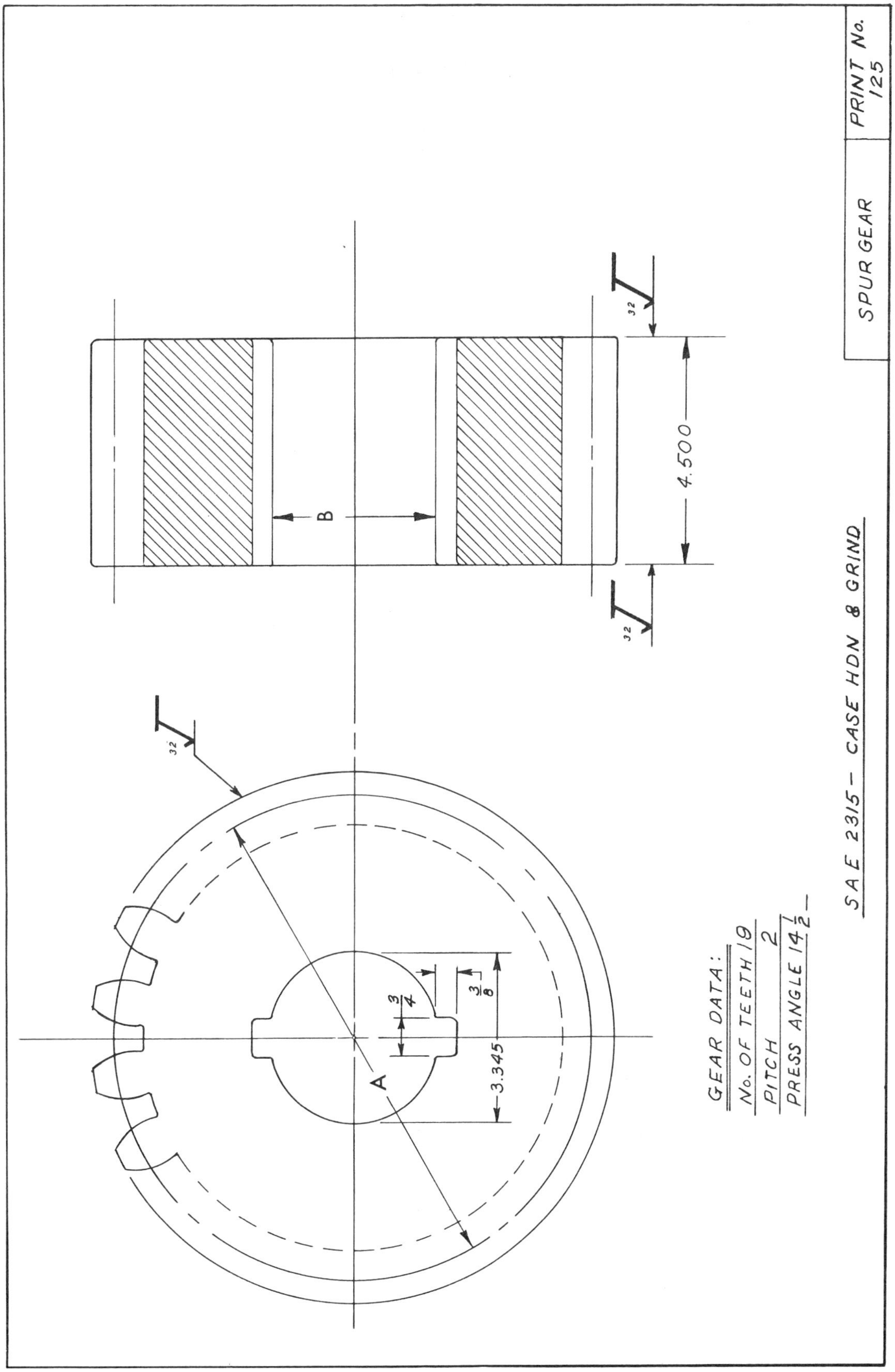
PRINT No. 125
SPUR GEAR
SAE 2315 - CASE HDN & GRIND
GEAR DATA:
No. OF TEETH 19
PITCH 2
PRESS ANGLE 14 1/2
4.500
B
A
3.345
3/4
3/8
32

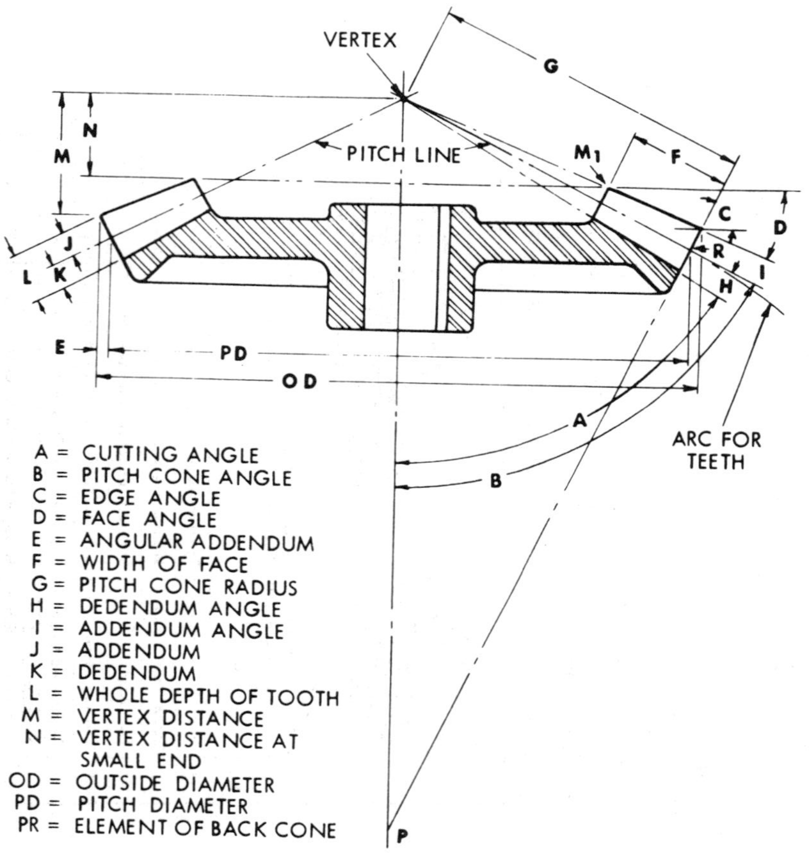

A = CUTTING ANGLE
B = PITCH CONE ANGLE
C = EDGE ANGLE
D = FACE ANGLE
E = ANGULAR ADDENDUM
F = WIDTH OF FACE
G = PITCH CONE RADIUS
H = DEDENDUM ANGLE
I = ADDENDUM ANGLE
J = ADDENDUM
K = DEDENDUM
L = WHOLE DEPTH OF TOOTH
M = VERTEX DISTANCE
N = VERTEX DISTANCE AT SMALL END
OD = OUTSIDE DIAMETER
PD = PITCH DIAMETER
PR = ELEMENT OF BACK CONE

PRINT NO. S-35

Rules and Formulas for Calculating Bevel Gears with Shafts at Right Angles*

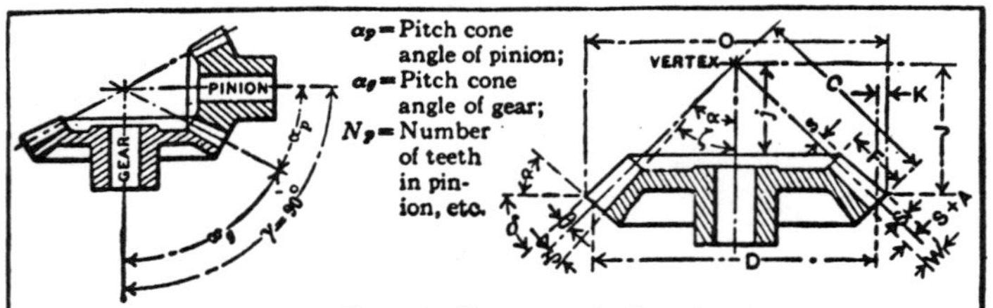

Use Rules and Formulas Nos. 1 to 21 in the order given.

No.	To Find	Rule	Formula
1	Pitch Cone Angle (or Edge Angle) of Pinion.	Divide the number of teeth in the pinion by the number of teeth in the gear to get the tangent.	$\tan \alpha_p = \frac{N_p}{N_g}$
2	Pitch Cone Angle (or Edge Angle) of Gear.	Divide the number of teeth in the gear by the number of teeth in the pinion to get the tangent.	$\tan \alpha_g = \frac{N_g}{N_p}$
3	Proof of Calculations for Pitch Cone Angles.	The sum of the pitch cone angles of the pinion and gear equals 90 degrees.	$\alpha_p + \alpha_g = 90°$
4	Pitch Diameter.	Divide the number of teeth by the diametral pitch; or multiply the number of teeth by the circular pitch and divide by 3.1416.	$D = \frac{N}{P} = \frac{NP'}{\pi}$
5	Addendum.	Divide 1.0 by the diametral pitch; or multiply the circular pitch by 0.318.	$S = \frac{1.0}{P} = 0.318\,P'$
6	Dedendum.	Divide 1.157 by the diametral pitch; or multiply the circular pitch by 0.368.	$S + A = \frac{1.157}{P} = 0.368\,P'$
7	Whole Depth of Tooth Space.	Divide 2.157 by the diametral pitch; or multiply the circular pitch by 0.687.	$W = \frac{2.157}{P} = 0.687\,P'$
8	Thickness of Tooth at Pitch Line.	Divide 1.571 by the diametral pitch; or divide the circular pitch by 2.	$T = \frac{1.571}{P} = \frac{P'}{2}$
9	Pitch Cone Radius.	Divide the pitch diameter by twice the sine of the pitch cone angle.	$C = \frac{D}{2 \times \sin \alpha}$
10	Addendum of Small End of Tooth.	Subtract the width of face from the pitch cone radius, divide the remainder by the pitch cone radius and multiply by the addendum.	$s = S \times \frac{C - F}{C}$
11	Thickness of Tooth at Pitch Line at Small End.	Subtract the width of face from the pitch cone radius, divide the remainder by the pitch cone radius and multiply by the thickness of the tooth at the pitch line.	$t = T \times \frac{C - F}{C}$
12	Addendum Angle.	Divide the addendum by the pitch cone radius to get the tangent.	$\tan \theta = \frac{S}{C}$
13	Dedendum Angle.	Divide the dedendum by the pitch cone radius to get the tangent.	$\tan \phi = \frac{S + A}{C}$

These dimensions are the same for both gear and pinion. (Nos. 5–13)

*Reprinted from MACHINERY'S Handbook.

See Print No. S-35 for Terms

Rules and Formulas for Calculating Bevel Gears with Shafts at Right Angles*

No.	To Find	Rule	Formula
14	Face Angle.	Subtract the sum of the pitch cone and addendum angles from 90 degrees.	$\delta = 90° - (\alpha + \theta)$
15	Cutting Angle.	Subtract the dedendum angle from the pitch cone angle	$\zeta = \alpha - \phi$
16	Angular Addendum.	Multiply the addendum by the cosine of the pitch cone angle.	$K = S \times \cos\alpha$
17	Outside Diameter.	Add twice the angular addendum to the pitch diameter.	$O = D + 2K$
18	Vertex or Apex Distance	Multiply one-half the outside diameter by the tangent of the face angle.	$J = \frac{O}{2} \times \tan\delta$
19	Vertex Distance at Small End of Tooth	Subtract the width of face from the pitch cone radius; divide the remainder by the pitch cone radius and multiply by the apex distance.	$j = J \times \frac{C - F}{C}$
20	Number of Teeth for which to Select Cutter	Divide the number of teeth by the cosine of the pitch cone angle.	$N' = \frac{N}{\cos\alpha}$
21	Proof of Calculations by Rules Nos. 9, 12, 14, 16 and 17.	The outside diameter equals twice the pitch cone radius multiplied by the cosine of the face angle and divided by the cosine of the addendum angle.	$O = \frac{2C \times \cos\delta}{\cos\theta}$

*Reprinted from MACHINERY'S Handbook.
Copyrighted by The Industrial Press, New York.

Symbols Used in Bevel Gear Calculations

P = Diametral pitch
P^1 = Circular pitch
α = Pitch cone angle and edge angle
D = Pitch diameter
S+A = Dedendum
A = Clearance
T = Thickness of tooth at pitch line
s = Addendum at small end of tooth
ϕ = Dedendum angle
θ = Addendum angle
K = Angular addendum
J = Vertex distance
N′ = Number of teeth for which to select cutter, also called "number of teeth in equivalent spur gear".

N = Number of teeth
π = 3.1416
γ = Center angle
S = Addendum
F = Width of face
C = Pitch cone radius
W = Whole depth of tooth space
t = Thickness of tooth at small end on pitch line
δ = Face angle
ζ = Cutting angle
O = Outside diameter (edge diameter for internal gears)
j = Vertex distance at small end

Rules and Formulas for Calculating Miter-Bevel Gearing*

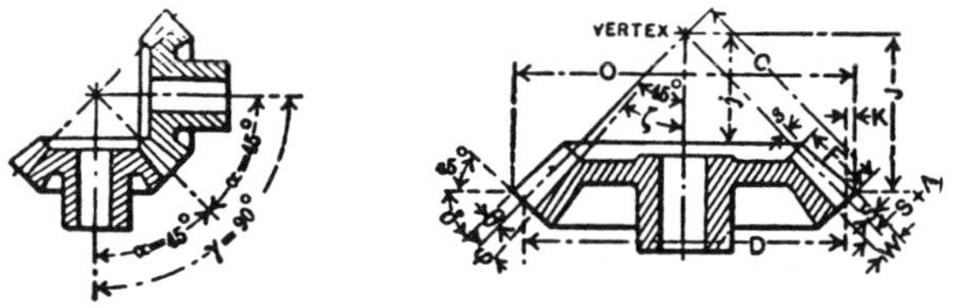

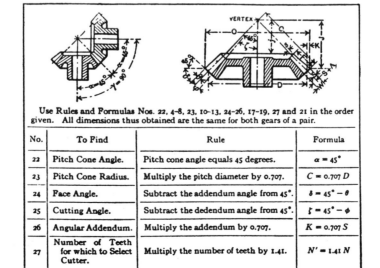

Use Rules and Formulas Nos. 22, 4–8, 23, 10–13, 24–26, 17–19, 27 and 21 in the order given. All dimensions thus obtained are the same for both gears of a pair.

No.	To Find	Rule	Formula
22	Pitch Cone Angle.	Pitch cone angle equals 45 degrees.	$\alpha = 45°$
23	Pitch Cone Radius.	Multiply the pitch diameter by 0.707.	$C = 0.707\,D$
24	Face Angle.	Subtract the addendum angle from 45°.	$\delta = 45° - \theta$
25	Cutting Angle.	Subtract the dedendum angle from 45°.	$\zeta = 45° - \phi$
26	Angular Addendum.	Multiply the addendum by 0.707.	$K = 0.707\,S$
27	Number of Teeth for which to Select Cutter.	Multiply the number of teeth by 1.41.	$N' = 1.41\,N$

*Reprinted from MACHINERY'S Handbook.
Copyrighted by The Industrial Press, New York.

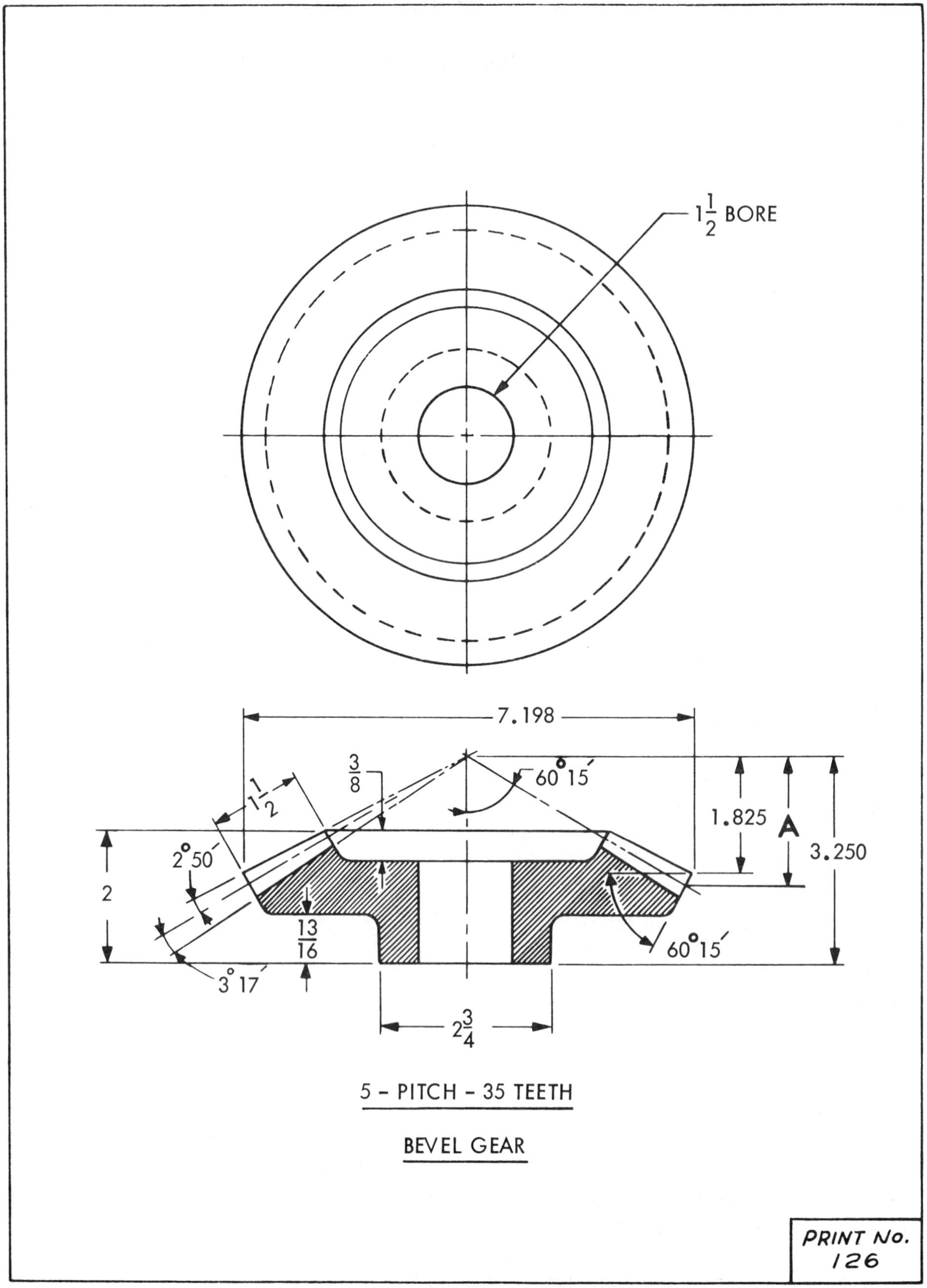
1 1/2 BORE
7.198
3/8
60° 15′
1 1/2
1.825
A
3.250
2° 50′
2
13/16
60° 15′
3° 17′
2 3/4
5 - PITCH - 35 TEETH
BEVEL GEAR
PRINT No. 126

TRADE COMPETENCY TEST

For Print No. 126

Student's Name ______________________ Instructor's Name ______________________

TEAR OFF HERE

Questions	Answers
1. Give the pitch diameter of gear.	1. ________
2. Give the outside diameter of gear.	2. ________
3. Give the width of face.	3. ________
4. Give the cutting angle.	4. ________
5. Give the pitch cone angle.	5. ________
6. Give the vertex distance at the large end of the tooth.	6. ________
7. Give the addendum at large end.	7. ________
8. Give the dedendum at large end.	8. ________
9. If the 35-tooth gear had 7 teeth, what would the pitch diameter be?	9. ________
10. What type of gear is shown?	10. ________
11. What is the outside diameter of the hub of the gear?	11. ________
12. What is the whole depth of tooth space at the large end?	12. ________
13. What would be the pitch cone angle of a mating gear mounted at a right angle with this gear?	13. ________
14. Dimension A is 2.000″. What is the pitch diameter of the mating pinion?	14. ________
15. How many teeth are in the mating pinion?	15. ________

TRADE COMPETENCY TEST

For Print No. 127

Student's Name________________________ Instructor's Name________________________

The hand wheel on Print 127 is used to operate the screw that controls the clamping device on a testing fixture. The spinning of the hand wheel to move the clamp in an open and closed position is a relatively quick and easy operation.

Questions	Answers
1. Give the dimensions A, B, C, D, and E.	1. A______ B______ C______ D______ E______
2. Give the amount of taper of the spokes.	2. ______
3. What is the number of the pattern?	3. ______
4. Give the maximum diameter of the bore.	4. ______
5. How may spokes does it have?	5. ______
6. Are the spokes round, flat, or oval?	6. ______
7. What material is the part made of?	7. ______
8. How much below the top surface of the rim is the top surface of the hub?	8. ______
9. What do the dashed lines in the section represent?	9. ______
10. What is the outside diameter of the hub?	10. ______

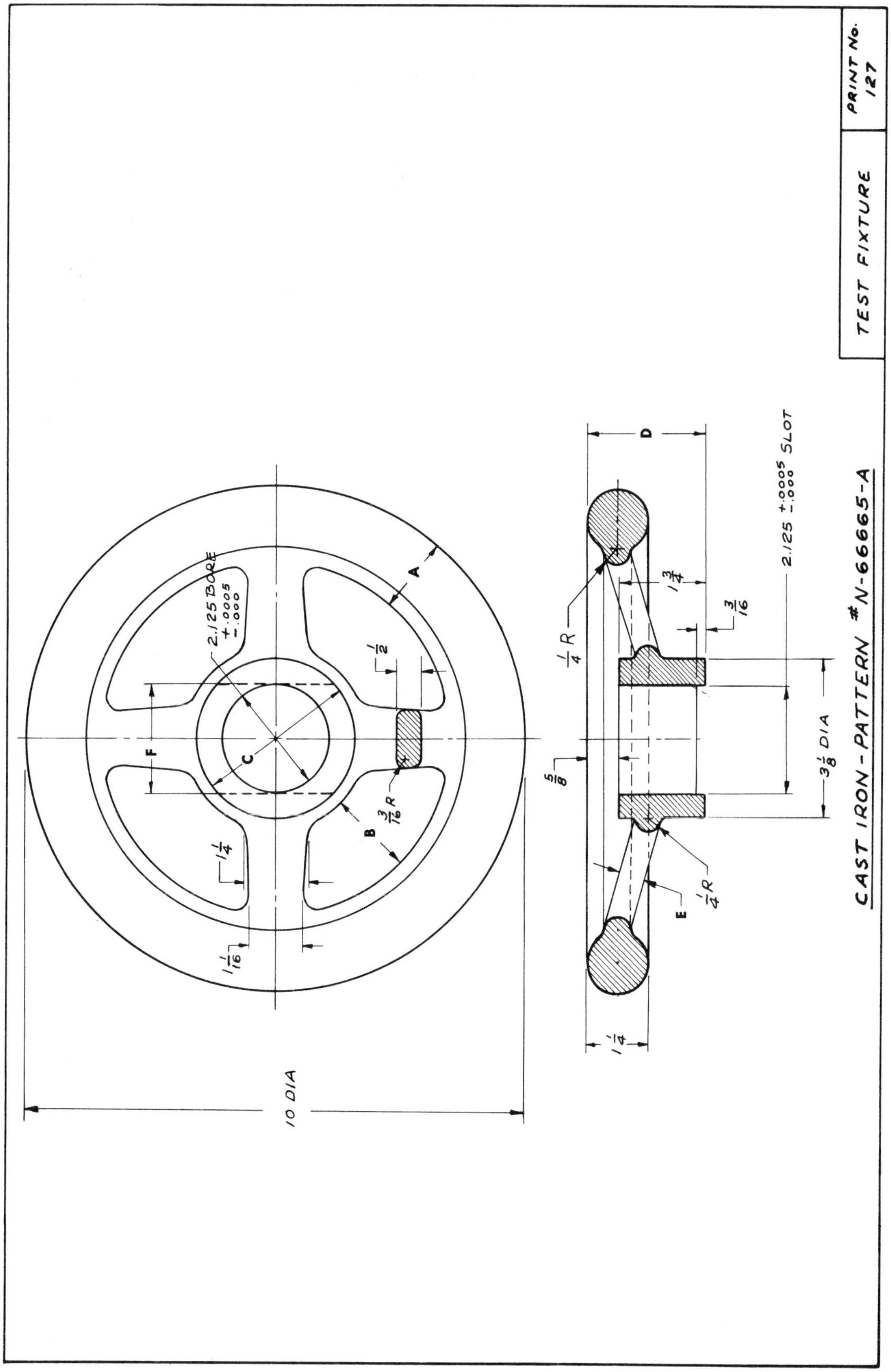
PRINT No. 127
TEST FIXTURE
CAST IRON - PATTERN #N-66665-A
10 DIA
2.125 BORE +.0005 -.000
2.125 +.0005 -.000 SLOT
3 1/8 DIA
1/4 R
3/16 R
1/2
1 1/4
1 1/16
1 3/4
3/16
5/8
A
B
C
D
E
F

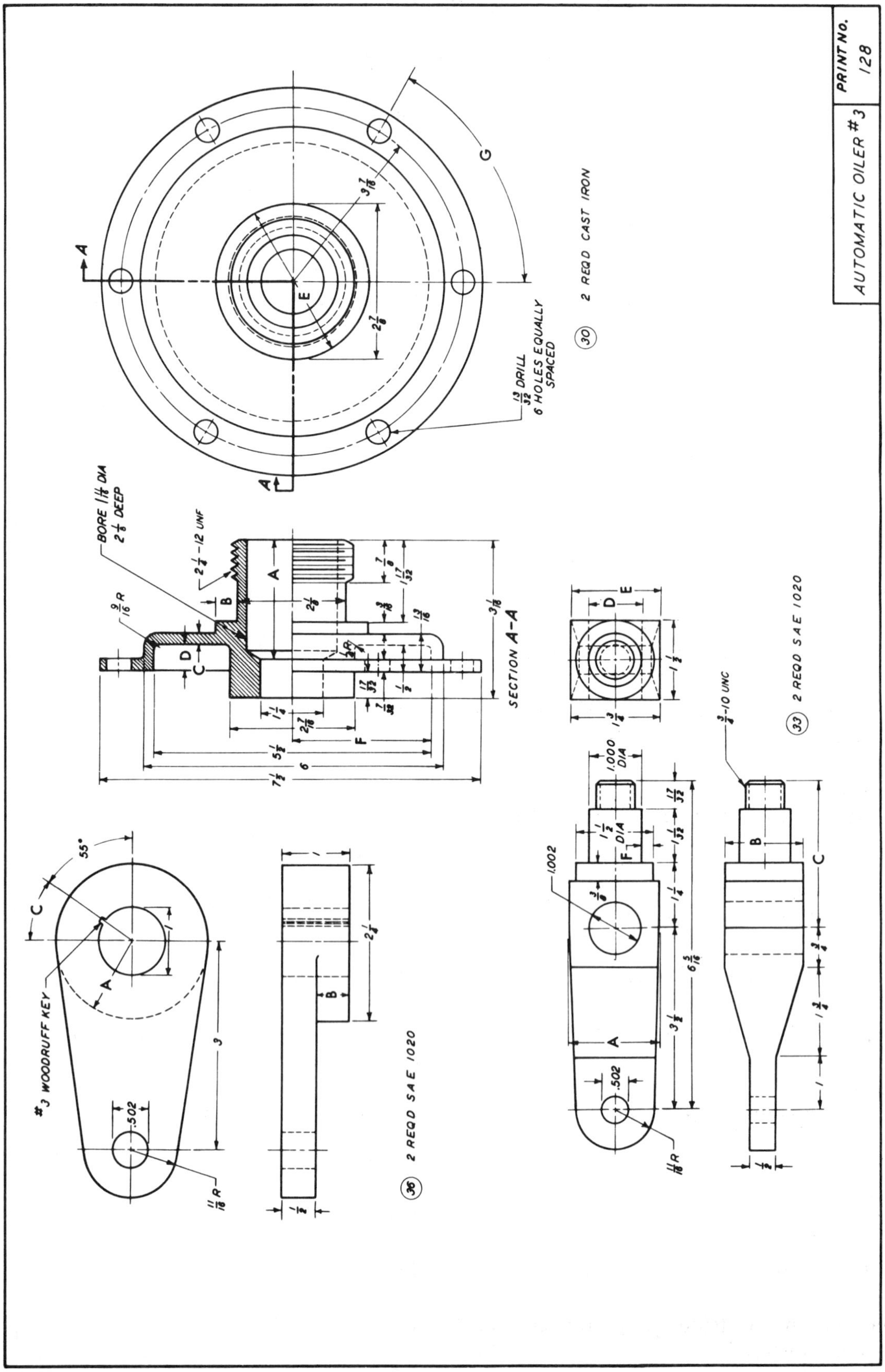
AUTOMATIC OILER #3
PRINT NO. 128
2 REQD CAST IRON
6 HOLES EQUALLY SPACED
SECTION A-A
BORE
2 REQD SAE 1020
#3 WOODRUFF KEY
2 REQD SAE 1020

TRADE COMPETENCY TEST

For Print No. 128

Student's Name________________ Instructor's Name________________

Prototypes, or the initial working models of machines, are frequently made with simplified designs. This practice facilitates the construction of the machine insofar as it eliminates the need for pattern and die equipment. The parts, though sometimes crude in shape, are functionally correct in design. Detail 30 is an exception to this, however, since its shape allows it to be easily machined to the contour of the finished casting. Details 33 and 36 will be designed as cast levers for the production models and their shape will be substantially altered.

T E A R O F F H E R E

Questions	Answers
Detail No. 36	
1. Give dimensions A and B.	1. A________ B________
2. Give angle C.	2. ________
3. Give the length, width, and height dimensions of lever.	3. ________
4. What is the ℄ dimension between holes?	4. ________
5. How could you make the .502″ hole in this piece?	5. ________
Detail No. 33	
6. Give dimensions A, B, C, D, E, and F.	6. A________ B________ C________ D________ E________ F________
7. How many pieces are required?	7. ________
8. What kind of material is to be used?	8. ________
9. Give the over-all size of this part.	9. ________
Detail No. 30	
10. Give dimensions A, B, C, D, E, and F.	10. A________ B________ C________ D________ E________ F________
11. Give angle G.	11. ________
12. What is the bolt circle diameter?	12. ________
13. What is the outside diameter of the threaded portion?	13. ________

TRADE COMPETENCY TEST

For Print No. 129

Student's Name____________________ Instructor's Name____________________

The same parts are sometimes required for the left and right hand sides of many machines. The automotive tie rod is a good example and also shows how parts are designed so that both the left and right-hand can be machined from the same rough forging, casting, or stamping. Section A-A shows that identical bosses are forged on either side of the large boss. The .373/.380 hole is drilled on one side only for the right hand tie rod and on the other side for the left hand. The 1-16 Am-NAT right-hand thread is also made left hand for the opposite tie rod.

Questions	Answers
1. Give the dimensions A, B, C, D, E, F, G, H, I, and J.	1. A____ B____ C____ D____ E____ F____ G____ H____ I____ J____
2. What do lines indicated by L represent?	2. ____
3. What does dashed line M indicate?	3. ____
4. Does the 5/32″ slot go through to the threaded hole?	4. ____
5. Is any part of the 17/32″ hole threaded?	5. ____
6. Note the oval shaped, dashed line construction in left end center of front view. Is this a hole or a projection?	6. ____
7. What view shows the dimension of G?	7. ____
8. What draft angle is specified?	8. ____

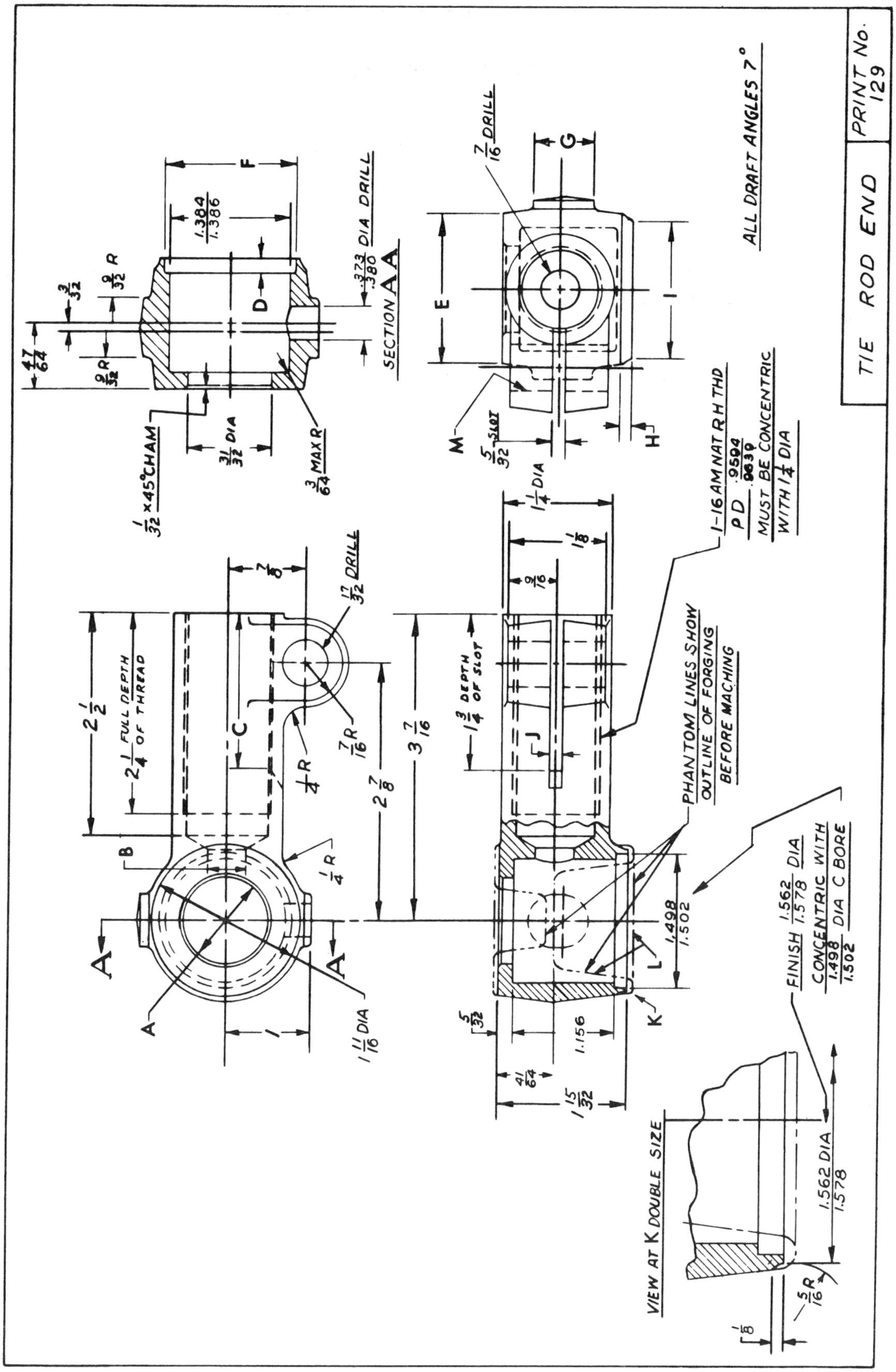
TIE ROD END
PRINT No. 129
ALL DRAFT ANGLES 7°
SECTION AA
7/16 DRILL
17/32 DRILL
PHANTOM LINES SHOW OUTLINE OF FORGING BEFORE MACHING
1-16 AM NAT R H THD
MUST BE CONCENTRIC WITH 1 1/4 DIA
FINISH 1.562/1.578 DIA CONCENTRIC WITH 1.498/1.502 DIA C BORE
VIEW AT K DOUBLE SIZE
1.562 DIA 1.578

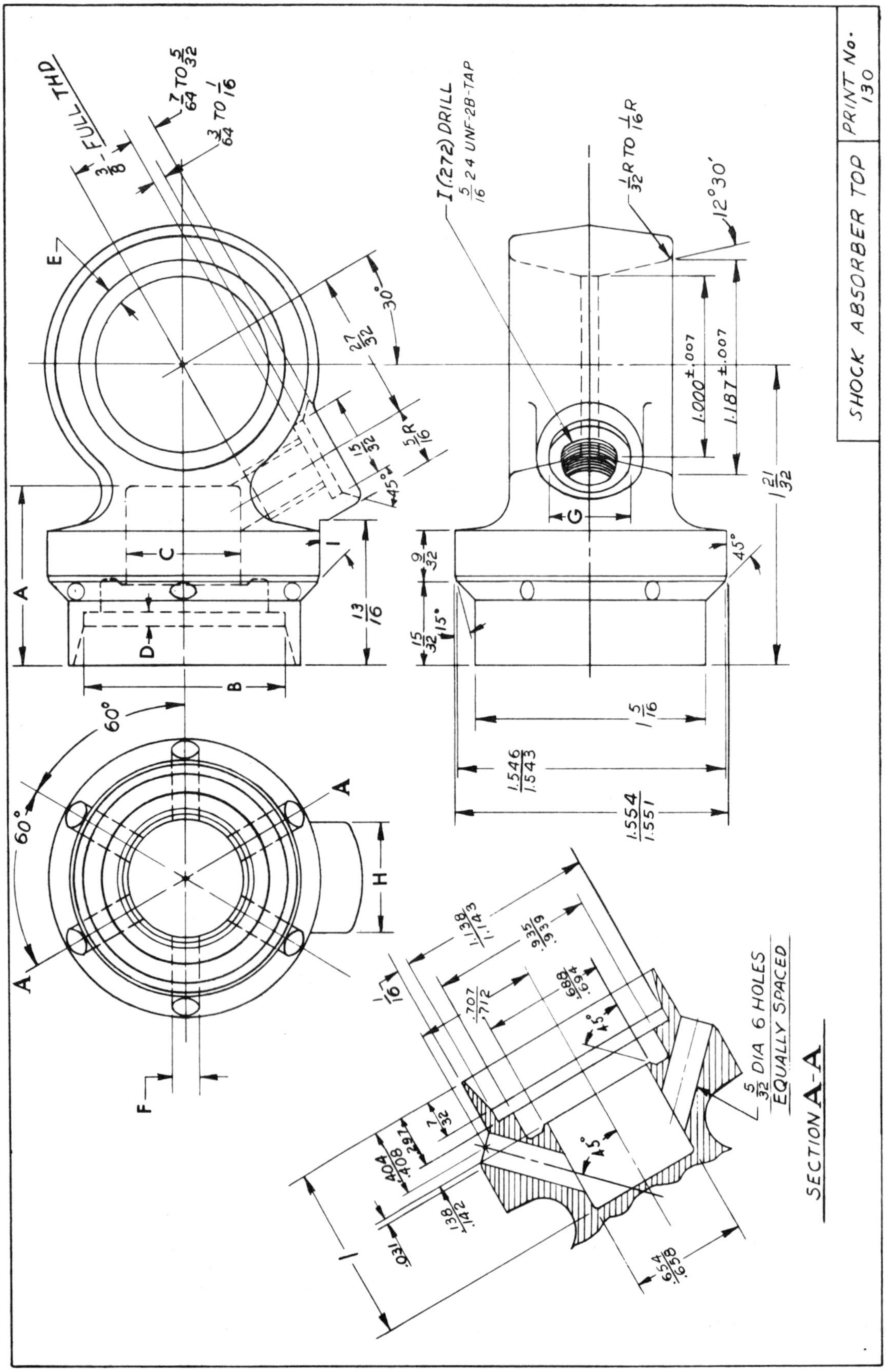
SHOCK ABSORBER TOP
PRINT No. 130
SECTION A-A
5/32 DIA 6 HOLES EQUALLY SPACED
I (.272) DRILL
5/16 24 UNF-2B-TAP
3/8 - FULL THD
7/64 TO 5/32
3/64 TO 1/16
1/32 R TO 1/16 R
12°30′
1.000 ±.007
1.187 ±.007
1 21/32
27/32
30°
15/32
5/16 R
45°
9/32
13/16
15/32
15°
1 5/16
1.546
1.543
1.554
1.551
60°
60°
1.138
1.143
.935
.939
.707
.712
.688
.694
1/16
45°
45°
7/32
.404
.408
.297
.138
.142
.031
.654
.658
A
B
C
D
E
F
G
H
I

TRADE COMPETENCY TEST

For Print No. 130

Student's Name________________ Instructor's Name________________

Incomplete drawings are frequently made when the processes of manufacture are not closely related. An example of this practice is shown in the Shock Absorber Top. This is a drawing of the finished part and contains only the information necessary to perform the finish operations. You will note the absence of the dimensions of those surfaces that are not machined. Conversely, the machining dimensions are not included on the forging drawing.

TEAR OFF HERE

Questions	Answers
1. Give the dimensions A, B, C, D, E, F, G, and H.	1. A____ B____ C____ D____ E____ F____ G____ H____
2. What is angle I?	2. ________
3. Are all invisible surfaces shown in the front view?	3. ________
4. Why can they be omitted without making the description of the object incomplete?	4. ________
5. Is there a line in the front view shown solid that should be shown dashed?	5. ________
6. How many threads per inch are used in the threaded hole?	6. ________
7. What is the angular spacing of the 5/32 holes?	7. ________
8. At what angle to the ℄ are the 5/32 holes drilled?	8. ________
9. At what angle to the horizontal ℄ is the threaded hole drilled?	9. ________
10. What is the linear distance that locates the center of the threaded hole from the center of the boss?	10. ________

TRADE COMPETENCY TEST

For Print No. 131

Student's Name____________________ Instructor's Name____________________

Phantom lines are used to show the finish machined flange. Only the critical dimensions, as they relate from the rough to the finished part, are added to the drawing. These dimensions are 3/32″, 1/32″ and 1/16″ dimensions in the top view.

Questions	Answers
1. Give dimensions A and B.	1. A________ B________
2. What is radius F?	2. ________
3. The phantom lines indicated by G represent the over-all length of the machined or finished flange. What is the length of the machined or finished flange?	3. ________
4. What two things does view H describe about the object which couldn't be described in the other view?	4. ________
5. Give the draft angles specified.	5. ________
6. Give the over-all length.	6. ________
7. Give the height dimension.	7. ________
8. Give the width dimension.	8. ________
9. Give dimension C.	9. ________
10. Give dimension E.	10. ________

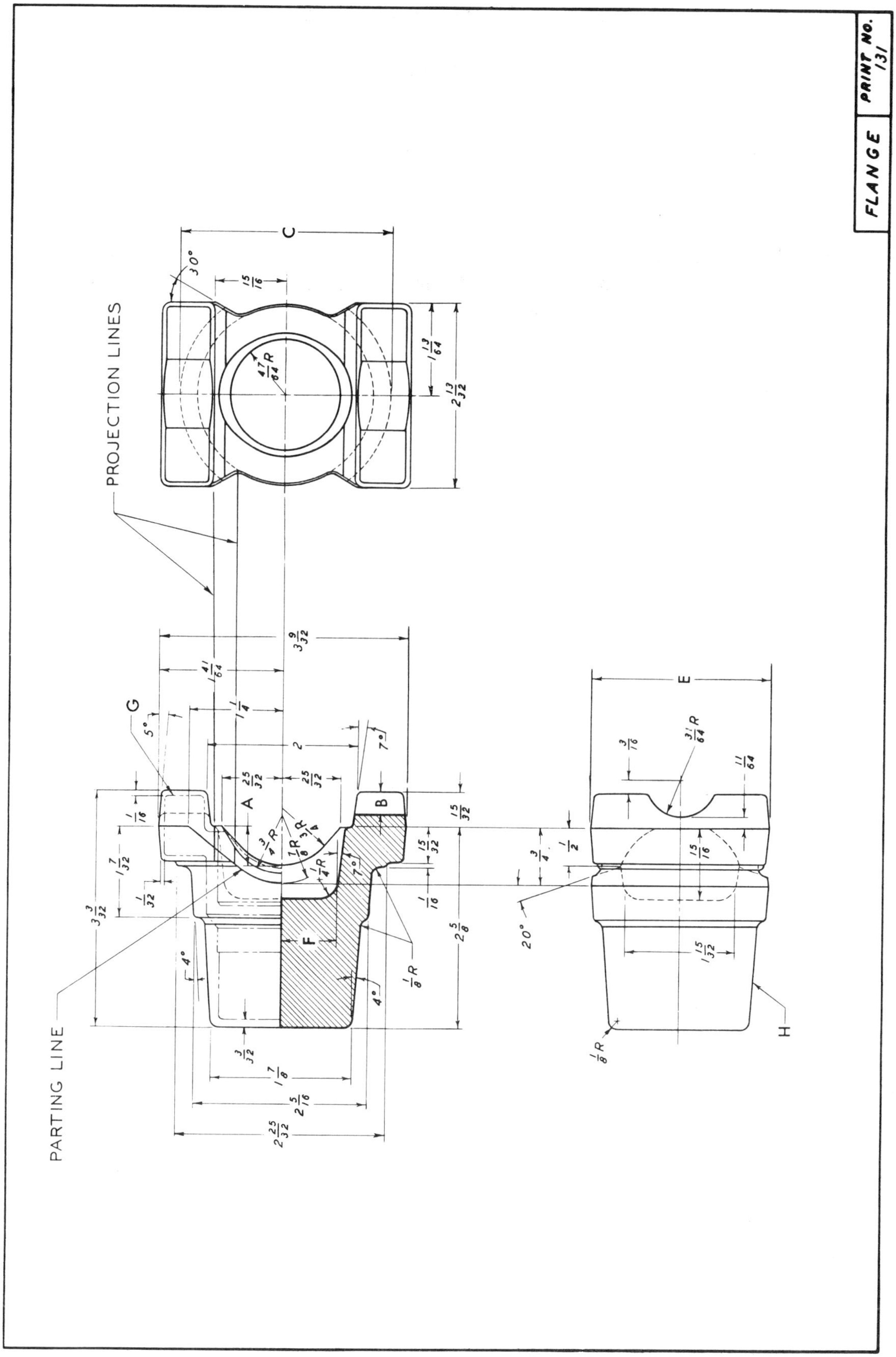
PRINT NO. 131
FLANGE
PROJECTION LINES
PARTING LINE
A
B
C
E
F
G
H

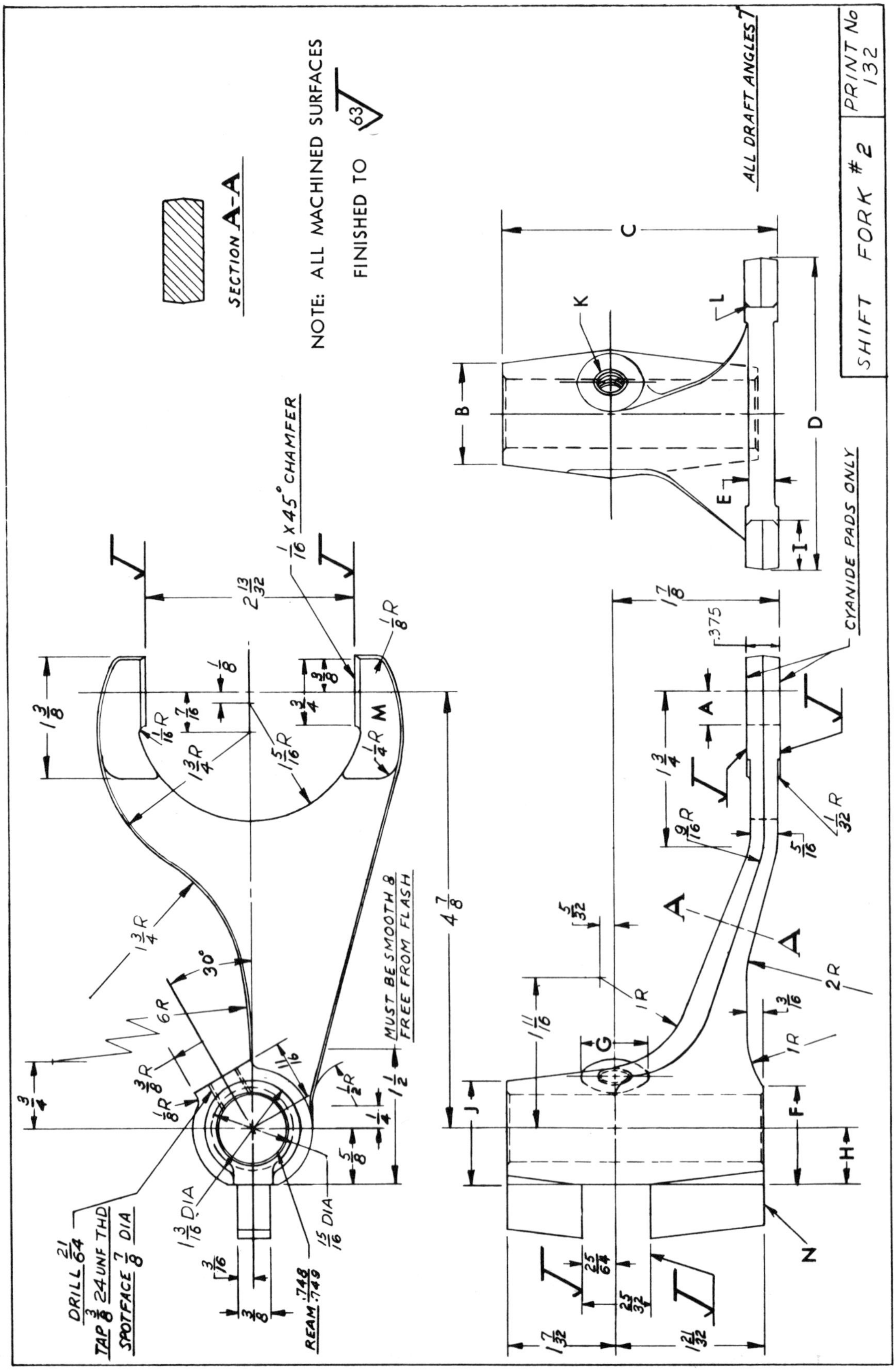
SECTION A-A
NOTE: ALL MACHINED SURFACES
FINISHED TO 63
ALL DRAFT ANGLES 7°
PRINT No
132
SHIFT FORK #2
1/16 X 45° CHAMFER
MUST BE SMOOTH & FREE FROM FLASH
CYANIDE PADS ONLY
DRILL 21/64
TAP 3/8 24 UNF THD
SPOTFACE 7/8 DIA
REAM .748/.749

TRADE COMPETENCY TEST

For Print No. 132

Student's Name________________ Instructor's Name________________

TEAR OFF HERE

Questions	Answers
1. Give dimensions A, B, C, D, E, F, G, H, I, and J.	1. A____ B____ C____ D____ E____ F____ G____ H____ I____ J____
2. Give the information necessary to machine hole indicated by K.	2. ____
3. What is the size of chamfer indicated by L?	3. ____
4. Is surface N above or below surface M?	4. ____
5. Is the .748/.749 reamed hole shown in the side view?	5. ____
6. What is the dimension between the pads?	6. ____
7. Give the parallel distance from surface M to the surface represented by line N in the front view.	7. ____
8. How would you describe the shape of Section A-A?	8. ____
9. What is the shape of the vertical part of the fork through which the reamed hole passes?	9. ____
10. Does the note "Must be smooth and free from flash" indicate that this is a forging?	10. ____
11. What is the draft angle specified?	11. ____

TRADE COMPETENCY TEST

For Print No. 133

Student's Name______________________ Instructor's Name______________________

Questions	Answers
1. Give dimensions A, B, C, D, E, F, G, H, I, J, K, L, M, N, O P, Q, R, and S.	1. A_____ B_____ C_____ D_____ E_____ F_____ G_____ H_____ I_____ J_____ K_____ L_____ M_____ N_____ O_____ P_____ Q_____ R_____ S_____
2. Is surface X in front of or behind surface W as seen in the front view?	2. __________
3. Is surface Y in front of or behind surface Z as seen in the right side view?	3. __________
4. How many degrees are in angle V in Section E-E?	4. __________
5. How many sectional views are shown?	5. __________
6. How many tapered holes are to be machined in the Steering Arm?	6. __________

Print 133 illustrates the method of dimensioning an object that has an irregular elevation around a curved contour. These dimensions are interpreted in the layout shown in Fig. 1. This layout is similar to that which is made by the skilled craftsman in preparing templates, or following gages, used in machining the cavities in the forging dies.

The centerline of the front view is laid out to exact scale, usually on sheet metal with a metal scriber. Small increments are measured along the curve and then transferred to a straight line extending from the tangential point of the 3-11/16″ radius. The center of the boss at #1 increment is then projected perpendicular to the base line. From this line the 5¼″ dimension determines the start of the curved elevation. The elevation of all increments can now be found by projecting them from the extension line perpendicular to the base line. For example, the elevation of increment #4 from the base line is equal to "X" and increment #9 is equal to "Y."

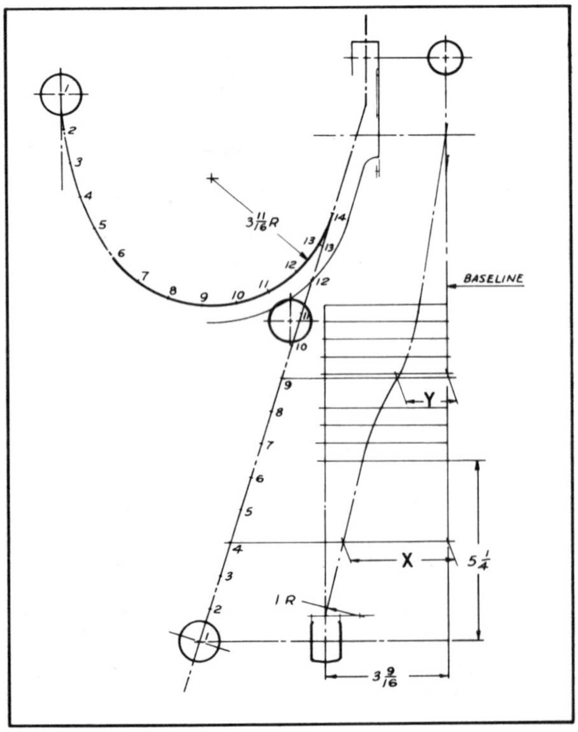

Fig. 1

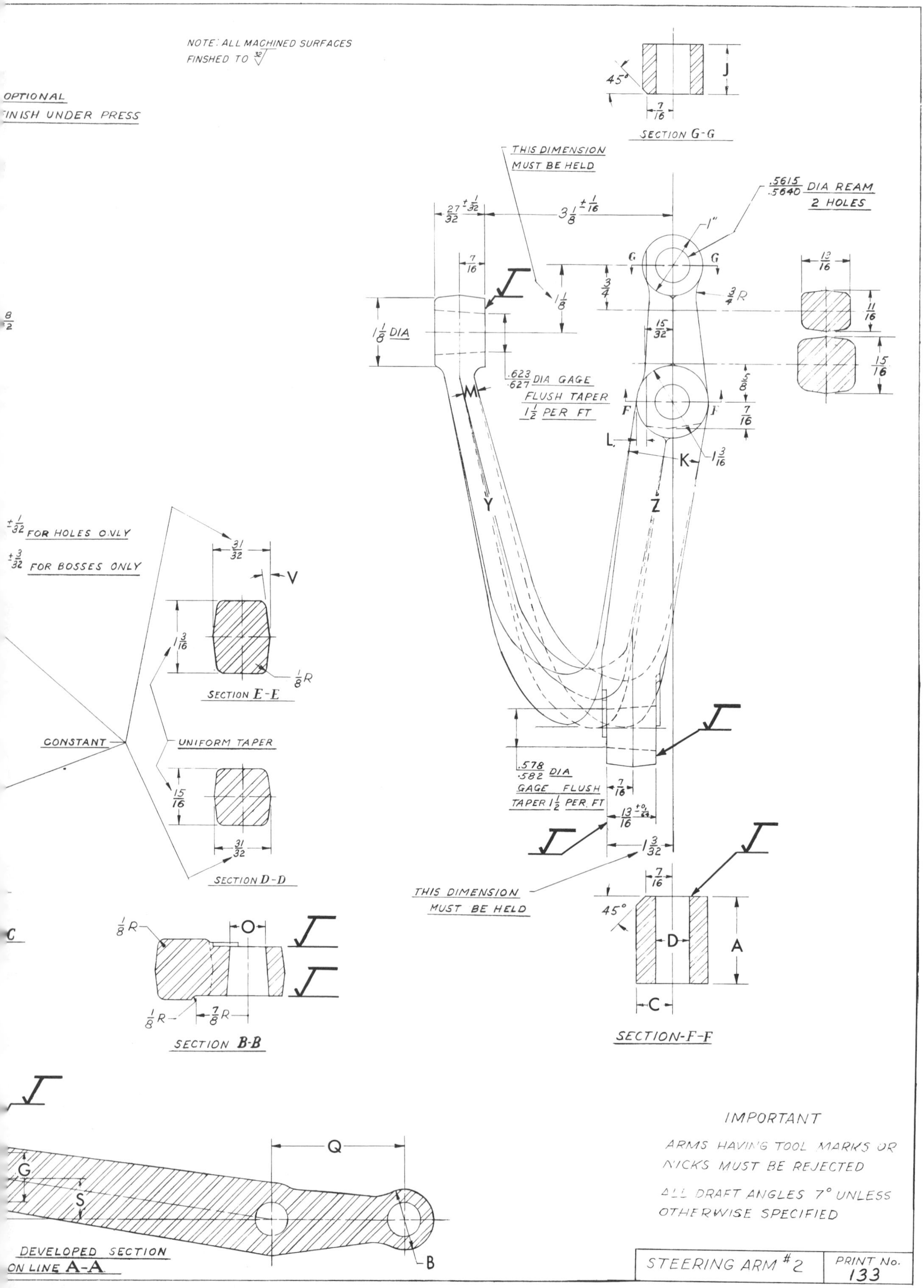
NOTE: ALL MACHINED SURFACES FINSHED TO 32
OPTIONAL
FINISH UNDER PRESS
SECTION G-G
THIS DIMENSION MUST BE HELD
.5615 .5640 DIA REAM 2 HOLES
1 1/8 DIA
.623 .627 DIA GAGE FLUSH TAPER 1 1/2 PER FT
± 1/32 FOR HOLES ONLY
± 3/32 FOR BOSSES ONLY
SECTION E-E
CONSTANT
UNIFORM TAPER
SECTION D-D
.578 .582 DIA GAGE FLUSH TAPER 1 1/2 PER. FT
THIS DIMENSION MUST BE HELD
SECTION B-B
SECTION-F-F
IMPORTANT
ARMS HAVING TOOL MARKS OR NICKS MUST BE REJECTED
ALL DRAFT ANGLES 7° UNLESS OTHERWISE SPECIFIED
DEVELOPED SECTION ON LINE A-A
STEERING ARM #2
PRINT No. 133

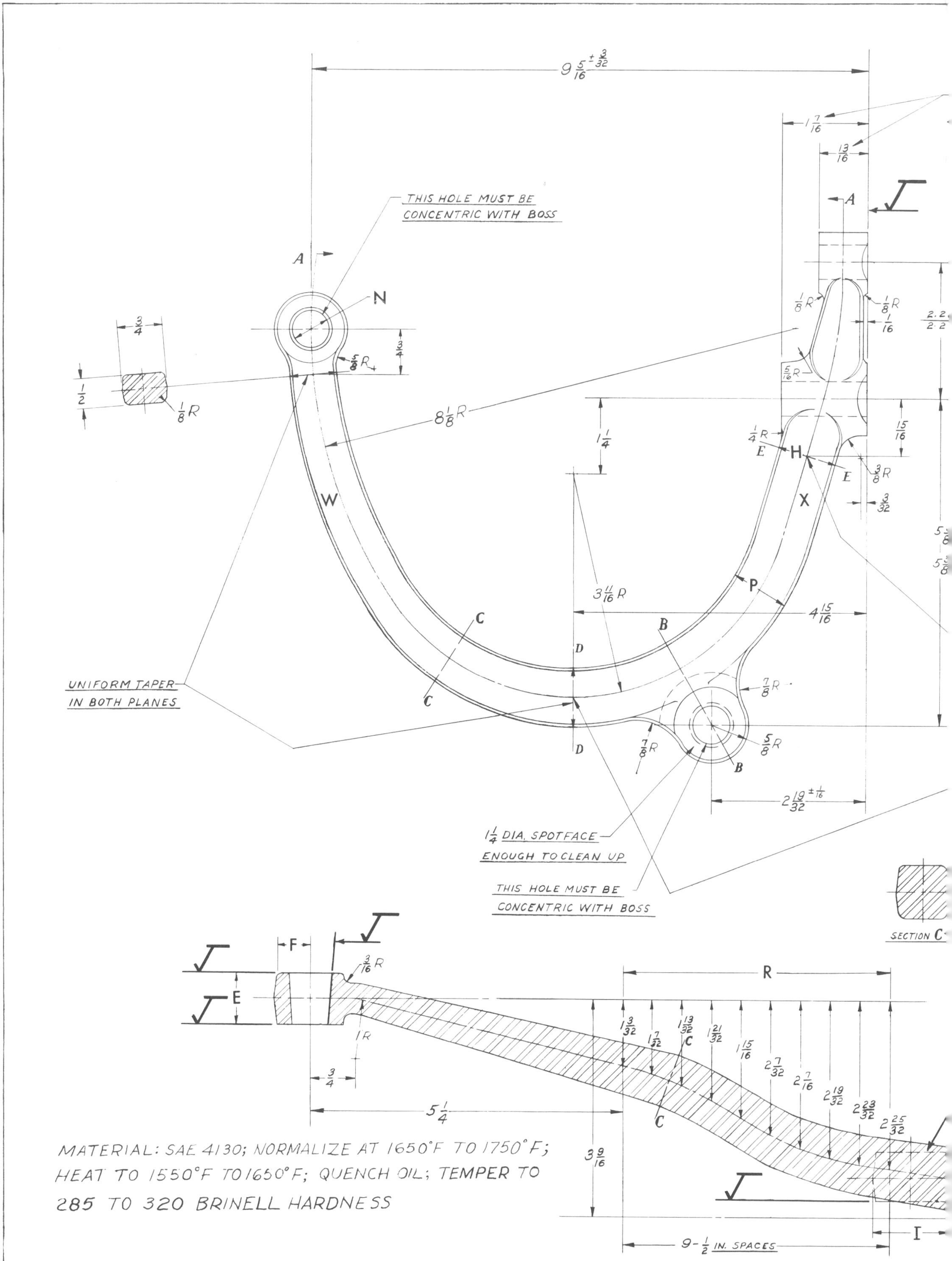
THIS HOLE MUST BE CONCENTRIC WITH BOSS
UNIFORM TAPER IN BOTH PLANES
1¼ DIA. SPOTFACE ENOUGH TO CLEAN UP
THIS HOLE MUST BE CONCENTRIC WITH BOSS
SECTION C
9-½ IN. SPACES
MATERIAL: SAE 4130; NORMALIZE AT 1650°F TO 1750°F; HEAT TO 1550°F TO 1650°F; QUENCH OIL; TEMPER TO 285 TO 320 BRINELL HARDNESS

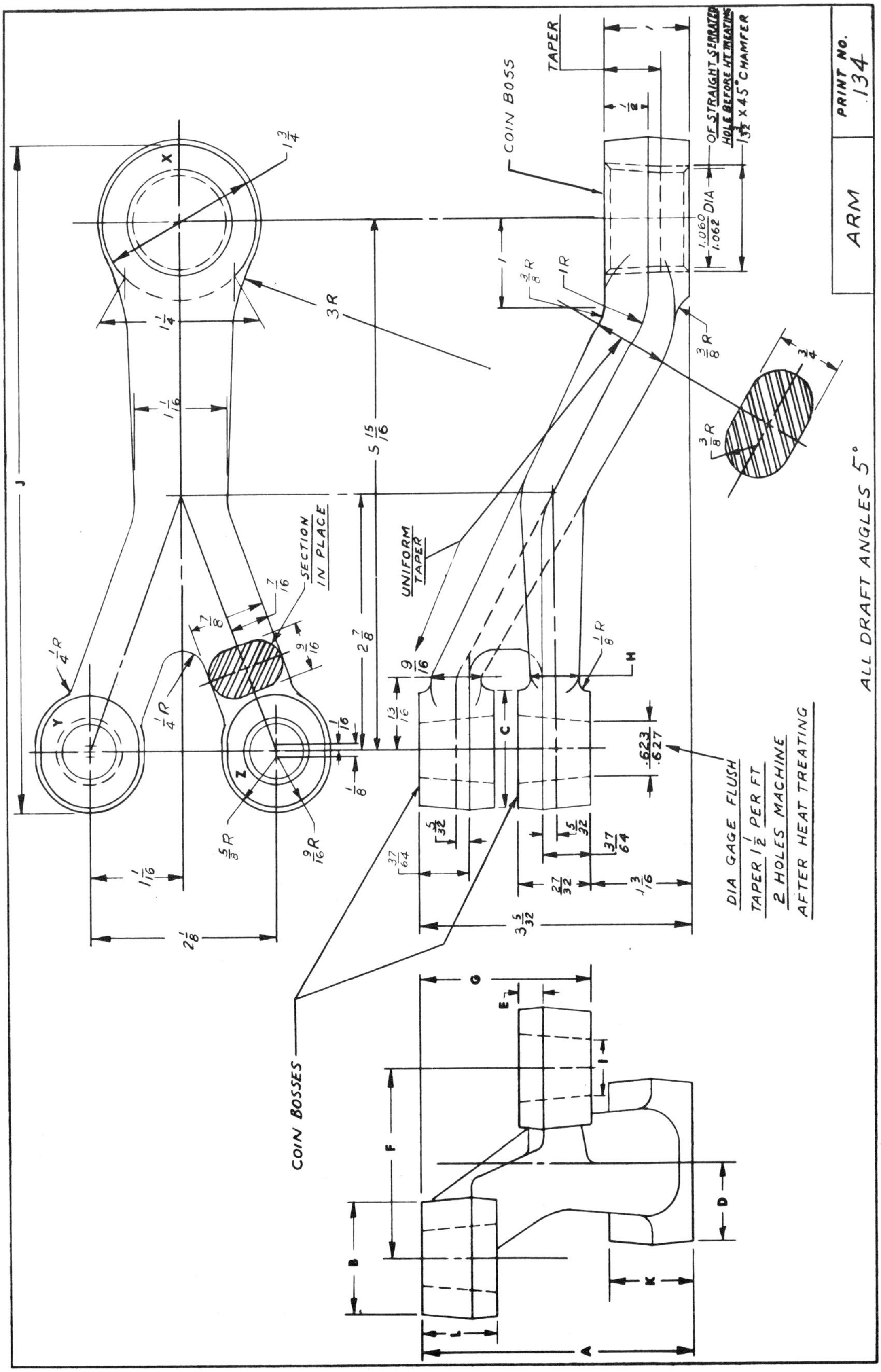
PRINT NO. 134
ARM
ALL DRAFT ANGLES 5°
COIN BOSS
COIN BOSSES
TAPER
OF STRAIGHT SERRATED HOLE BEFORE HT TREATING
1/32 X 45° CHAMFER
1.060/1.062 DIA
SECTION IN PLACE
UNIFORM TAPER
DIA GAGE FLUSH
TAPER 1 1/2 PER FT
2 HOLES MACHINE
AFTER HEAT TREATING
.623/.627
3R
5 15/16
2 7/8
2 1/8
3 5/32
X
Y
Z

TRADE COMPETENCY TEST

For Print No. 134

Student's Name______________________ Instructor's Name______________________

Questions	Answers
1. Give dimensions A, B, C, D, E, F, G, H, I, J, K, and L.	1. A______ B______ C______ D______ E______ F______ G______ H______ I______ J______ K______ L______
2. How many surfaces are to be coined?	2. ______________________
3. Is surface Y above or below surface Z?	3. ______________________
4. Is surface X above or below surface Z?	4. ______________________
5. Where is boss Y shown in the front view and left side view?	5. ______________________
6. Why does boss Y show a dashed circle?	6. ______________________
7. What is the difference in thickness between the largest and smallest sections of arm connecting the bosses?	7. ______________________
8. Which of the bosses in the front view is nearest to you, Y or Z?	8. ______________________
9. What is the tolerance on the small diameter of the holes in boss Z?	9. ______________________

TEAR OFF HERE

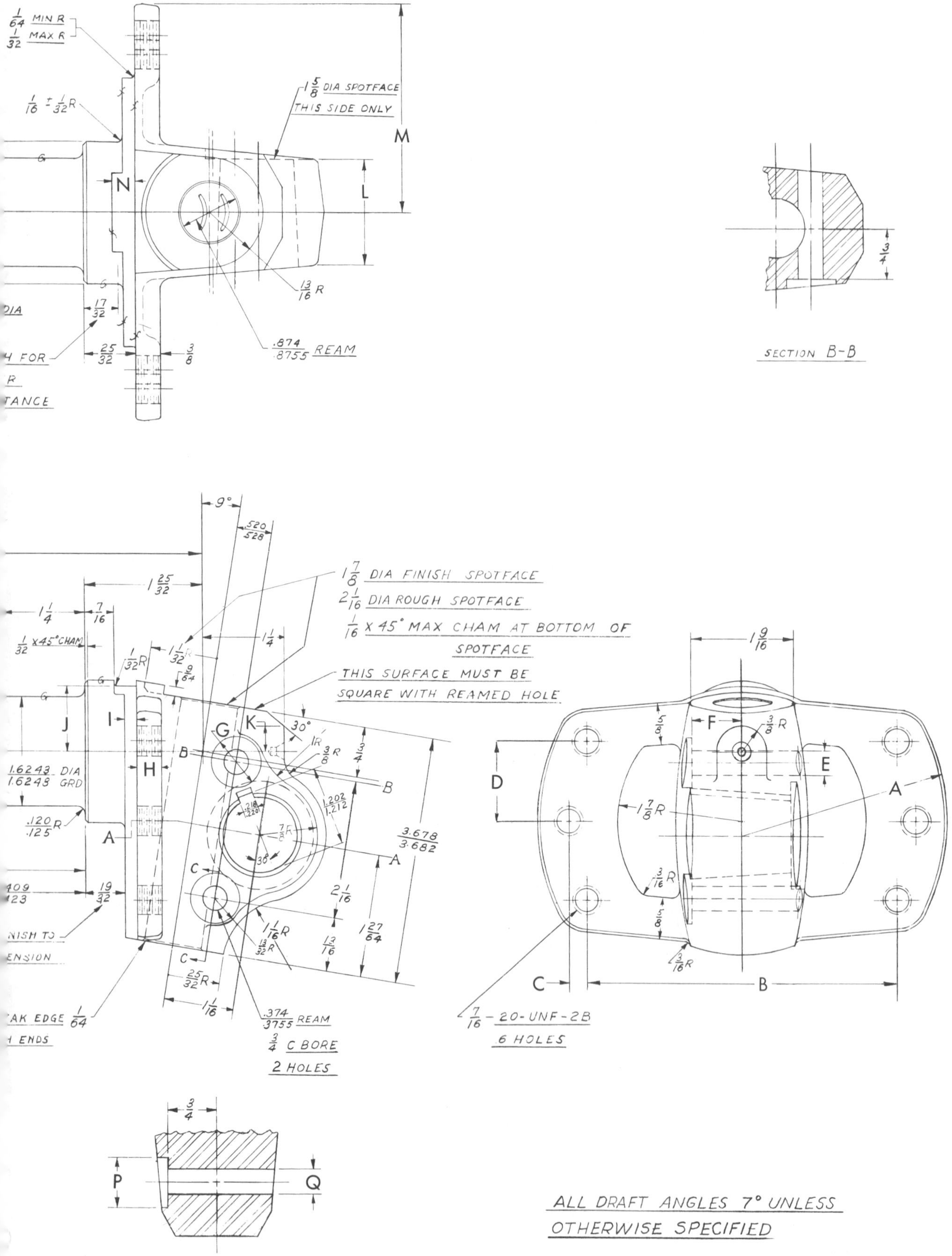

ALL DRAFT ANGLES 7° UNLESS OTHERWISE SPECIFIED

STEERING KNUCKLE	PRINT No 135

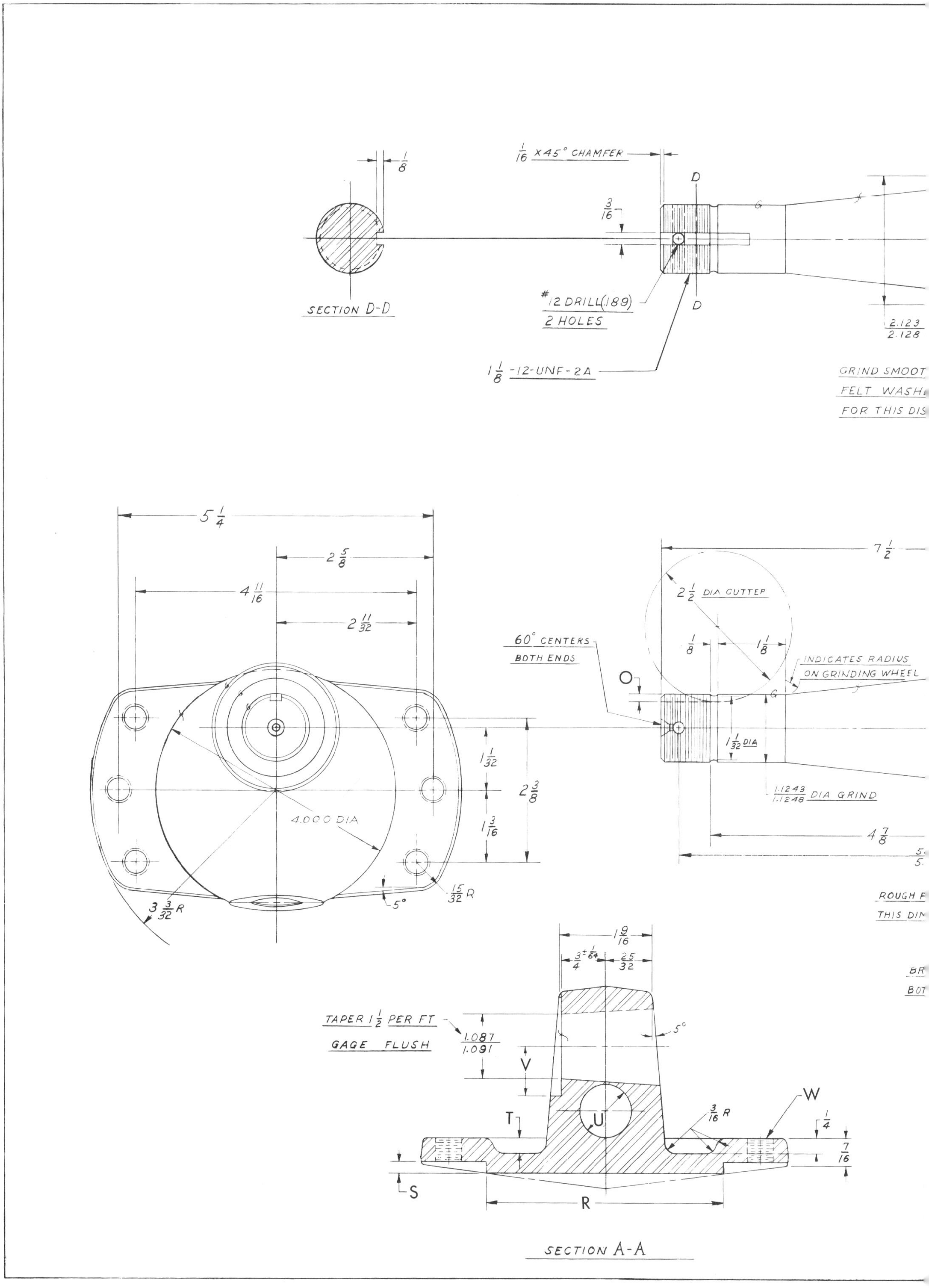

1/8
1/16 X 45° CHAMFER
D
3/16
SECTION D-D
#12 DRILL (.189)
2 HOLES
D
1 1/8 -12-UNF-2A
2.123
2.128
GRIND SMOOT
FELT WASH
FOR THIS DIS
5 1/4
2 5/8
4 11/16
2 11/32
1 1/32
2 3/8
1 3/16
4.000 DIA
3 3/32 R
15/32 R
5°
7 1/2
2 1/2 DIA CUTTER
60° CENTERS
BOTH ENDS
1/8
1 1/8
INDICATES RADIUS
ON GRINDING WHEEL
O
1 1/32 DIA
.1243
.1248 DIA GRIND
4 7/8
ROUGH F
THIS DIM
BR
BOT
1 9/16
3/4 ± 1/64
25/32
TAPER 1 1/2 PER FT
GAGE FLUSH
1.087
1.091
5°
V
W
T
U
3/16 R
1/4
7/16
S
R
SECTION A-A

TRADE COMPETENCY TEST

For Print No. 135

Student's Name________________ Instructor's Name________________

TEAR OFF HERE

Questions	Answers
1. Give dimensions A, B, C, D, E, F, G, H, I, J, K, L, M, N, O, P, Q, R, S, T, U, and V.	1. A____ B____ C____ D____ E____ F____ G____ H____ I____ J____ K____ L____ M____ N____ O____ P____ Q____ R____ S____ T____ U____ V____
2. Give necessary information to drill and tap hole indicated by W. (See chart in Appendix B.)	2. ____
3. How many sectional views are shown?	3. ____
4. How many holes go through the object?	4. ____
5. How many tapered holes go through the object?	5. ____
6. The king pin is a pin on which the steering knuckle pivots on the front axle. The angle formed between the wheel spindle and the king pin is not a right angle. Give the angular difference between this angle and a right angle.	6. ____
7. The steering knuckle arm is fastened to the steering knuckle with a tapered section. What is the lowest limit of the smallest diameter of the hole in which the arm is inserted?	7. ____

TRADE COMPETENCY TEST

For Print No. 136

Student's Name____________________ Instructor's Name____________________

Questions	Answers
1. Give the length of the following dimensions: G, H, I, J, K, L, M, N, O, P, R, S, W, and Z.	1. G_____ H_____ I_____ J_____ K_____ L_____ M_____ N_____ O_____ P_____ R_____ S_____ W_____ Z_____
2. Is surface T in front of or behind surface U as you look at the front view?	2. _____
3. Is surface V in front of or behind surface X as you look at the front view?	3. _____
4. Does dimension Y show the true length of pedal pad? Why?	4. _____
5. How many auxiliary views are shown?	5. _____
6. How many sectional views are shown?	6. _____
7. How is the recess in channel section shown in auxiliary view?	7. _____
8. Is the projection at W a true circle?	8. _____
9. Which view gives the answer to #8?	9. _____
10. In section F-F note extension with 11/32 R. Where is this extension located in the side view?	10. _____
11. What is the diameter of the opposite end to M?	11. _____
12. What is the thickness of extension at S?	12. _____

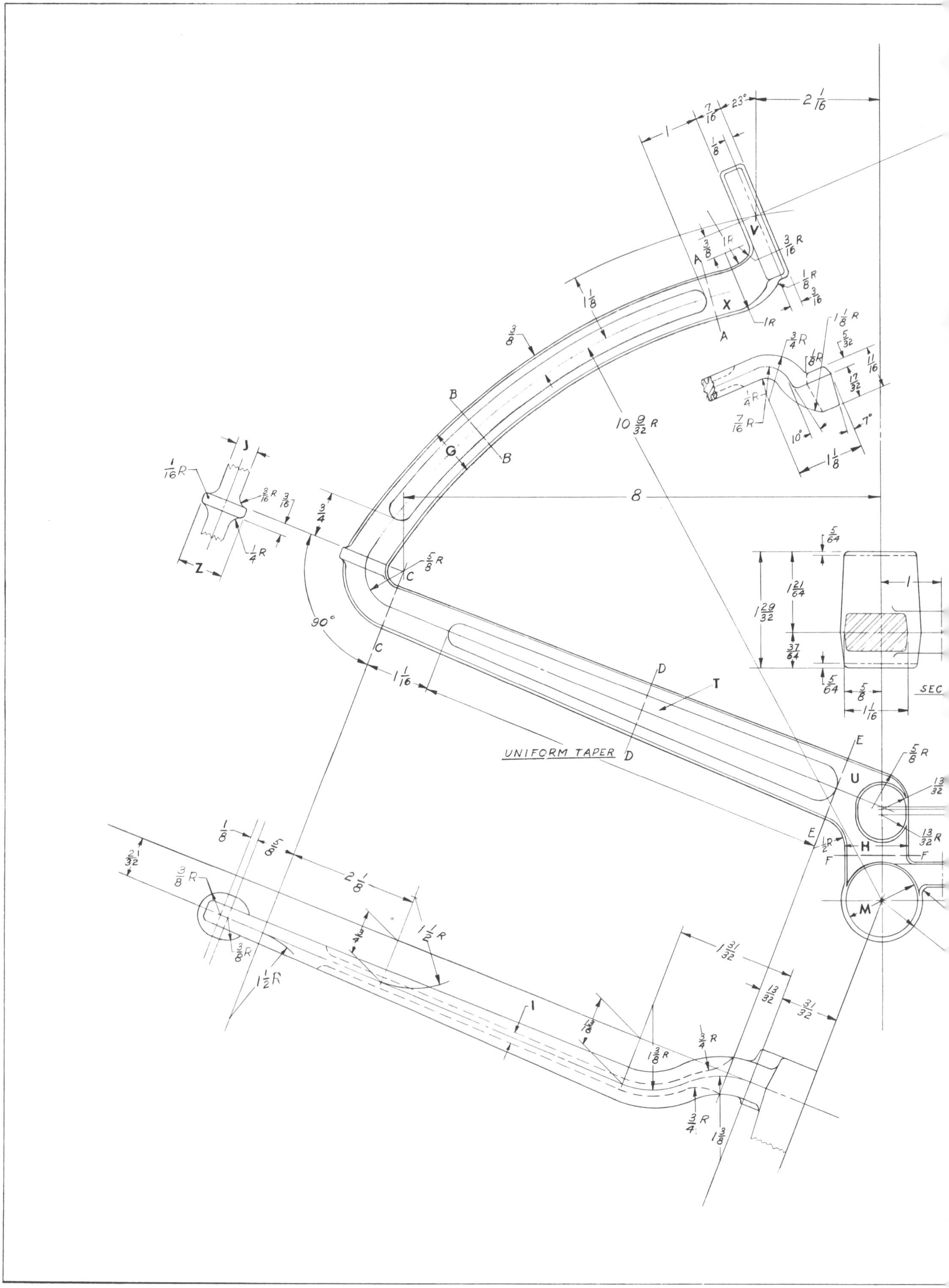

UNIFORM TAPER
SEC
90°
23°
10°
7°
A
B
C
D
E
F
G
H
J
M
T
U
V
X
Z

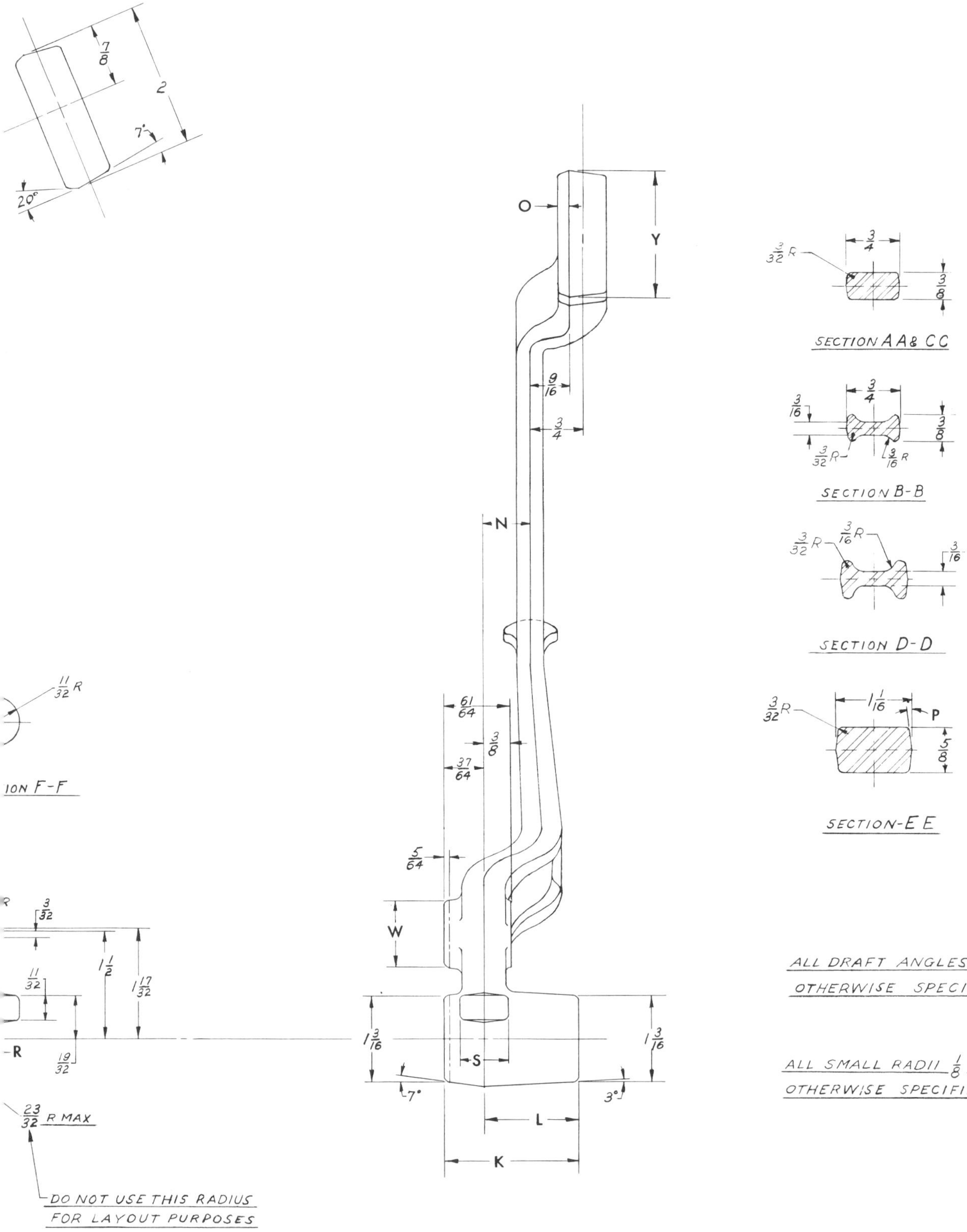

ALL DRAFT ANGLES 7° UNLESS OTHERWISE SPECIFIED

ALL SMALL RADII 1/8 UNLESS OTHERWISE SPECIFIED

BRAKE PEDAL	PRINT No. 136

TRADE COMPETENCY TEST

For Print No. 137

Student's Name______________________ Instructor's Name______________________

Questions	Answers
1. The detail numbers have been replaced by letters on the detail sheet. Give the numbers for details as they are listed on the assembly drawing.	1. A_____ B_____ C_____ D_____ E_____ F_____ G_____ H_____ I__________ J__________

2. When a die, fixture or other mechanism requires several parts in its construction, the design engineer prepares a Bill of Material. The Bill of Material gives the material required and other necessary information for each detail of the assembly.

The Bill of Materials for the Combination Trim Die is only partly completed. Fill in the material sizes necessary to make each detail, adding ⅛″ material to each surface to provide material for finishing (except where material sizes are specified). Do not list material sizes in fractions smaller than ⅛″. On cylindrical surfaces of less than 5″ diameter add only ⅛″ to diameter. For flat objects of less than ½″ thickness add only ⅛″ over-all (1/16″ each side). For all screws give diameter, number of threads and length. Screw lengths on Print 137 are measured under the head, except for flat head screws which are measured over-all.

For the spring give the outside diameter, free length and wire diameter. (Make free length ⅛″ longer than its *maximum* working height.) List standard parts, such as screws, under material as "purchase." Give material specifications if they are shown on the detail sheet. Wherever necessary on those parts that are not detailed, *scale* the assembly drawing.

TEAR OFF HERE

BILL OF MATERIALS — PRINT 137

DET NO.	DETAIL NAME	MATERIAL SIZE	MATERIAL	NO. REQD
1	GUIDE PIN	2 1/8 DIA x 8 3/4	SAE 1045	2
2	TOP DIE HOLDER			
3	SQ HD SET SCR			
4	TOP PUNCH			
5	TRIM DIE			
6	DOWEL PIN			
7	SLEEVE			
8	STRIPPER PCH			
9	RETAINER			
10	PIERCING PCH			
11	SPRING			
12	PEDESTAL			
13	BASE			
14	FL HD SCR			
15	SOC HD SCR			
16	SOC HD SCR	1/2 - 13 x 1 1/4	PURCHASE	4
17	SQ HD SET SCR			
18	SOC HD SCR	1/2 - 13 x 1 1/2	PURCHASE	4
19	SQ HD SET SCR			
20	SQ HD SET SCR			

THIS PAGE FOR STUDENT NOTES

CHAPTER 9

Reading Industrial Blueprints

Monarch Machine Tool Co.

INDUSTRIAL PRINTS

In the previous chapters you have found variations in the manner in which blueprints are drawn. Notes and symbols do not always appear in the same form on all drawings. Don't let variations from standard procedures throw you. Interpret the information intelligently and when in doubt don't be afraid to ask questions. Ask your instructor or, on the job, ask your foreman. They'll know the answer, and they'll be impressed that you can read blueprints so well that you know the right questions to ask.

You have learned the basic principles of blueprint reading and in searching for the answers to the problems in the previous chapters you have gained a substantial amount of valuable experience in reading industrial blueprints.

In Chapter 9 we have included blueprints now being used by five well-known companies. These include a print of a zinc die casting from a gas meter manufacturer; aluminum forgings and machining prints from an aircraft company; a print from a manufacturer of boilers, illustrating methods of drawing weldments and using welding symbols; a drawing from a tool company showing a plastic part used in the manufacture of aluminum levels; and prints illustrating parts made on numerically controlled machines.

The print on numerical control illustrates a new departure in drawing techniques. This drawing also teaches a greater lesson in demonstrating that objects that are not complex in structure can now be made automatically. Does this eliminate or reduce the need for blueprints reading? Quite the contrary. The skilled mechanic who performed the accurate drilling operation on the flange prior to automation is now assigned to more complex work for which automated procedures have not yet been devised. His new assignment is not a routine one and requires greater skill and a broader knowledge of shop practices. The automated machine operator who has replaced him was not required on his previous production job to read blueprints. This knowledge was needed only by the setup man and the inspector who checked his work. On his new assignment, he must now be able to interpret the setup chart and with automation he is provided with time during which he must check the work he produces. To do this, he too must now be able to interpret blueprints.

In studying this chapter we suggest that, before reading the explanation which accompanies these prints, you examine the blueprint thoroughly. To do this, first read the information on the title block. You will learn what the part is and also have other information necessary to make a complete interpretation of the blueprint. Next, visualize the shape and size of the object shown and then examine its vital statistics as given in the notes, symbols and dimensions. Now make a pencil check mark by those elements of the drawing which are not clear. After this, read the explanation accompanying the prints. Check your interpretation and learn the meaning of those of which you were not sure.

ZINC DIE CASTING: DRAWING 44328

This is a typical drawing of a zinc die casting. It is possible to hold relatively close tolerances by this method of fabrication thereby eliminating many machining operations.

In the upper right corner of the drawing is a block which shows the part number, item number and material specification. You will note that FO 14 has been added to the drawing number to make the part number. "FO" is a code used by this company to identify forgings and castings. "PO" is used to identify finished components and "GO" identifies assemblies. Item 1 gives this part an identity in an assembly. This item number is also repeated in a circle above the title block.

Zamak per 2003 is the company's identification of this zinc alloy and it is comparable to ASTM b-86-52.

Section A–A – This company uses this general pattern for crosshatching all material. Note the close tolerance of ± .001″ on the cored hole. This restriction is imposed to eliminate the drilling operation prior to tapping.

(REF) – (Section A–A). This notation under the angles 2°-22′ and 2°-12′ indicates that these are reference dimensions (that is, dimensions used for informational purposes only and not essential for manufacturing). The angle 2°-22′ is controlled by diameters .347 ± .003″ and .377
+ .000″
— .003″
— .003″. The 2°-12′ is controlled by dimensions .301 ± .003″ and .330″ + .000″
— .003″

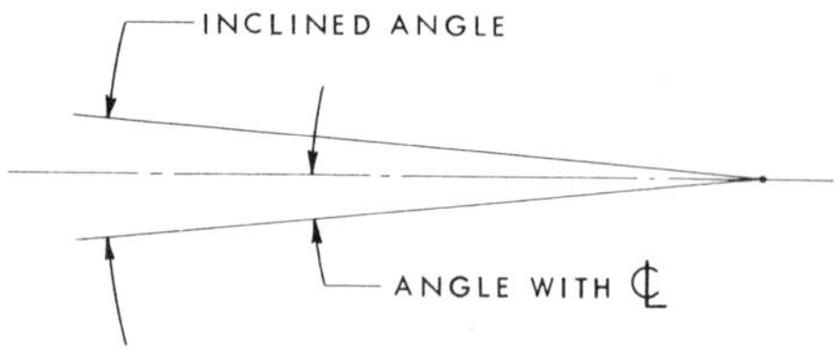

Fig. 1

500 DIA + DFT (1° INCL) – (Front View) This note means that draft equal to 1° included angle is added to the .500 diameter. Draft angles can be specified with the centerline or included angle. See Fig. 1.

You will note that the flat portion of the spindle on the 54° ± ½° angle is not shown in the front view. This is good practice because it would not be shown in its true shape and it is thoroughly described in the right side view and in Section A–A.

Title block – The title block has provisions for separate tolerances of fractions, decimals and angles.

MP11354 – (Title block) This is a release number. This company has a memo system that describes the need or advantages of new designs. This release number on the drawing makes it possible to refer to a memo of the same number to obtain this information.

REF AL-5000 (Title block) This is the model number of the meter on which this crank is used.

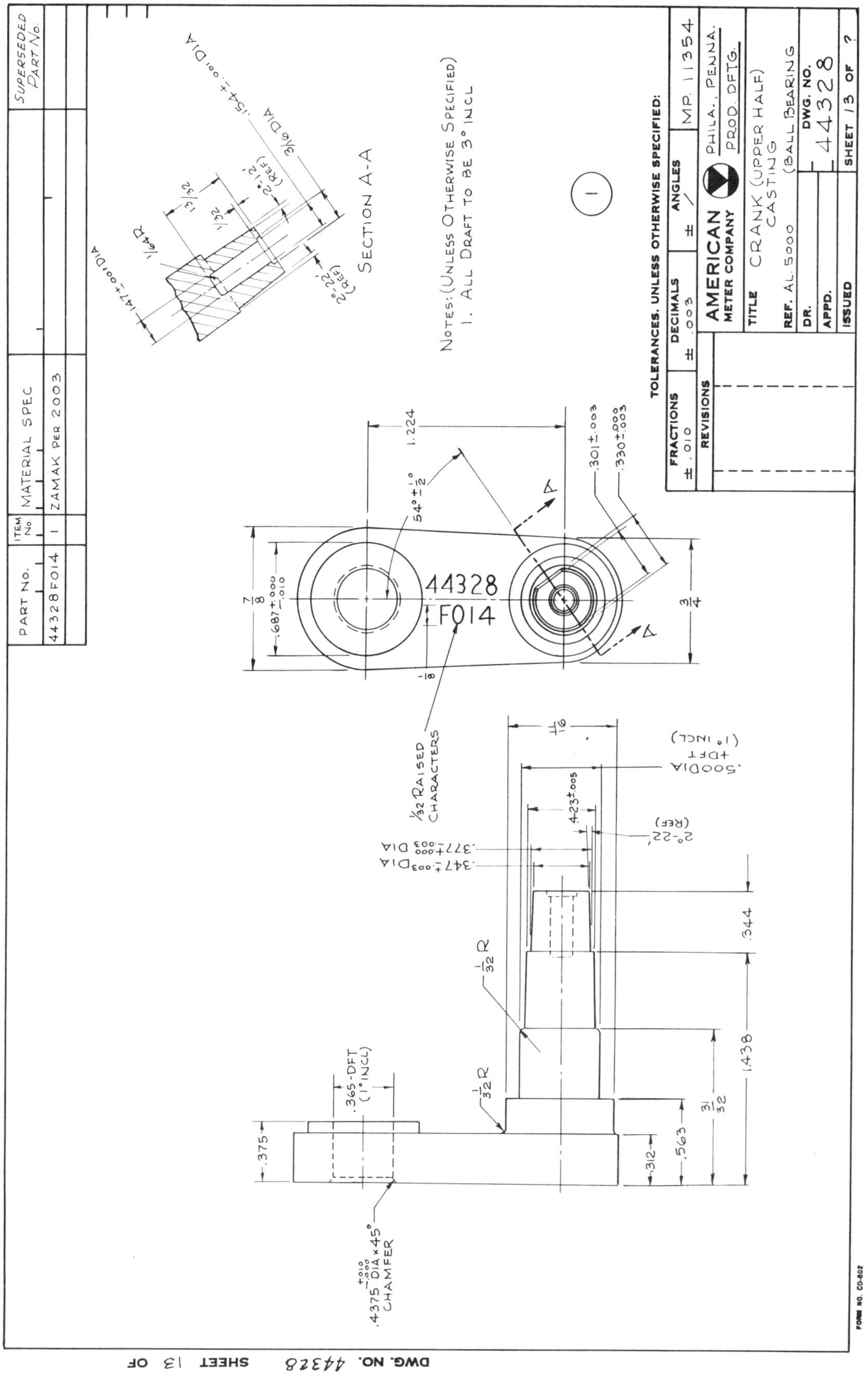
SUPERSEDED PART No.
PART No.
44328F014
ITEM No
1
MATERIAL SPEC
ZAMAK PER 2003
SECTION A-A
NOTES: (UNLESS OTHERWISE SPECIFIED)
1. ALL DRAFT TO BE 3° INCL
TOLERANCES, UNLESS OTHERWISE SPECIFIED:
FRACTIONS ± .010
DECIMALS ± .003
ANGLES
REVISIONS
AMERICAN METER COMPANY
PHILA., PENNA.
PROD. DFTG.
MP 11354
TITLE CRANK (UPPER HALF) CASTING (BALL BEARING)
REF. AL 5000
DR.
APPD.
ISSUED
DWG. NO. 44328
SHEET 13 OF ?
44328
F014
1/32 RAISED CHARACTERS
1.224
.344
1.438
.375
.563
.312
.365-DFT (1° INCL)
CHAMFER
DWG. NO. 44328 SHEET 13 OF
FORM NO. CO-802

ZINC DIE CASTNG: FINISH DRAWING 48459

This drawing does not repeat any information shown on the casting drawing except that specifically needed for machining and inspection purposes. You will note in the upper right hand corner that PO34 has been added to the drawing number to make the part number.

Under material specifications you are referred to the casting drawing.

In the front view the purpose of the reference .563 dimension is to show that the machining on the .4731 diameter portion of the spindle does not include facing the shoulder adjacent to it.

63 (Front view). This denotes that this diameter is machine finished to a smoothness of 63 AA (63 microinches, arithmetical average). A microinch (MU in.) is one millionth of an inch. The arithmetical average refers to the deviation from the mean surface.

.003 T.I.R. - (Notes). The note "eccentricity not to exceed .003 T.I.R." requires that the diameters of the spindle be held concentric within .003″ total indicator reading.

8–32 UNC–2B - (Side view). This indicates the size, the number of threads per inch and a Unified National Coarse, class 2, internal thread.

X-PRESS TAP - (Side view). This is a trade name for a tap that is used to form threads by rolling instead of cutting. This is possible in ductile metal and particularly suited for tapping blind holes because no chips are developed.

You will note the casting and finished drawings of this part are not drawn to actual size. No provisions are made for this information on the drawing and it is an apparent policy of the company to omit this information.

To assist in filing the blueprints, the part numbers are repeated in the upper left border.

Fig. 2 illustrates a typical meter. Notice the cranks shown in the upper part of the figure. These are similar to the crank shown in drawings #44328 and #48459.

Fig. 2

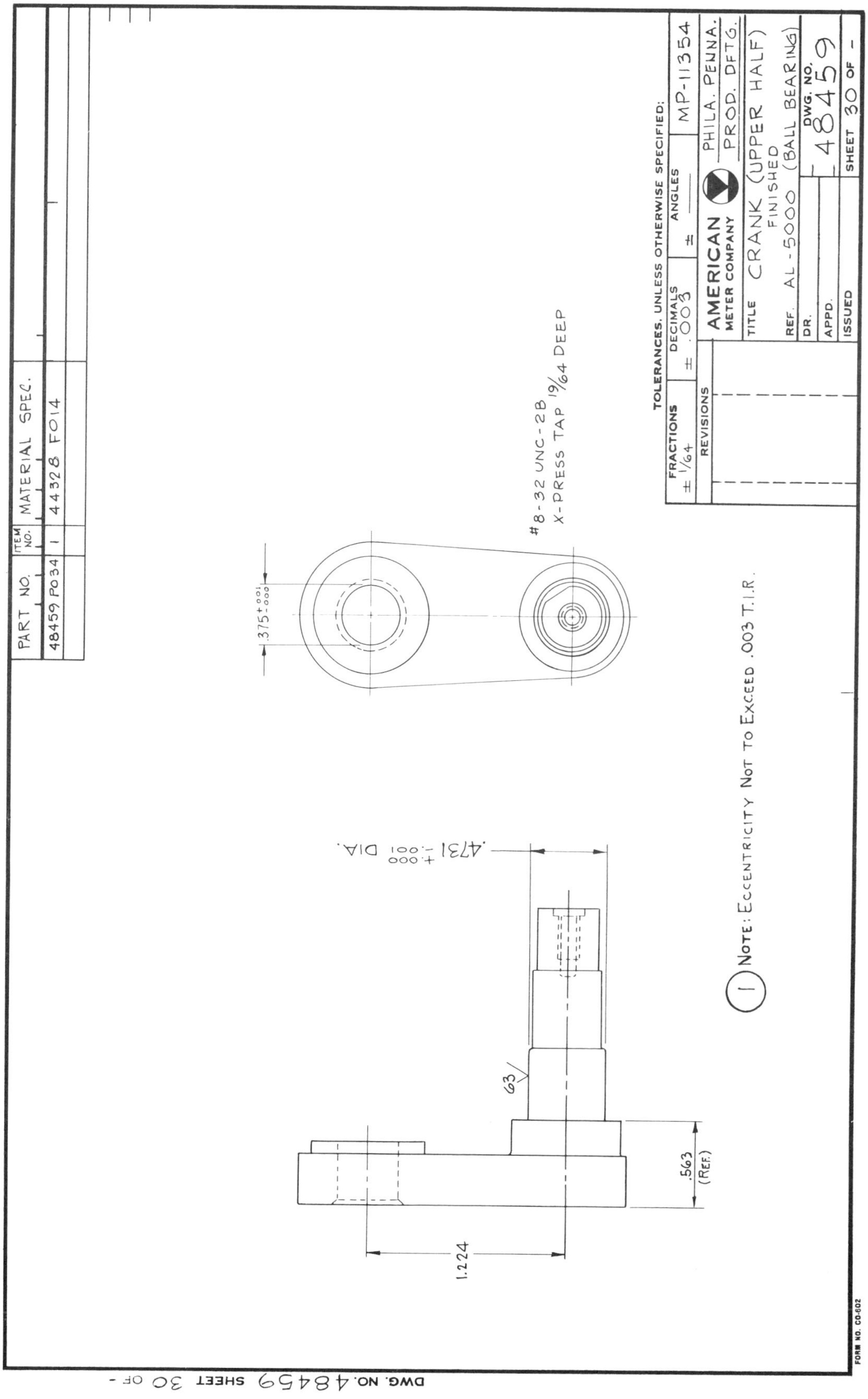
TOLERANCES, UNLESS OTHERWISE SPECIFIED:
FRACTIONS ± 1/64
DECIMALS ± .003
ANGLES ± —
MP-11354
AMERICAN METER COMPANY
PHILA. PENNA.
PROD. DFTG.
TITLE CRANK (UPPER HALF) FINISHED
REF. AL-5000 (BALL BEARING)
DR.
APPD.
ISSUED
DWG. NO. 48459
SHEET 30 OF -
REVISIONS
PART NO.
ITEM NO.
MATERIAL SPEC.
48459 P034
1
44328 F014
#8-32 UNC-2B
X-PRESS TAP 19/64 DEEP
.375 +.001 -.000
.4731 +.000 -.001 DIA.
63
.563 (REF.)
1.224
1 NOTE: ECCENTRICITY NOT TO EXCEED .003 T.I.R.
DWG. NO. 48459 SHEET 30 OF -
FORM NO. C0-602

ALUMINUM FORGING

The two drawings #40283 and #40284 show the rough forging and the finished assembly of a Landing Gear Link. These drawings serve to demonstrate accepted dimensioning practices and other prevalent drawing methods now in use in the aircraft industry.

First you will note that all dimensions are in decimals. The same tolerances are allowed for dimensions in hundredths of an inch as for fractional dimensions on other drawings. This is a relatively new trend in dimensioning. Its basic advantage is that machine tools are calibrated in decimals of an inch. Therefore, dimensioning in this manner eliminates the need for converting fractions to decimals for machine operation.

ALUMINUM FORGING: DRAWING 40283

On the forging drawing notice that all tolerances and other specific information are given in the notes.

DRAFT ANGLES – (Notes). The draft angle is the slant surface around the object that permits easy removal from the dies during the forging operation. Its extent is shown by the double lines in the front view.

MISMATCH – (Notes). While forging a piece, if the two dies are not in perfect alignment when they are forced together, the resultant forging is said to be mismatched. This condition is shown in Fig. 3. The tolerance specified is .018″ maximum.

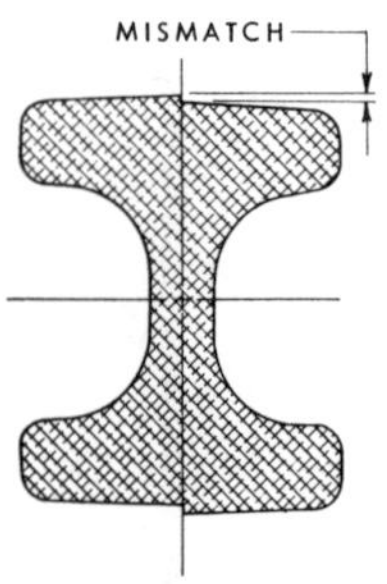

Fig. 3

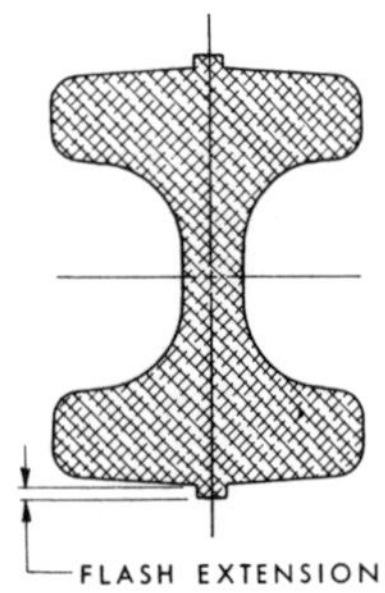

Fig. 4

STRAIGHTNESS – (Notes). The straightness of forgings, castings, and stampings is important. There is always a possibility of production error and some materials have a tendency to warp during heat treating operations. Straightness must be controlled to allowable limits to facilitate machining operations and proper dimensional control of the finished piece. The straightness tolerance of .015″ means that the center line of the forging can vary up or down this amount from end to end.

DIE CLOSURE – (Notes). If forging dies do not have proper closure, the resultant forging can be either over– or undersize. This can complicate the machining operations to be performed on it.

LENGTH AND WIDTH – (Notes). Die wear or the useful life of the forging dies is limited by this tolerance. In large forgings or castings, variations due to shrinkage irregularities would also be governed by this tolerance.

FLASH EXTENSION – (Notes). This is a limit imposed on trimming and finishing operations at the forge plant. Fig. 4 shows the flash extension for this part.

SPEC. AMS 4135G – (Notes). AMS is an abbreviation for Aeronautical Materials Specification. 4135G specifies 2014 aluminum alloy with a T6 treatment. The physical requirements for these specifications are 65,000 tensile, 55,000 yield, 10% elongation.

.030 (TYP.) – (Section A–A). This states that a .030″ variance of the center location of the .090″ radius is typical in all places of similar shape.

.56 (REF) – (Section A–A). The dimension .56 (REF) is the distance from the center line of the part to the mold line. This dimension is also shown by the .56 radius in the front view. Usually when a dimension is repeated in this manner, it is designated as a reference dimension.

℄ SYM. – (Section A–A and Top View). This symbol means that the part is the same on both sides of the center line or symmetrical about the center line.

MOLD LINE – (Section A–A). The mold line indicates the intersection of a line at

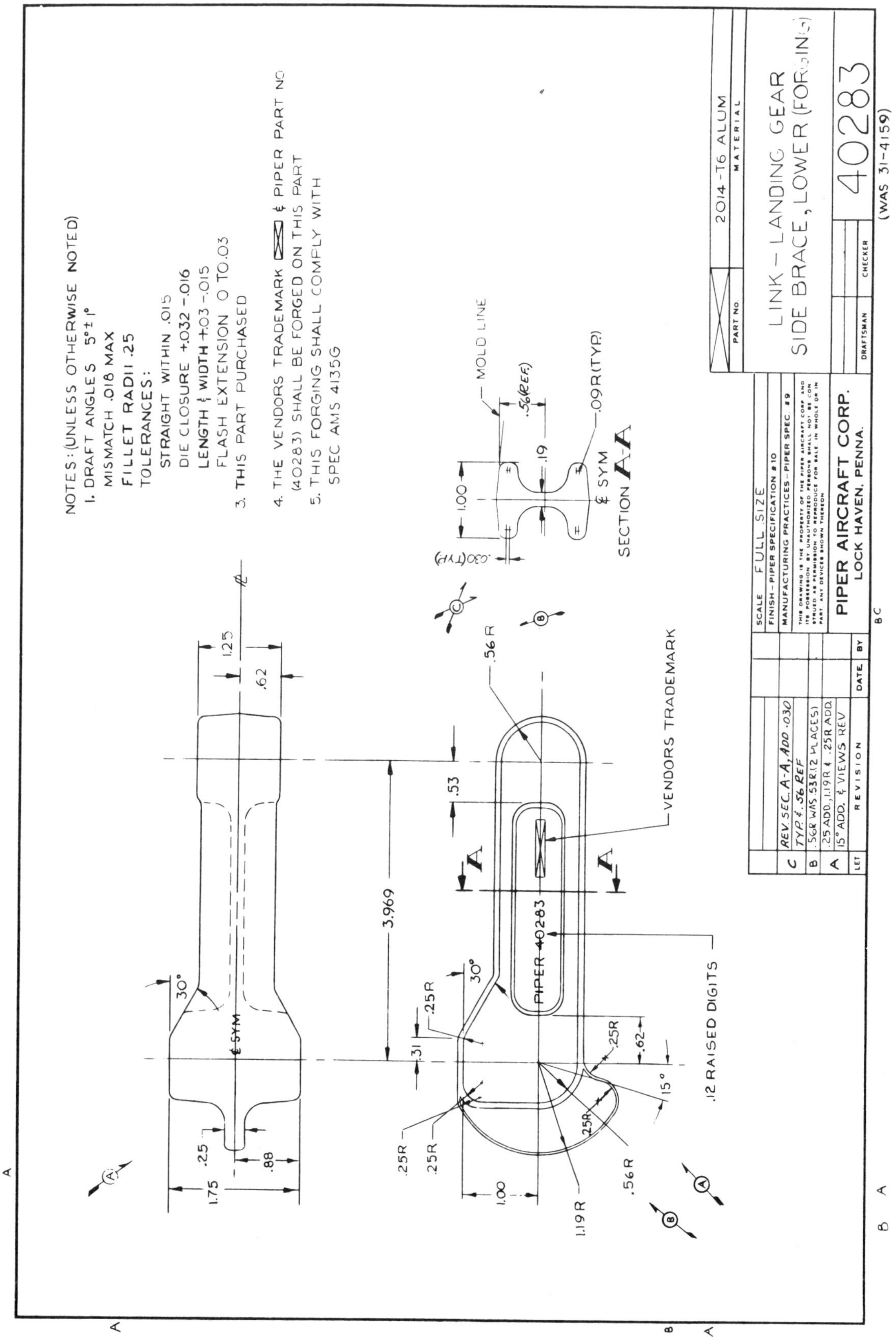
NOTES: (UNLESS OTHERWISE NOTED)
1. DRAFT ANGLES 5° ± 1°
MISMATCH .018 MAX
FILLET RADII .25
TOLERANCES:
STRAIGHT WITHIN .015
DIE CLOSURE +.032 -.016
LENGTH & WIDTH +.03 -.015
FLASH EXTENSION 0 TO .03
3. THIS PART PURCHASED
4. THE VENDORS TRADEMARK & PIPER PART NO (40283) SHALL BE FORGED ON THIS PART
5. THIS FORGING SHALL COMPLY WITH SPEC AMS 4135G
MOLD LINE
.56(REF)
.09R(TYP)
.19
1.00
.030(TYP)
℄ SYM
SECTION A-A
VENDORS TRADEMARK
PIPER 40283
.12 RAISED DIGITS
3.969
1.25
.62
.53
.56 R
30°
.25R
.31
.62
15°
1.19R
1.00
1.75
.25
.88
2014-T6 ALUM
MATERIAL
PART NO
LINK - LANDING GEAR
SIDE BRACE, LOWER (FORGING)
40283
(WAS 31-4159)
SCALE FULL SIZE
FINISH - PIPER SPECIFICATION #10
MANUFACTURING PRACTICES - PIPER SPEC #9
PIPER AIRCRAFT CORP.
LOCK HAVEN, PENNA.
DRAFTSMAN
CHECKER
C REV. SEC. A-A, ADD .030 TYP & .56 REF
B .56R WAS .53R (2 PLACES)
A .25 ADD, 1.19R & .25R ADD, 15° ADD & VIEWS REV
LET REVISION DATE BY

the lower surface of the part with the .56 dimension. Fig. 5 (scale 4″ equals 1″) shows the relation of this line to the part.

CROSS-HATCHING – (Section A–A). Cross-hatching of Section A–A has been omitted on both the forging and the finish drawings. This is a common practice in some companies for this type of section.

℄ – (Top view). This is an abbreviation of parting line. This line designates the face of each die or the parting line of the dies.

CHANGE FLAGS – The change flags embellished with arrows point out the changes and make reference to them in the revision space of the title block. To aid in locating changes, they are also noted at border of drawing.

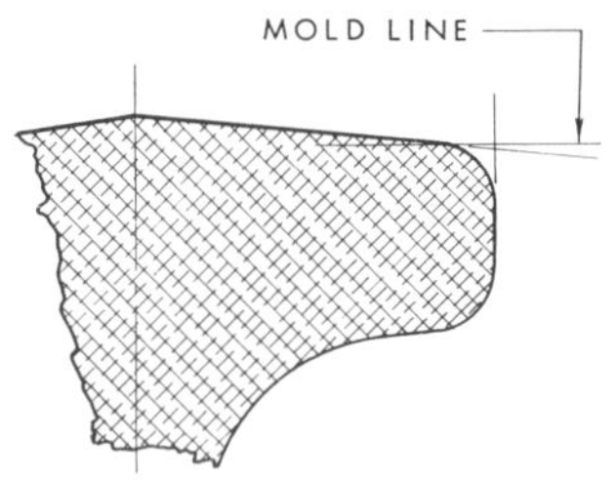

Fig. 5

ALUMINUM FORGING: FINISH DRAWING 40284

This drawing describes the machining and assembly operations necessary to make the finished part. In the title block a bill of materials has been added which lists the items required to complete the part.

FLARE PIN ENDS – (Top View). The Lock Pin, #40376, is secured by means of flaring each end. To accomplish this, a 30° countersink (see Front View) is made to widen the .2495 diameter hole to .281 at the mouth. The pin is then inserted and by means of a tool it is deformed at each end to the shape of the hole. The widening of the ends of the pin is called flaring.

LINE REAM – (Front View). You will note that the Bushings #NAS77-8-50, are pressed in from either side of the boss. It is possible that this might make them undersize or out of alignment. This can be corrected by line reaming with a .4990 reamer.

.213 HOLE THRU – (Front view). .213 or a #3 drill is used to drill a hole through to the .6245 hole prior to tapping.

TAP ¼ 28 UNF-3B THRU – (Front view). Use a ¼ diameter tap, 28 threads per inch, unified fine thread; 3B designates an internal thread with a class 3 fit.

ALODINE OR ANODIZE FINISH – (Notes). These are alternate methods of coating aluminum to prevent corrosion.

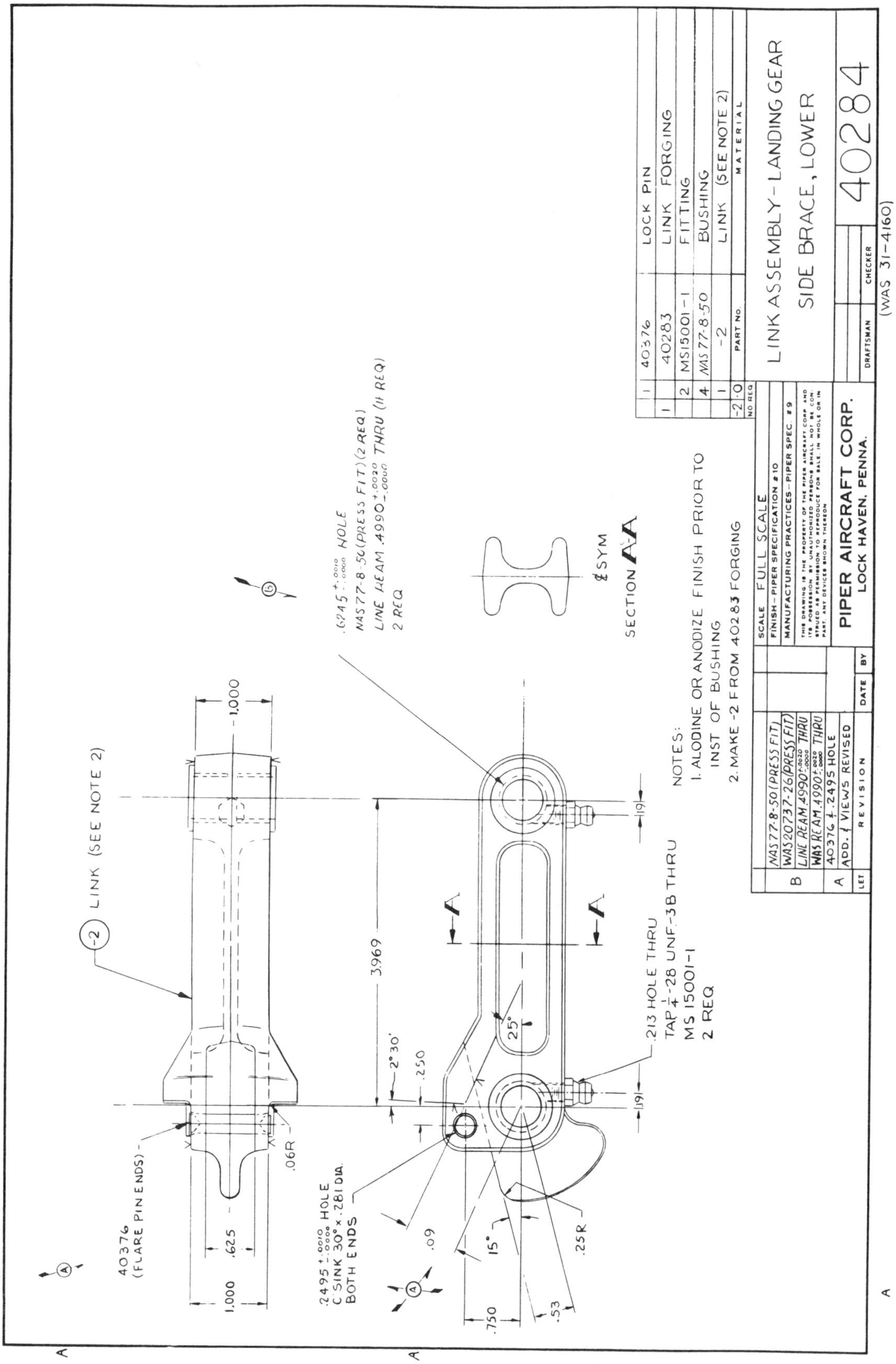

1 | 40376 | LOCK PIN
1 | 40283 | LINK FORGING
2 | MS15001-1 | FITTING
4 | NAS77-8-50 | BUSHING
1 | -2 | LINK (SEE NOTE 2)
-2 | -0 |
NO REQ | PART NO. | MATERIAL
LINK ASSEMBLY - LANDING GEAR
SIDE BRACE, LOWER
40284
(WAS 31-4160)
DRAFTSMAN
CHECKER
SCALE FULL SCALE
FINISH - PIPER SPECIFICATION #10
MANUFACTURING PRACTICES - PIPER SPEC #9
PIPER AIRCRAFT CORP.
LOCK HAVEN, PENNA.
B NAS77-8-50(PRESS FIT) WAS 20737-26(PRESS FIT) LINE REAM .4990 THRU WAS REAM .4990 THRU
A 40376 ± .2495 HOLE ADD. ℄ VIEWS REVISED
LET REVISION DATE BY
NOTES:
1. ALODINE OR ANODIZE FINISH PRIOR TO INST OF BUSHING
2. MAKE -2 FROM 40283 FORGING
SECTION A-A
℄ SYM
.6245 +.0010 -.0000 HOLE
NAS77-8-56(PRESS FIT)(2 REQ)
LINE REAM .4990 +.0020 -.0000 THRU (11 REQ)
2 REQ
-2 LINK (SEE NOTE 2)
1.000
3.969
.19
2°30'
.250
25°
.213 HOLE THRU
TAP 1/4-28 UNF-3B THRU
MS 15001-1
2 REQ
40376
(FLARE PIN ENDS)
.06R
.625
1.000
.2495 +.0010 -.0000 HOLE
C SINK 30° x .281 DIA
BOTH ENDS
.09
15°
.25R
.750
.53

WELDMENT DRAWING ID-437

Steel shapes joined together by welding are known as weldments. This method of fabrication is widely used in some industries that require a large number of various shapes to construct their products. A weldment is shown on Erie City Iron Works Drawing ID-437. Four of these saddles serve as a mounting base for a large drum attached to a boiler. You will note this weldment drawing gives the overall sizes of each plate used in this fabrication and the number required.

Dimensions are given in feet and inches and, therefore, each dimension is followed with the foot or inch symbol (3′ – 5″). Dimensioning in feet and inches is standard practice with most companies when their constructions range into exceptionally large work..

PL – (Front view). This symbol is used to denote plate. The type of plate is not specified on the drawing because in this industry steel plate is universally used. The steel plate is hot rolled from low carbon steel. It welds easily and its ductility allows it to be formed into shapes as shown on the contoured plate of the Drum Saddle. (The symbol used for plate is a duplication of that used to denote parting lines of castings, forgings, etc.)

ROLL TO SUIT DRUM – (Front view). This note indicates that this plate is to be formed prior to assembly to conform to the contour of the drum.

SHOP NOTE – (Front view). Certain fabricating operations on a boiler might be performed in the field during erection. Should there be any question as to where the operation will be performed, it is specified "shop" or "field."

WELDING SYMBOL – (Section B–B). This symbol specifies a 3⁄8″ fillet weld on each side of joining plates and this same weld is typical on all joints. (See Figs. 6 and 7 for welding symbols.) The circular flag specifies a weld all around. The welding symbols are not shown in the front view. In place of the symbols, the drawing indicates where welds are to be made by a solid triangular mark depicting each weld. Specifications for these particular welds are not given on this view, so the 3⁄8″ fillet weld accompanied by TYP on the welding symbol (Section B–B) applies.

ESS CURVE – (Front view). A curved line placed at the end of the arrow pointing to the 1″ x 2′ x 3′3″ plate. Note that the arrows pointing to all other plates touch on the edge of the plate. This plate does not have an exposed edge that is in common with one of the other plates. Therefore, this symbol is used to denote that the arrow points to the surface of the plate.

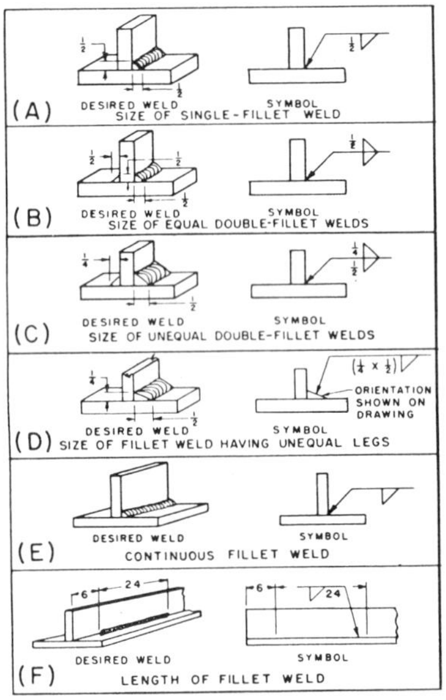

Fig. 6

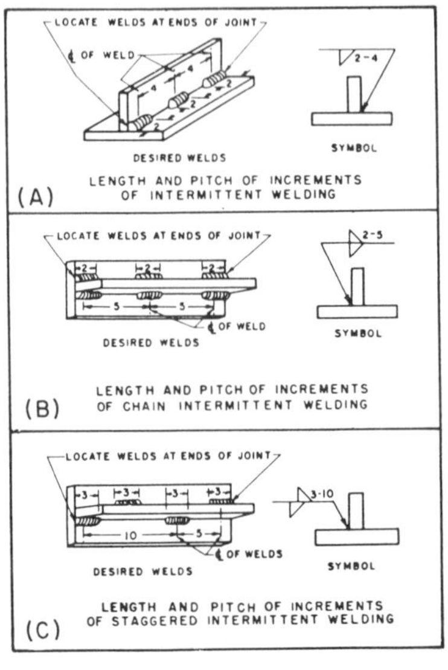

Fig. 7

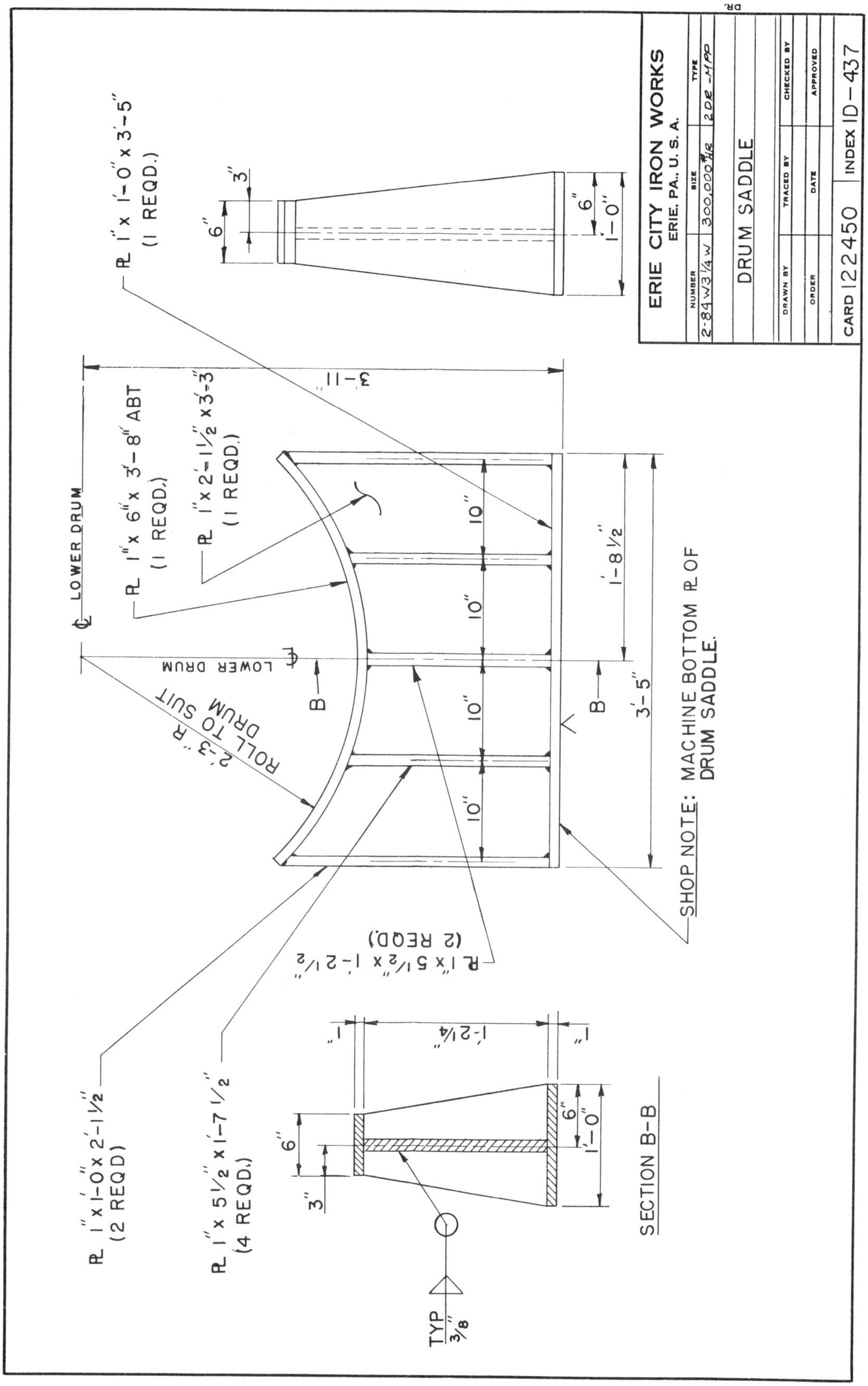
ERIE CITY IRON WORKS
ERIE, PA., U. S. A.
NUMBER 2-84W3 1/4W
SIZE 300,000# HR
TYPE 2DE-MPP
DRUM SADDLE
DRAWN BY
TRACED BY
CHECKED BY
ORDER
DATE
APPROVED
DR.
CARD 122450
INDEX ID-437
PL 1" x 1'-0" x 3'-5" (1 REQD.)
PL 1" x 6" x 3'-8" ABT (1 REQD.)
PL 1" x 2'-1 1/2" x 3'-3" (1 REQD.)
PL 1" x 5 1/2" x 1'-2 1/2" (2 REQD.)
PL 1" x 1'-0" x 2'-1 1/2" (2 REQD)
PL 1" x 5 1/2" x 1'-7 1/2" (4 REQD.)
LOWER DRUM
2'-3" R ROLL TO SUIT DRUM
SHOP NOTE: MACHINE BOTTOM PL OF DRUM SADDLE.
SECTION B-B
TYP 3/8"
3'-11"
1'-8 1/2"
3'-5"
1'-2 1/4"

INJECTION MOLDED PLASTIC: DRAWING NO. L-2467

Injection molded plastic parts are now used in the construction of a wide variety of products. Their strength, appearance and economy of manufacture have accelerated their prominence in the industrial field. Fig. 8 illustrates the molding process: an injection tree is being removed from the die. The drawing practices used for showing plastic parts are the same as those used for similar shapes of other materials.

The drawing L-2467 of the Vial Support Sleeves is made to a scale of 4″ = 1″. This scale is used because of the complexity of design, the thin wall sections, and the difficulty of clearly dimensioning a part so small.

You will note in the upper right corner of the print there is an actual size view shown. This is good practice because it would be difficult to visualize the true size of the object from a drawing made four times its shape. (Note, however, that the print has been reduced to fit the page). Fig. 9 shows the Vial Support Sleeve placed in the aluminum level.

TOLERANCES – Tolerances that are not specified are shown with two sets of limits. Dimensions in thousandths of an inch (.000″) are limited to ± .002″. Dimensions in hundredths of an inch (.00″) are more liberal and allow ± .005″.

Photograph by Ron Norman

Fig. 8

Mayes Brothers Tool Manufacturing Co.

Fig. 9

CHANGE BLOCK – Change "A" shows reference dimensions added in three places. This indicates that after the drawing was made, it was decided that additional dimensions would perhaps aid in inspecting the part. The reference dimensions were added for checking purposes only, and they are followed by the note "REF" to show they are not the controlling dimensions. For example: The outside diameter of 1.400 (REF) is controlled by the 1.391 dimension + 1° draft for a distance of .265″. The 10° (REF) dimension is controlled by the draft angle necessary to match the 1.3 diameter with 1° draft angle on the opposite side.

PL – (Section B–B). There are two parting line symbols given since the object has a parting line in a different plane for the inside and outside diameters.

1° TYP – (Section B–B). This refers to the external draft added to the 1.340 diameter and the 1.391 diameter core.

.340 + .005 – .000″ DIAMETER CORE – (Front view). This specifies a molding operation. This is performed by a cam operated mechanism during the molding cycle.

MISMATCH – (Section B–B). Appearance is important. Therefore, no mismatch is allowed and the two conical surfaces must meet without showing a visual offset.

MATERIAL – (Notes). Polystyrene is a type of plastic material that is available in several colors. Therefore, specifications must include the color required.

DRAFT ANGLES – (Notes). Though most draft angles are specified, those for the ribs at a 20° angle and the .06 and .12 ribs are not specified and are therefore 2°.

MANUFACTURERS' NUMBER – (Notes). Manufacturers' numbers and trade marks are prohibited because they would detract from the appearance of the sleeve. Sharp corners on small plastic parts improve the lines and provide a better appearance.

EJECTOR PINS – (Notes). Ejector pins are used to push the part from the die after molding. If they are not flush, they would leave either a depression or a projection on the surface of the piece. Either condition is objectionable.

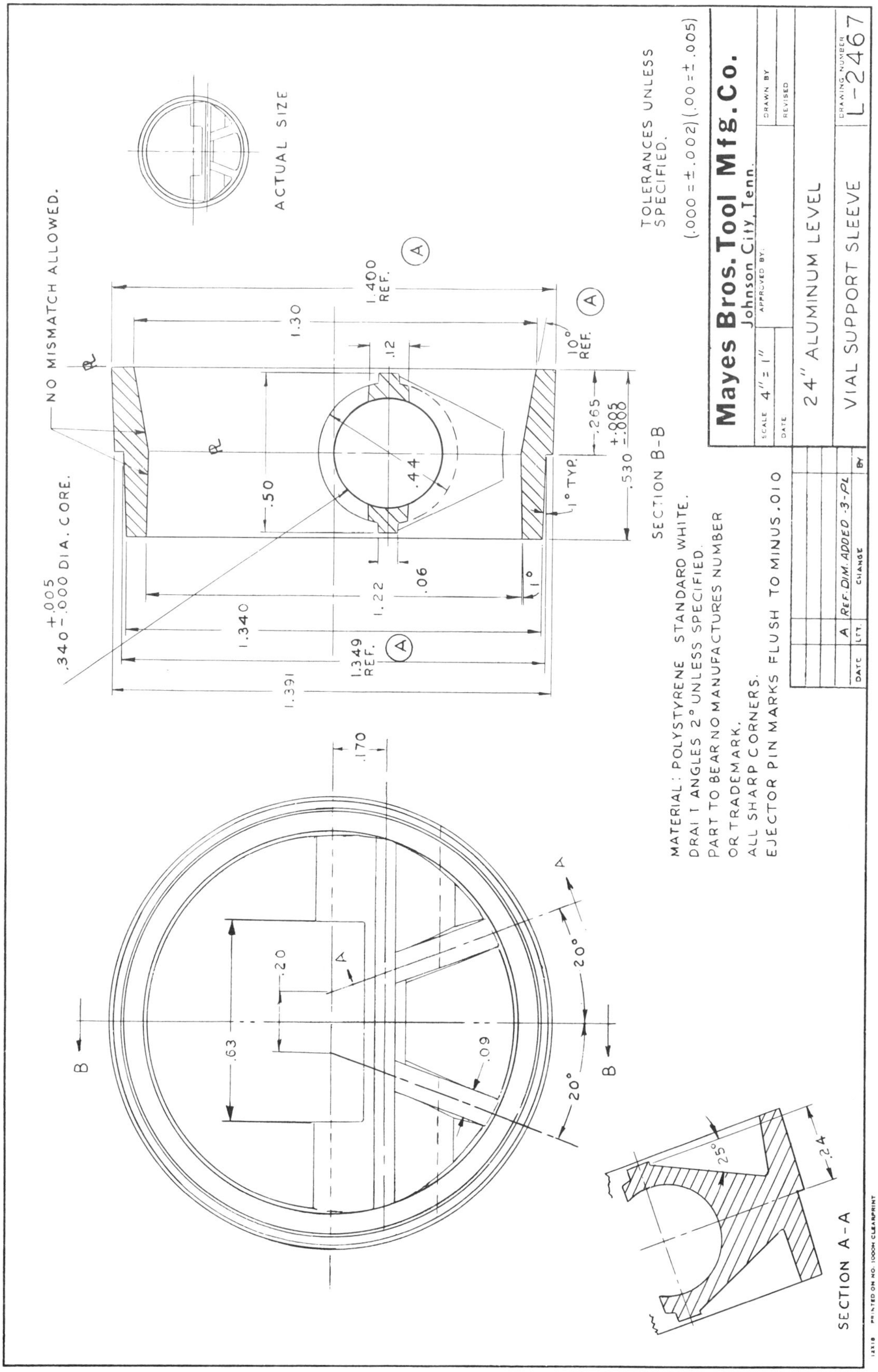
Mayes Bros. Tool Mfg. Co.
Johnson City, Tenn.
DRAWN BY
APPROVED BY
REVISED
SCALE 4" = 1"
DATE
24" ALUMINUM LEVEL
VIAL SUPPORT SLEEVE
DRAWING NUMBER L-2467
TOLERANCES UNLESS SPECIFIED.
(.000 = ±.002) (.00 = ±.005)
MATERIAL: POLYSTYRENE STANDARD WHITE.
DRAI T ANGLES 2° UNLESS SPECIFIED.
PART TO BEAR NO MANUFACTURES NUMBER
OR TRADEMARK.
ALL SHARP CORNERS.
EJECTOR PIN MARKS FLUSH TO MINUS .010
A REF. DIM. ADDED -3-PL
DATE LTR. CHANGE BY
NO MISMATCH ALLOWED.
.340 +.005 -.000 DIA. CORE.
ACTUAL SIZE
SECTION B-B
SECTION A-A
1.400 REF.
1.30
.12
10° REF.
.265
.530 +.008 -.000
1° TYP.
.50
.44
.06
1.22
1°
1.340
1.349 REF.
1.391
.170
.20
.63
.09
20°
20°
25°
.24
PRINTED ON NO. 1000H CLEARPRINT

AUTOMATION THROUGH NUMERICAL CONTROL

Major progress in automation has been made in the mechanical industry by the use of electronic tape to control the movements and operation of machine tools.

Many advantages are claimed for this method including greater accuracy, less scrap, increased production, greater tool life, reduced inspection and lower labor costs.

A machine tool designed for this type of control is shown in Fig. 10. This machine has an eight-spindle turret with vertical travel. The table moves either from left to right or front to back. To the right of this machine is shown an electronic control instrument called a "reader." As a prepared tape is fed through the reader, commands are given to the machine that will cause the table on which the work is located to move into proper position, index the turret selecting the proper tool and then perform the work required.

The information necessary to set up the piece to be machined and prepare the tape is contained on the Setup Sheet for Part No. 40991-1-40 and on the blueprint #40991.

Section A–A on the Setup Sheet describes the method of locating and holding the piece. The jig with the two angle plates on the table of the machine center each piece in an exact location.

Burgmaster Corp.

Fig. 10

The movement of the table from left to right, or horizontally, is called the "X" axis. The movement from front to back, or vertically, (in relation to the blueprint) is the "Y" axis. The turret moves up or down in a fixed location and this is called the "Z" axis. The X and Y axis lines are the horizontal and vertical center lines of the part (as viewed on the print). The table of the machine tool on which the part to be machined is fixed must first be moved to a zero reference point. The Setup Sheet gives the X axis zero reference point as —11.6250 (left from sleeve center line); the Y axis "O" reference point as —11.6250″ (up from sleeve center line). These lines are shown on the Print 40991 as the horizontal and vertical lines tangent to the bolt circle. The "O" reference lines are now the means of measuring the vertical and horizontal travel of the machine table necessary to locate the work for each operation. The Z axis "O" reference is 1.000″ above the work surface. This requires that the turret height be pre-adjusted so that the tip of the drills or other tools in the turret are set 1.000″ above surface of the work.

TAPE PREPARATION

A typewriter with a special attachment is used to punch the tape that will control the machining operation. Fig. 11 shows a section of the prepared tape. The part of the typewritten sheet made in preparing the tape for Part #40991 is shown in Fig. 12. Across the top of the first sheet is a list of the operations to be performed. The first column (N001, N002, etc.) is the code number of each hole to be made in the piece. The next column, G80, clears the machine of previous commands. G81 in this column will cause the machine to go into a drilling cycle. The column prefixed by "X" locates the table in the "X" or horizontal axis and the next column locates it in the vertical axis. The number preceded by "R" controls the distance of rapid traverse. As explained in the setup of the machine, the tip of each tool is present 1.000″ above the surface of the work. R0067 commands the turret to lower at a rapid rate for a distance of .670″

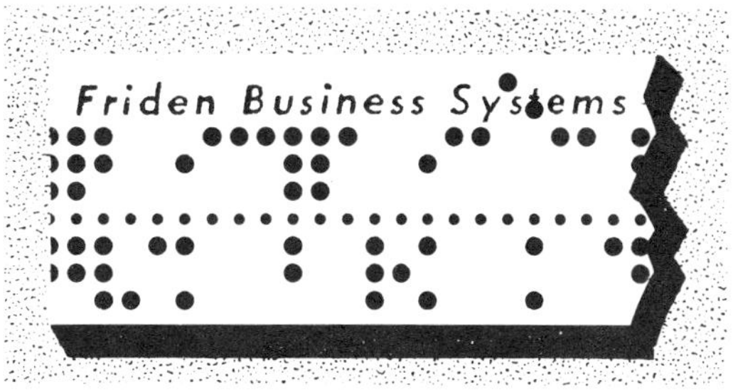

Fig. 11

CTR DR, L DR, DR, REAM & CHF 18 HOLES CTR DR, L DR, DR, CHF & TAP 2 HOLES
40991/1 10/5/65 CJ

	G80	X116250	Y116250			FO		T01	M00
	G80	X156010	Y007011			F07	S14	T01	
N001	G81			R0067	Z017000				M08
N002		X190974	Y027198						
N003		X216926	Y058125						
N004		X230734	Y096063						
N005		X230734	Y136437						
N006		X216926	Y174375						
N019		X205302	Y190974						
N007		X190974	Y205302						
N008		X156010	Y225489						
N009		X116250	Y232500						
N010		X076490	Y225489						
N011		X041526	Y205302						

Fig. 12

and then proceed at the feed rate specified by F07. F07 is a code number and indicates a feed as shown in the setup sheet of 3½″ per minute. S14 is the code that sets the speed at a rate of 665 revolutions per minute. The next column prefixed by "T" rotates the turret to position the proper drill or tap. The last column gives auxiliary commands to the machine. For example, M00 stops and holds the machine at the bottom of the cycle to allow for chip cleaning. This command is used during operation #8 because of possible damage due to clogging of chips in the Cogsdill chamfer operation. M08 turns on the coolant pump.

The following will explain the sequence of commands for the operations listed on the Setup Sheet to drill, chamfer and ream, and drill and tap the 20 holes shown on the flange of Part #40991.

#1 – G80 clears machine of previous commands.

#2 – X116250 moves table 11.625″ horizontally from "O" axis to vertical centerline of part.

#3 – Y116250 moves table 11.625″ vertically from "O" on the Y axis to exact center of the part.

#4 – T01 selects #1 turret which holds the #8 center drill.

#5 – M00 holds machine in position.

#6 – Now that the machine is located in proper position, the G80 command is repeated to erase all previous commands.

#7 – X156010 and Y007011 place the hole #1 hole, or the first hole to the right of 12 o'clock, in proper location.

#8 – the feed (F07) and speed (S14) are set.

#9 – the turret is again selected by T01 because this command had been cleared.

#10 – G81 causes the machine to perform a drilling cycle.

#11 – the rapid traverse is controlled by R0067.

#12 – Z01700 commands the turret to lower 1.700″ causing the center drill to drill to a depth of .700″.

LOAD CENTER #1601 BURGMASTER 3 BHTL SETUP SHEET ZURN INDUSTRIES INC. ERIE, PA.

OPER No.	OPERATION NAME	MATERIAL	PART No.
4	CTR. DR., 4 DR., DR., CHF, TAP (2) CTR. DR. 4DR. DR. REAM & CHF (18)	1045	40991-1-40

P/C	BY	DATE	REVISIONS:	PART NAME
83				SLEEVE 109

TURRET No.	TOOLING	DEPTH "Z" FROM O REF. POINT	SPINDLE FEED	SPINDLE SPEED
1	CTR DR #8 (1-20)	1.7000	F07 3.5	S14 665
2	LEAD DR $\frac{9}{16}$ (1-20)	2.7260	F10 5.0	S12 420
3	DR $1\frac{1}{4}$ (1-18)	2.9750	F07 3.5	S06 150
4	DR $1\frac{11}{64}$ (19, 20)	2.9520	F07 3.5	S06 150
5	CHF 2"	1.1600	F30 15.0	S05 110
6	TAP $1\frac{1}{4}$-12	3.0170	F13 6.5	S04 75
7	REAM $1\frac{1}{4}$	2.9750	F09 4.5	S05 110
8	CHF (COGSDILL $1\frac{1}{4}$)	3.6875	F15 7.5	S07 220

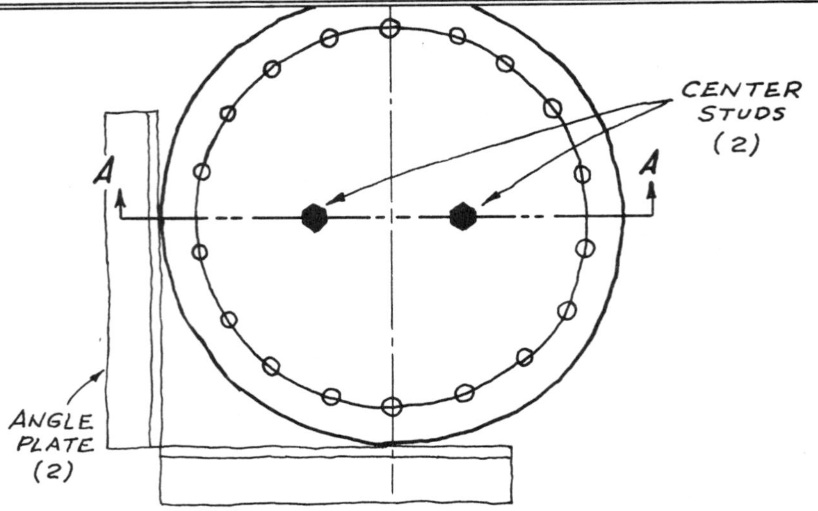

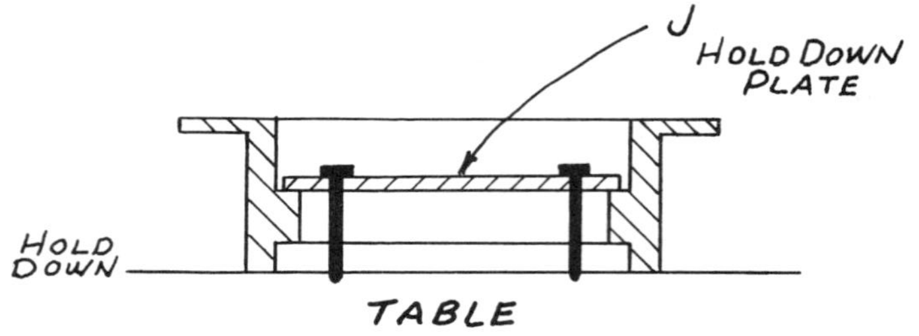

SECTION A-A

RISER BLOCK No. NONE FIXTURE No. J
DESCRIPTION:

REGISTRATION: REGISTER EACH PIECE FROM ANGLE PLATES FOR APPROXIMATE POSITION. INDICATE OUTSIDE DIAMETER TO DETERMINE EXACT CENTER & ADJUST TABLE TO COMPENSATE FOR ERROR.

X AXIS "O" REFERENCE - SUBTRACT 11.6250" FROM ℄
Y AXIS "O" REFERENCE - SUBTRACT 11.6250" FROM ℄
Z AXIS "O" REFERENCE - 1.000" ABOVE SURFACE TO BE DRILLED

NOTE: ___ INCHES HAS BEEN ADDED TO Y AXIS TO BRING TABLE FORWARD AT END OF CYCLE

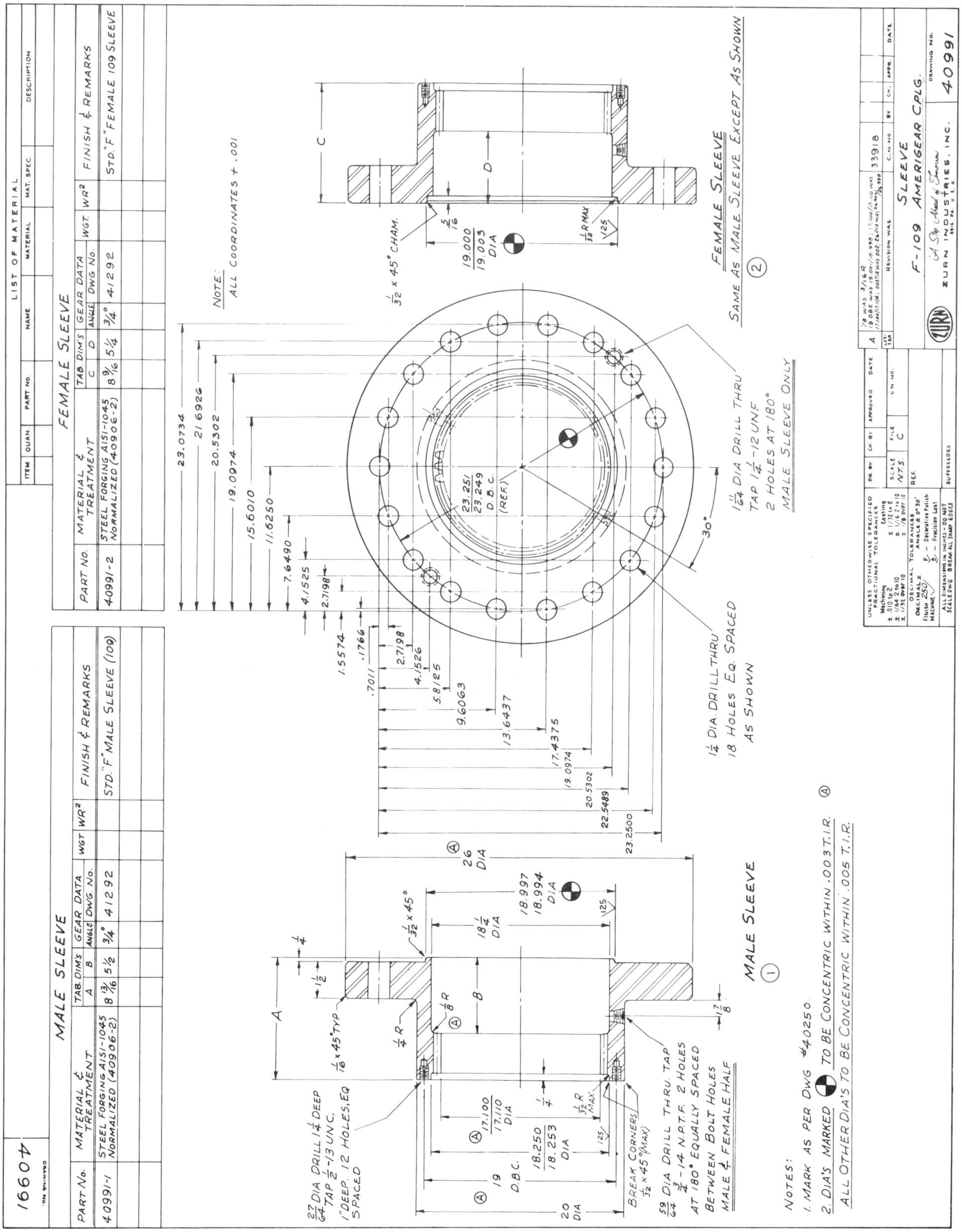
MALE SLEEVE
PART No. | MATERIAL & TREATMENT | TAB. DIM'S A B | GEAR DATA ANGLE DWG. No. | WGT | WR² | FINISH & REMARKS
40991-1 | STEEL FORGING AISI-1045 NORMALIZED (40906-2) | 8 13/16 5 1/2 | 3/4° 41292 | | | STD. "F" MALE SLEEVE (109)
FEMALE SLEEVE
PART No. | MATERIAL & TREATMENT | TAB. DIMS C D | GEAR DATA ANGLE DWG No. | WGT. | WR² | FINISH & REMARKS
40991-2 | STEEL FORGING AISI-1045 NORMALIZED (40906-2) | 8 9/16 5 1/4 | 3/4° 41292 | | | STD. "F" FEMALE 109 SLEEVE
LIST OF MATERIAL
NOTE: ALL COORDINATES ± .001
FEMALE SLEEVE
SAME AS MALE SLEEVE EXCEPT AS SHOWN
1 1/4 DIA DRILL THRU 18 HOLES EQ. SPACED AS SHOWN
11/64 DIA DRILL THRU TAP 1/4-12 UNF 2 HOLES AT 180° MALE SLEEVE ONLY
MALE SLEEVE
NOTES:
1. MARK AS PER DWG #40250.
2. DIA'S MARKED TO BE CONCENTRIC WITHIN .003 T.I.R.
ALL OTHER DIA'S TO BE CONCENTRIC WITHIN .005 T.I.R.
SLEEVE
F-109 AMERIGEAR CPLG.
ZURN INDUSTRIES, INC.
40991

#13 – M08 is punched to start the coolant pump.

When the first hole is center drilled, the turret returns to a position that places the drill 1″ above the surface of the work. The table is commanded by the reader to move to position X190974 and Y027198, and the cycle is continued until the twenty holes shown on the part are completed.

The automatic operation of a machine of this type makes is possible for one mechanic to supervise the operation of several similar machines.

NUMERICAL CONTROL: DRAWING NO. 40991

DATUM LINES – (Front view). Location of points on the part is made from two datum lines. The horizontal datum line is designated "X"; the vertical datum line is designated "Y".

23.251/23.249 D. B. C. – (Front view). This refers to the diameter of bolt circle and is termed "reference" because all holes on this circle are to be located by the co-ordinates given from the datum lines.

LETTERS: A, B, C, and D – (Section views and part description). These are "Tab" dimensions. "Tab" is an abbreviation for "tabulated"; it is a normal procedure to have couplings made in various lengths. This drawing could take care of different length couplings by simply changing these dimensions.

REGISTRATION – (Setup Sheet).. Any variation on the outside diameter of the sleeve will cause a change in the location of its center on the table, since it is located by the two angle plates. To correct this variation, it is necessary to indicate the outside surface of each flange after it is fastened to the table and find the exact center. When this is done, a correction is made on the machine setting to compensate for this variation.

METRIC DIMENSIONING DRAWING NO. 8910

By an act of Congress in 1866 the United States made the meter the legal measure of length. The United States' yard is defined in relation to the meter as 1 yard equals 3600/3937 meter. The length of the meter was fixed by law in July of 1866 as equal to 39.37 inches. The method of converting inches to millimeters is based on this relationship and, accordingly, one inch equals 25.4000508 millimeters. From this, the American Standard Association has approved the use of 25.4 as the practical equivalent of one inch.

Drafting rooms and machine shops do not generally use the metric system for dimensioning drawings. There are United States companies, however, that have operations in areas where the metric system is used, and they have provided a dual system of dimensioning.

Drawing No. 8910 illustrates how this is accomplished. The metric equivalent is given in millimeters shown in parentheses following the inch dimensions. For example, the top dimension has 14.70 inches followed by 373,4 millimeters in parenthesis. Note that the millimeter dimensions use a comma for the decimal point, following European practice.

The method used to obtain the equivalent is shown in the following example:

$$14.70 \times 25.4 = 373{,}4$$

Clark Equipment Co. has followed the usual practice in showing the millimeters in one place decimals only. The unit of length of a millimeter is slightly over 1/32 of an inch; so, except for very precise work, it is not necessary to show decimal divisions less than one-tenth of a millimeter.

Table 1 is convenient for making conversions to fourth place decimal accuracy by simple addition, and is particularly useful where a high degree of precision is required. Its use is illustrated in an example.

Example: Find the millimeter equivalent of 6.3847 in.

6.0000 in.	=	152.4000 mm
.3000 in.	=	7.6200 mm
.0800 in.	=	2.0320 mm
.0040 in.	=	.1016 mm
.0007 in.	=	.0178 mm
6.3847 in.	=	162.1714 mm

TABLE 1.

Inch—Millimeter and Inch—Centimeter Conversion Table*

(Based on 1 inch = 25.4 millimeters, exactly)†

INCHES TO MILLIMETERS

In.	Mm.	In.	Mm.	In.	Mm.	In.	Mm.	In.	Mm.	In.	Mm.
10	254.00000	1	25.40000	.1	2.54000	.01	.25400	.001	.02540	.0001	.00254
20	508.00000	2	50.80000	.2	5.08000	.02	.50800	.002	.05080	.0002	.00508
30	762.00000	3	76.20000	.3	7.62000	.03	.76200	.003	.07620	.0003	.00762
40	1,016.00000	4	101.60000	.4	10.16000	.04	1.01600	.004	.10160	.0004	.01016
50	1,270.00000	5	127.00000	.5	12.70000	.05	1.27000	.005	.12700	.0005	.01270
60	1,524.00000	6	152.40000	.6	15.24000	.06	1.52400	.006	.15240	.0006	.01524
70	1,778.00000	7	177.80000	.7	17.78000	.07	1.77800	.007	.17780	.0007	.01778
80	2,032.00000	8	203.20000	.8	20.32000	.08	2.03200	.008	.20320	.0008	.02032
90	2,286.00000	9	228.60000	.9	22.86000	.09	2.28600	.009	.22860	.0009	.02286
100	2,540.00000	10	254.00000	1.0	25.40000	.10	2.54000	.010	.25400	.0010	.02540

MILLIMETERS TO INCHES

Mm.	In.	Mm.	In.	Mm.	In.	Mm.	In.	Mm.	In.	Mm.	In.
100	3.93701	10	.39370	1	.03937	.1	.00394	.01	.00039	.001	.00004
200	7.87402	20	.78740	2	.07874	.2	.00787	.02	.00079	.002	.00008
300	11.81102	30	1.18110	3	.11811	.3	.01181	.03	.00118	.003	.00012
400	15.74803	40	1.57480	4	.15748	.4	.01575	.04	.00157	.004	.00016
500	19.68504	50	1.96850	5	.19685	.5	.01969	.05	.00197	.005	.00020
600	23.62205	60	2.36220	6	.23622	.6	.02362	.06	.00236	.006	.00024
700	27.55906	70	2.75591	7	.27559	.7	.02756	.07	.00276	.007	.00028
800	31.49606	80	3.14961	8	.31496	.8	.03150	.08	.00315	.008	.00031
900	35.43307	90	3.54331	9	.35433	.9	.03543	.09	.00354	.009	.00035
1,000	39.37008	100	3.93701	10	.39370	1.0	.03937	.10	.00394	.010	.00039

* For inches to centimeters, shift decimal point in mm column one place to left and read centimeters, thus:

40 in. = 1016 mm = 101.6 cm

For centimeters to inches, shift decimal point of centimeter value one place to right and enter mm column, thus:

70 cm = 700 mm = 27.55906 inches

† USA Standard Practice for Industrial Use (ANS B48.1)

Machinery's Handbook

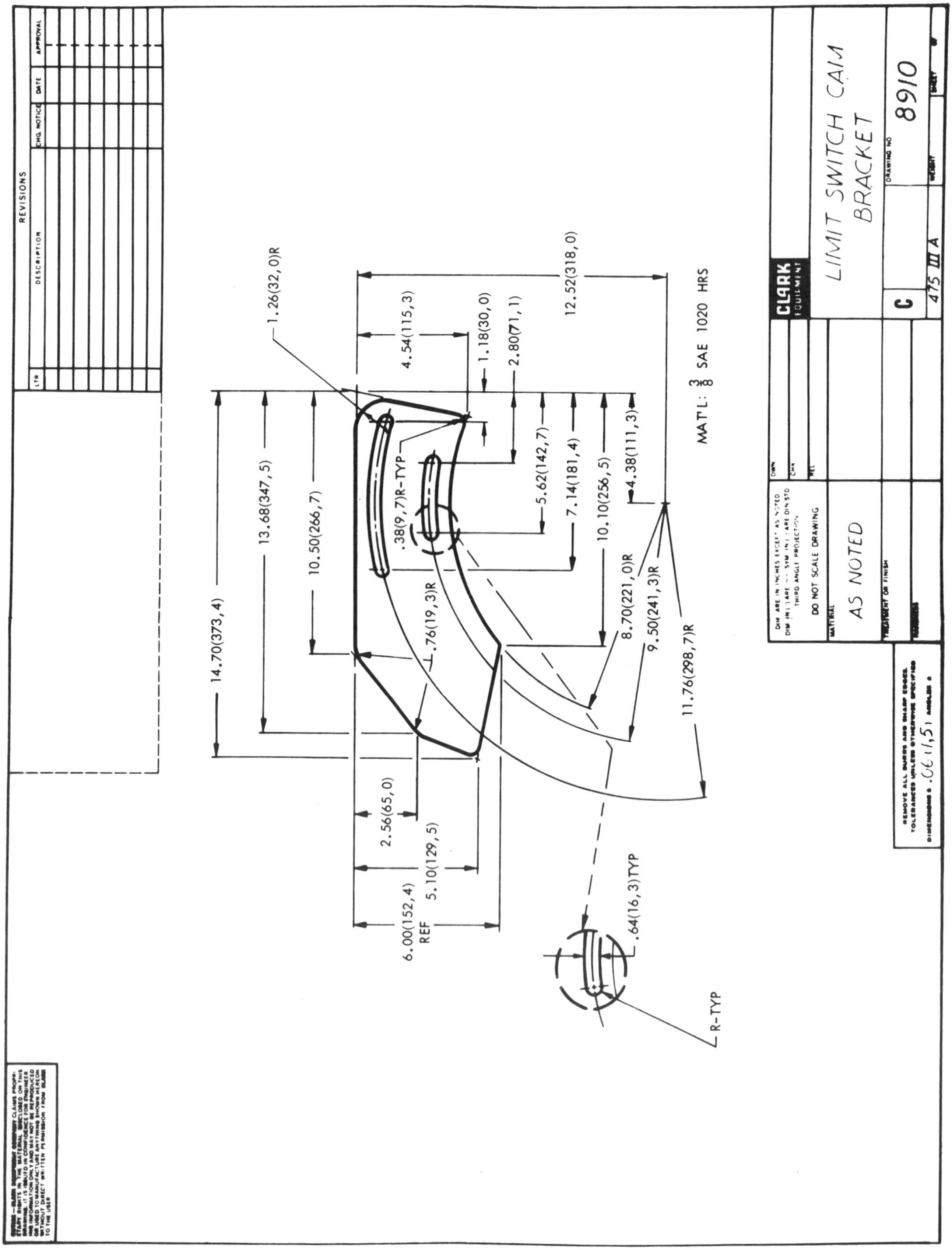
REVISIONS
LTR
DESCRIPTION
CHG. NOTICE
DATE
APPROVAL
1.26(32,0)R
4.54(115,3)
1.18(30,0)
2.80(71,1)
12.52(318,0)
14.70(373,4)
13.68(347,5)
10.50(266,7)
.38(9,7)R-TYP
5.62(142,7)
7.14(181,4)
10.10(256,5)
4.38(111,3)
.76(19,3)R
8.70(221,0)R
9.50(241,3)R
11.76(298,7)R
2.56(65,0)
5.10(129,5)
6.00(152,4) REF
.64(16,3)TYP
R-TYP
MAT'L: 3/8 SAE 1020 HRS
CLARK EQUIPMENT
LIMIT SWITCH CAM BRACKET
DRAWING NO 8910
C
475 III A
DO NOT SCALE DRAWING
THIRD ANGLE PROJECTION
MATERIAL
AS NOTED
TREATMENT OR FINISH
REMOVE ALL BURRS AND SHARP EDGES

APPENDIX A

GEOMETRIC DIMENSIONING AND TOLERANCING

Primarily, geometric dimensioning and tolerancing is a means of specifying dimensions and tolerances on a drawing with respect to the actual *function* or *relationship* of part features which can be most economically produced.[1]

Generally, it is the system of modern building blocks for good drawing practice which provides the means of stating almost *any necessary* dimensional requirement on the drawing not otherwise covered by implication or standard interpretation.

Geometric dimensioning and tolerancing should be used in cases where:

1) Part features are critical to function or interchangeability.
2) Functional gaging techniques are desirable.
3) Datum references are desirable to insure consistency between manufacturing and gaging operations.
4) Standard interpretation or tolerance is not already implied.

It should be noted at this point that geometric dimensioning and tolerancing does not replace conventional dimensioning. One method supplements the other, and the two

TABLE 1. GEOMETRIC CHARACTERISTICS

THE GEOMETRIC CHARACTERISTIC ELEMENTS AND SYMBOLS THAT ARE USED AS THE BUILDING BLOCKS FOR GEOMETRIC DIMENSIONING AND TOLERANCING ARE:

- ⏥ FLATNESS ♦
- — STRAIGHTNESS
- ∠ ANGULARITY
- ⊥ PERPENDICULARITY (SQUARENESS, NORMALITY)
- ∥ PARALLELISM
- ○ ROUNDNESS (CIRCULARITY)
- ⌭ CYLINDRICITY
- ⌓ PROFILE OF ANY SURFACE
- ⌒ PROFILE OF ANY LINE
- ↗ RUNOUT
- ⌖ TRUE POSITION
- ◉ CONCENTRICITY
- ⌯ SYMMETRY

MODIFIERS:

- Ⓜ MAXIMUM MATERIAL CONDITION
- Ⓢ REGARDLESS OF FEATURE SIZE

SPECIAL SYMBOLS: ★

- ⏥ (struck through) NEGATIVE NOTATION
- Ⓟ PROJECTED TOLERANCE ZONE

SPECIAL MODIFIERS:

- Ⓛ LEAST MATERIAL CONDITION (LMC)

(♦) Symbol formerly ⌒ under MIL-STD-8B, and — under MIL-STD-8C.

(★) Not per USASI Y14.5. Industry used symbols and modifiers not yet officially adopted. Advisory only, use at discretion.

[1] This material is extracted from *A Treatise on Geometric Dimensioning and Tolerancing,* pages 5 through 21, by Lowell W. Foster, Honeywill, Inc.

should be used in combinaton to best advantage for the application.

There are, of course, instances where the principles we are now discussing or the symbology simply are not necessary or perhaps do not apply, such as on an unusual application where a clearly written note is better. Geometric dimensioning and tolerancing is the *servant,* not the *master,* and should be applied whenever it can do something for us.

The geometric characteristic elements and symbols that are used as the building blocks for geometric dimensioning and tolerancing are shown in Table 1.

Using Symbols. The advantages of the use of symbols are reasonably obvious. Fig. 1 compares the same part using symbols above and notes below. Symbols are more quickly drawn and because of their compactness can be applied at or near the place on the drawing where they apply.

In combinations of these symbols, plus box tolerances, a requirement may be clearly stated which would otherwise require a long note. These notes, usually inconsistently written, take up drawing space, make for a cluttered drawing and, most important, necessitate visually skipping back and forth from the body of the drawing to the notes and vice versa. Notes require something close to memorization before the drawing can be properly interpreted. Symbols are in view and are instantly and uniformly interpretable at the place of application on the drawing. Note, too, how the symbol approach lends more order and preciseness to the general appearance of the drawing.

Maximum Material Condition. Perhaps the most important principle involved in geometric dimensioning and tolerancing is *maximum material condition.* It is well to be sure that there is a good understanding of its meaning.

Note on Fig. 2 that the maximum material size of the .250 ± .005 hole is .245 or at its *low* limit of size. The hole at its low limit

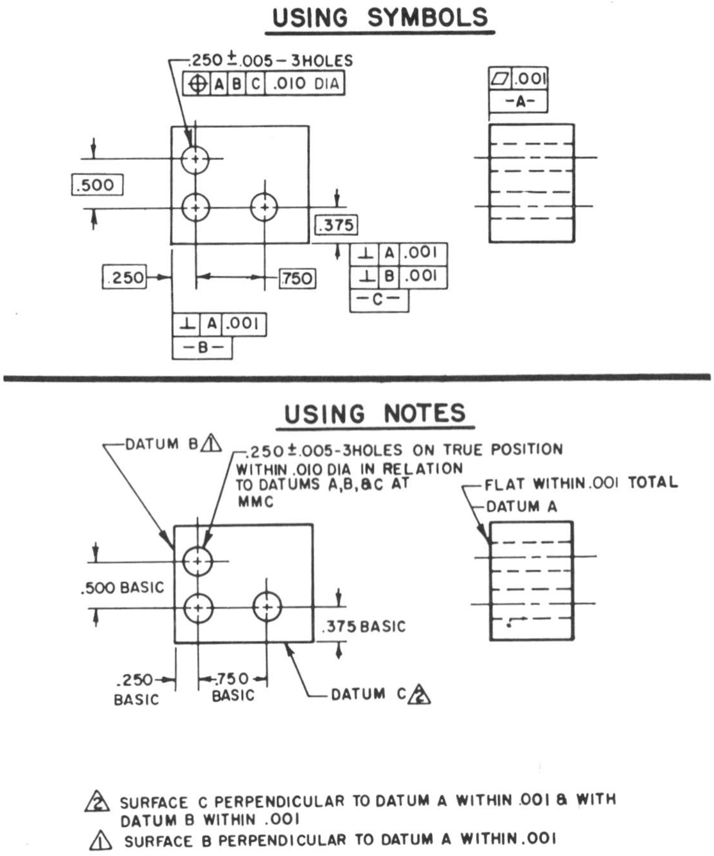

Fig. 1

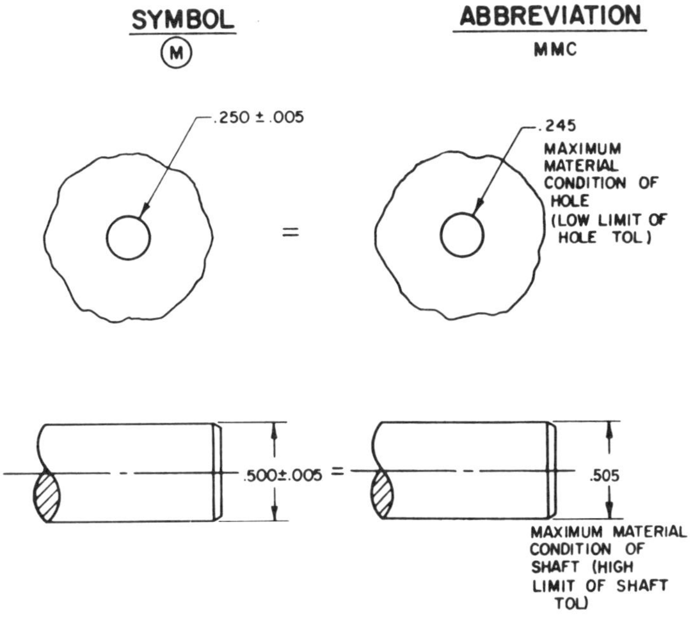

Fig. 2

obviously retains *more* part material than if it were at its *high* limit or larger size; thus, the term *maximum material condition* as it applies to a hole or similar feature.

The maximum material condition of the .500 ± .005 shaft is .505, its *largest* size, or the condition in which it contains the most material; thus, the term *maximum material condition* as it applies to a shaft or a similar feature.

The symbol for maximum material condition, the circle with the M inside, and the occasionally used abbreviation MMC, are shown at the top of Fig. 2. The symbol, or the abbreviation for maximum material condition, is known as a *modifier*.

Generally, the use of the maximum material condition principle permits greater possible tolerance as part sizes vary from their calculated limits. It also insures interchangeability and permits functional gaging techniques. It is the fundamental principle upon which the system of geometric dimensioning and tolerancing is based.

Regardless of Feature Size. This means the condition where the tolerance of form or position of a feature must be met irrespective of where the feature lies within its size tolerance.

Regardless of feature size is yet another principle involved in geometric dimensioning and tolerancing which should be well understood.

Unlike maximum material condition, under a "regardless of feature size" application no extra positional or form tolerance is available no matter to which size the related features are produced.

It is really an independent form of dimensioning and tolerancing which has always been used prior to introduction of the maximum material condition principle.

The symbol for the "regardless of feature size" is the circle with the S inside similar to that used for maximum material condition. The occasionally used abbreviation is RFS. The symbol or the abbreviation for "regardless of feature size" is also known as a *modifier*.

Basic and Datum. Two additional terms which are inherent with geometric dimensioning and tolerancing, and an understanding of which is required, are the terms *basic* and *datum*. Proper application of these provide additional necessary building blocks for effective dimensioning.

The term *basic* means a theoretically exact dimension used to describe size, shape, or location of a feature from which permissible variations are established by tolerances on other dimensions or notes.

It should be remembered, however, that a basic dimension tells only half the story. The tolerance which states the permissible variation from these exact locations must also be specified elsewhere on the drawing. This tolerance is specified with the affected feature or features.

A basic dimension on a drawing is identified by the word "basic" or "BSC" next to or below the dimension, by a general note on the drawing, or by enclosing each of the basic dimensions within a rectangular box. See Fig. 3.

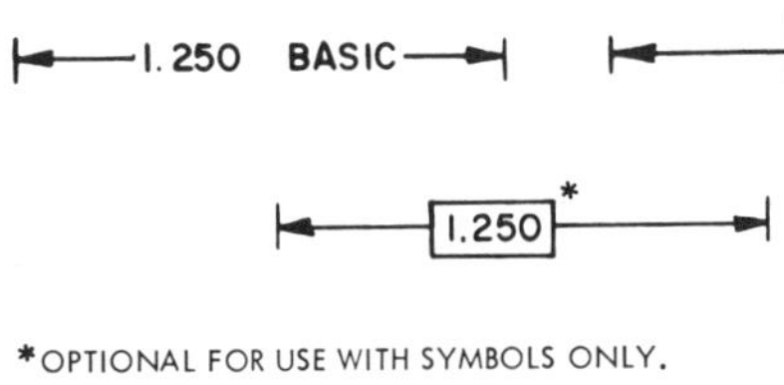

Fig. 3

Datums. Datums are points, lines, planes, cylinders, etc., assumed to be exact for purposes of computation or reference and from which the location or form of features of a part may be established. Datums are established by, or relative to, actual part features or surfaces.

The datum identification symbol used on the drawing is shown in Fig. 4. Note that each datum requiring identification on a drawing is assigned a different reference letter.

Datum surfaces and datum features are actual part surfaces or features used to establish datums and which include all the surface or feature innaccuracies.

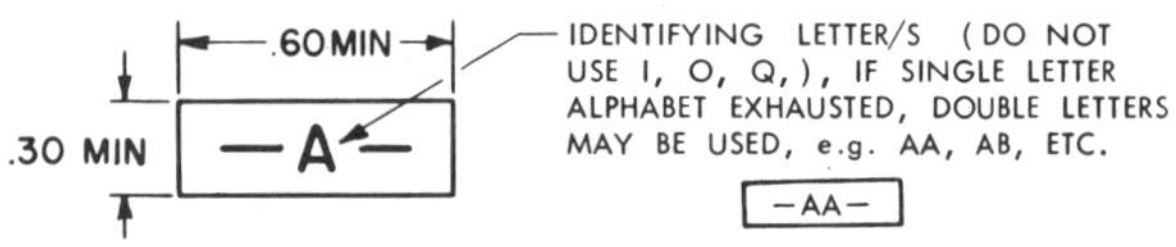

Fig. 4

The datum identification symbol is associated with the feature being identified as a datum by the following method:

1) If the datum identification symbol is shown on the *extension line* of a feature, the datum is for that feature only. See Figs. 5 and 6.
2) If the datum identification symbol is shown on a *dimension line* below or adjacent to a dimension, with a leader, note, or with a feature control symbol it applies to the entire dimension or feature with which it is associated. See Fig. 7.

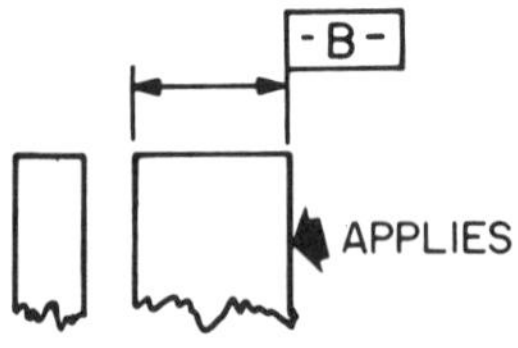

PLANE SURFACES

Fig. 5

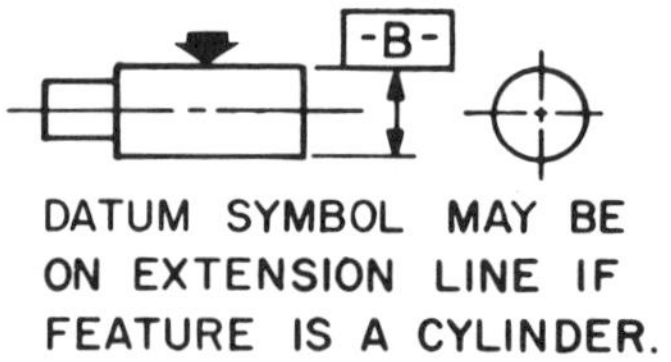

CYLINDERS

Fig. 6

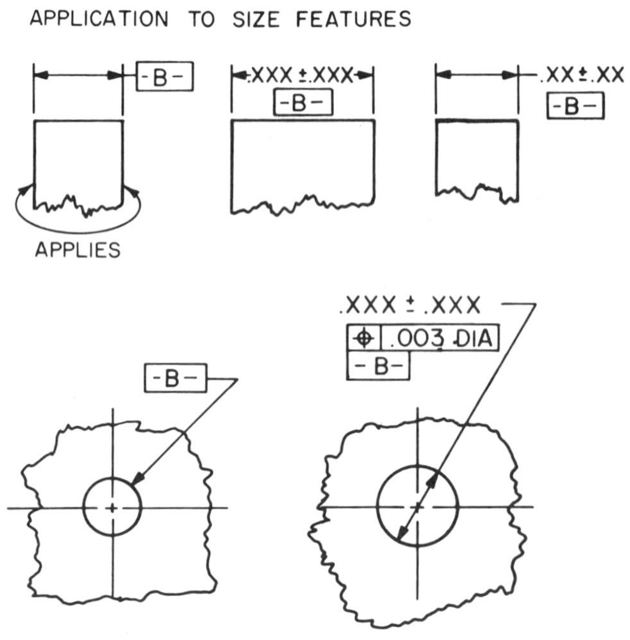

Fig. 7

Feature Control Symbol. The geometric characteristic symbols, plus the datum references, tolerances, or modifiers, form the *feature control symbol.* This is tied to the feature described. See Fig. 8. Fig. 9 shows this feature control symbol used on a part.

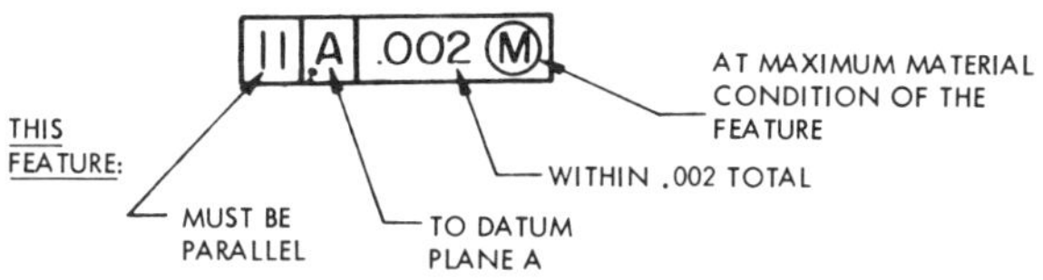

Fig. 8

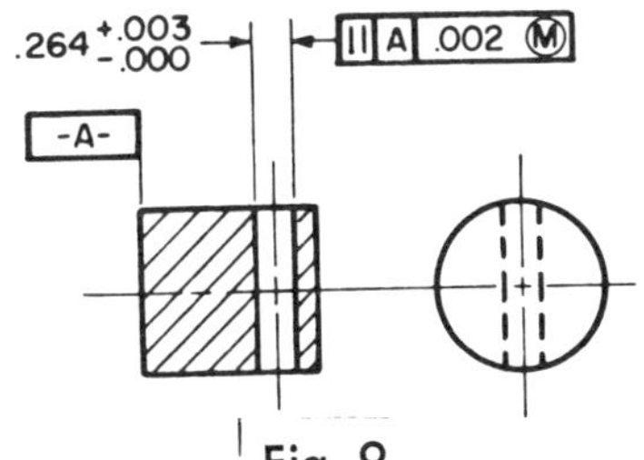

Fig. 9

The feature control symbols are associated with the feature(s) being toleranced by one of the following methods:

1) Attaching a side, end, or corner of the symbol box to an *extension line* from the feature. (Used on most form tolerances). See Fig. 10.

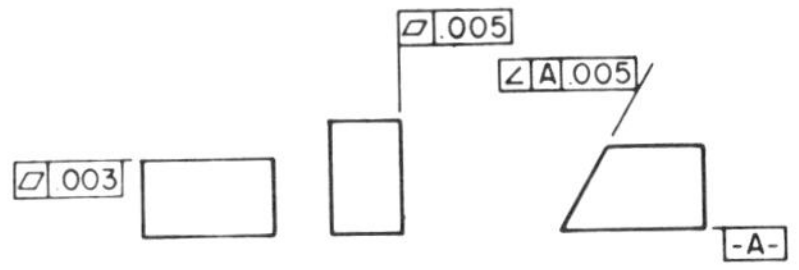

Fig. 10

2) Attaching a side or end of the symbol box to the *dimension line* pertaining to the feature when it is cylindrical, as in Fig. 11.

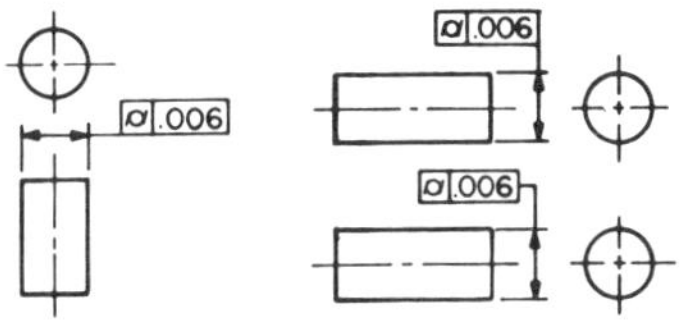

Fig. 11

3) Placing the symbol box below or closely adjacent to the dimension or note pertaining to the feature, as in Fig. 12.
4) Running a leader line from the symbol to the feature, as in Fig. 13.

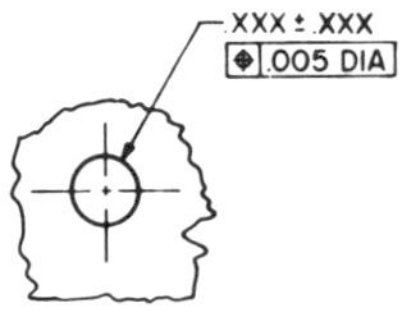

Fig. 12

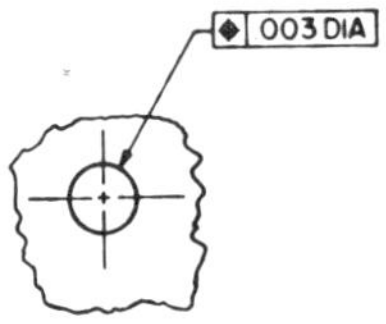

Fig. 13

Combined Feature Control Symbol and Datum Identifying Symbol. When a feature serves as a datum and is also controlled by a positional or form tolerance, the feature control symbol and the datum identifying the symbol should be combined as shown in Fig. 14.

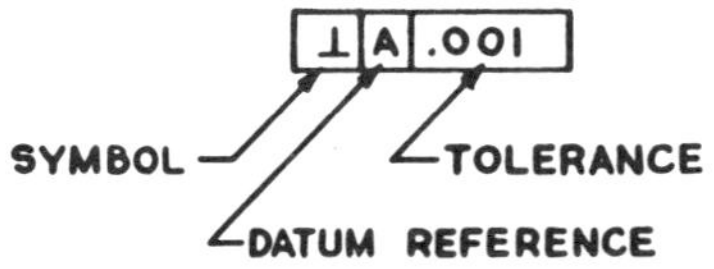

Fig. 14

Reference to Datum. When a positional or form tolerance must be related to a datum, this relationship is stated by placing the datum reference letter between the geometric characteristic symbol and the tolerance.

The following figures show additional examples of the feature control symbols with reference to datums.

Fig. 15 shows a feature control symbol with *two* datums. The symbol reads "This feature shall be located at true position with respect to both datums A and B within .002 diameter at maximum material condition."

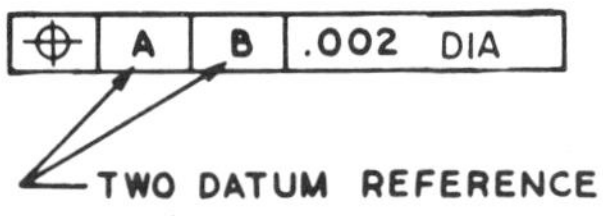

Fig. 15

Note that vertical lines are used to separate the characteristic symbols, the datum references, and the feature tolerance. These vertical lines are used on all feature control symbols to insure clarity. One reason for this is illustrated in Fig. 16, in which modifiers are used. The lines clarify that the modifiers apply only to those datums or tolerances within which they are enclosed in a separate area of the symbol box.

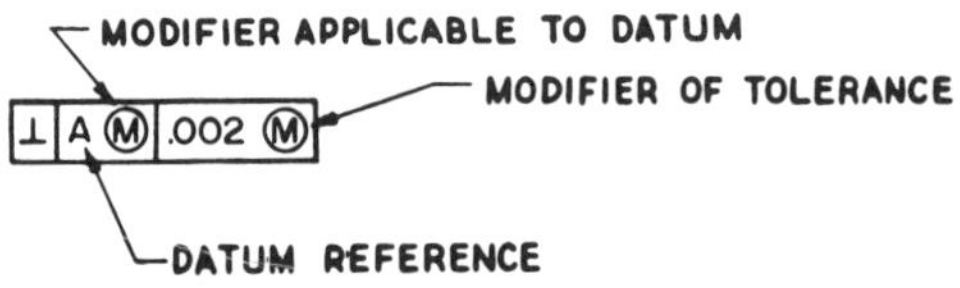

Fig. 16

Fig. 17 illustrates primary, secondary, and tertiary datums showing the order of precedence of the datums. When the order of precedence of datums is significant to function, datum references should be classified as primary, secondary, and tertiary. The datum precedence is shown by placing each datum reference letter in the proper order. The first datum letter (left to right) is considered the primary datum, the second letter secondary, etc. Thus the datum reference letters will not necessarily be in alphabetical order.

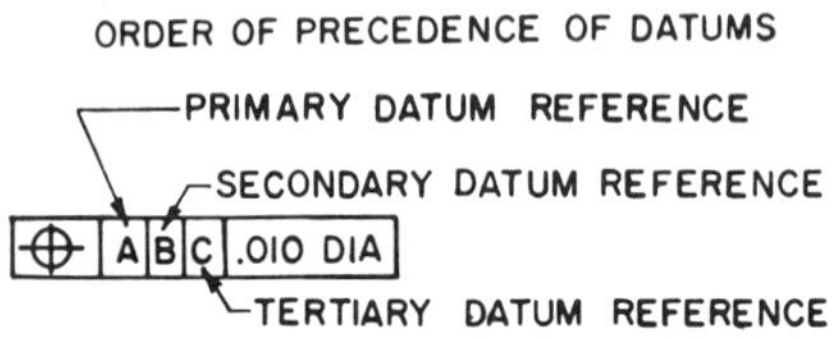

Fig. 17

Fig. 18 illustrates a feature control symbol where multiple datum features are used simultaneously to establish a single datum reference, e.g., to establish a common datum axis.

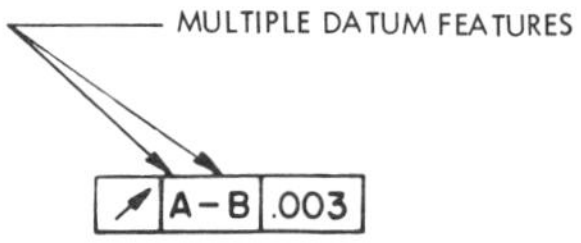

Fig. 18

Fig. 19 illustrates a normally impractical use of datum references. Note that datum A has been modified to MMC; whereas the feature controlled is RFS. This sets up a situation where the datum reference is subject to variation and cannot serve as a fixed reference for an RFS relationship.

Although there may be exceptions under certain special circumstances, generally wherever MMC is used on any datum the

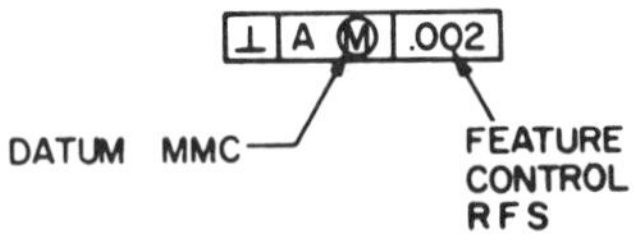

Fig. 19

feature controlled should also then be controlled at MMC.

Again, shown in Table 2 for review, are the basic elements or building blocks which form the language of geometric dimensioning and tolerancing. This includes the "geometric characteristics" and the "modifiers"; the terms "basic" and "datum"; and the "datum identification" and "feature control" symbols.

At this point we should recognize that the geometric characteristics are of two varieties: *form* and *position*.

The *form* elements relate to the conventional elements of geometry and generally suggest *magnitude* of the feature dimensioned. The *position* elements generally show relationships of centerlines and show *location* of the feature mentioned.

Occasionally a *form* element characteristic is used in a manner which, by the above definition, indicates that it is being used as a *positional* application. This is an exception to the common use of form tolerance tolerancing, but it is used when necessary and applicable to the situation. The use of position tolerance characteristics should be con-

TABLE 2. GEOMETRIC CHARACTERISTICS

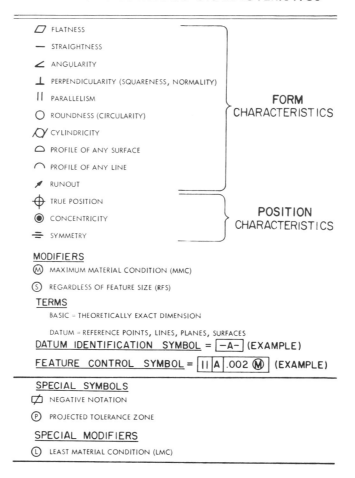

Symbol	Characteristic	Group
⏥	FLATNESS	FORM CHARACTERISTICS
—	STRAIGHTNESS	
∠	ANGULARITY	
⊥	PERPENDICULARITY (SQUARENESS, NORMALITY)	
∥	PARALLELISM	
○	ROUNDNESS (CIRCULARITY)	
⌭	CYLINDRICITY	
⌓	PROFILE OF ANY SURFACE	
⌒	PROFILE OF ANY LINE	
↗	RUNOUT	
⌖	TRUE POSITION	POSITION CHARACTERISTICS
◎	CONCENTRICITY	
⌯	SYMMETRY	

MODIFIERS

Ⓜ MAXIMUM MATERIAL CONDITION (MMC)

Ⓢ REGARDLESS OF FEATURE SIZE (RFS)

TERMS

BASIC = THEORETICALLY EXACT DIMENSION

DATUM = REFERENCE POINTS, LINES, PLANES, SURFACES

DATUM IDENTIFICATION SYMBOL = [-A-] (EXAMPLE)

FEATURE CONTROL SYMBOL = [∥ | A | .002 Ⓜ] (EXAMPLE)

SPECIAL SYMBOLS

⏥̸ NEGATIVE NOTATION

Ⓟ PROJECTED TOLERANCE ZONE

SPECIAL MODIFIERS

Ⓛ LEAST MATERIAL CONDITION (LMC)

sidered, however, instead of form tolerance characteristics if feasible under this situation.

From the preceding it will be seen, therefore, that the application of some form tolerance characteristic (namely ⊥, ∠, ∥) to a given situation will determine whether it is actually used as a pure *form* tolerance (normally related to plane surfaces) or whether it is a form tolerance characteristic used as a *positional* tolerance (normally involving features with size and a center plane or axis).

APPENDIX B

THREAD ELEMENTS AND TAP DRILL SIZES UNIFIED NATIONAL COARSE THREAD SERIES

Sizes	Threads Per Inch	DIAMETERS (Basic) Major Diameter —Inches	Pitch Diameter —Inches	Minor Diameter Inches Ext. Thds.	Minor Diameter Inches Int. Thds.	TAP DRILLS Tap Drill To Produce Approx. 75% Full Thread	Decimal Equivalent of Tap Drill —Inches
1	64	.073	.0629	.0538	.0561	No. 53	.0595
2	56	.086	.0744	.0641	.0667	No. 50	.0700
3	48	.099	.0855	.0734	.0764	No. 47	.0785
4	40	.112	.0958	.0813	.0849	No. 43	.0890
5	40	.125	.1088	.0943	.0979	No. 38	.1015
6	32	.138	.1177	.0997	.1042	No. 36	.1065
8	32	.164	.1437	.1257	.1302	No. 29	.1360
10	24	.190	.1629	.1389	.1449	No. 25	.1495
12	24	.216	.1889	.1649	.1709	No. 16	.1770
1/4	20	.2500	.2175	.1887	.1959	No. 7	.2010
5/16	18	.3125	.2764	.2443	.2524	Let. F	.2570
3/8	16	.3750	.3344	.2983	.3073	5/16	.3125
7/16	14	.4375	.3911	.3499	.3602	Let. U	.3680
1/2	13	.5000	.4500	.4056	.4167	27/64	.4219
9/16	12	.5625	.5084	.4603	.4723	31/64	.4844
5/8	11	.6250	.5660	.5135	.5266	17/32	.5312
3/4	10	.7500	.6850	.6273	.6417	21/32	.6562
7/8	9	.8750	.8028	.7387	.7547	49/64	.7656
1	8	1.0000	.9188	.8466	.8647	7/8	.8750
1 1/8	7	1.1250	1.0322	.9407	.9704	63/64	.9844
1 1/4	7	1.2500	1.1572	1.0747	1.0954	1 7/64	1.1093
1 1/2	6	1.5000	1.3917	1.2955	1.3196	1 11/32	1.4218
1 3/4	5	1.7500	1.6201	1.5046	1.5335	1 9/16	1.5625
2	4 1/2	2.0000	1.8557	1.7274	1.7594	1 25/32	1.7812
2 1/4	4 1/2	2.2500	2.1057	1.9774	2.0094	2 1/32	2.0312
2 1/2	4	2.5000	2.3376	2.1933	2.2294	2 1/4	2.2500
2 3/4	4	2.7500	2.5876	2.4433	2.4794	2 1/2	2.5000
3	4	3.0000	2.8376	2.6933	2.7294	2 3/4	2.7500

THREAD ELEMENTS AND TAP DRILL SIZES UNIFIED NATIONAL FINE THREAD SERIES

Sizes	Threads Per Inch	DIAMETERS (Basic) Major Diameter —Inches	Pitch Diameter —Inches	Minor Diameter Inches Ext. Thds.	Minor Diameter Inches Int. Thds.	TAP DRILLS Tap Drill To Produce Approx. 75% Full Thread	Decimal Equivalent of Tap Drill —Inches
0	80	.060	.0519	.0447	.0465	3/64	.0469
1	72	.073	.0640	.0560	.0580	No. 53	.0595
2	64	.086	.0759	.0668	.0691	No. 50	.0700
3	56	.099	.0874	.0771	.0797	No. 45	.0820
4	48	.112	.0985	.0864	.0894	No. 42	.0935
5	44	.125	.1102	.0971	.1004	No. 37	.1040
6	40	.138	.1218	.1073	.1109	No. 33	.1130
8	36	.164	.1460	.1299	.1339	No. 29	.1360
10	32	.190	.1697	.1517	.1562	No. 21	.1590
12	28	.216	.1928	.1722	.1773	No. 14	.1820
1/4	28	.2500	.2268	.2060	.2113	No. 3	.2130
5/16	24	.3125	.2854	.2614	.2674	Let. I	.2720
3/8	24	.3750	.3479	.3239	.3299	Let. Q	.3346
7/16	20	.4375	.4050	.3762	.3834	25/64	.3906
1/2	20	.5000	.4675	.4387	.4459	29/64	.4531
9/16	18	.5625	.5264	.4943	.5024	33/64	.5156
5/8	18	.6250	.5889	.5568	.5649	37/64	.5781
3/4	16	.7500	.7094	.6733	.6823	11/16	.6875
7/8	14	.8750	.8286	.7874	.7977	13/16	.8125
1	12	1.0000	.9549	.8978	.9098	59/64	.9219
1 1/8	12	1.1250	1.0709	1.0228	1.0348	1 3/64	1.0469
1 1/4	12	1.2500	1.1959	1.1478	1.1598	1 11/64	1.1719
1 1/2	12	1.5000	1.4459	1.3978	1.4098	1 27/64	1.4219

APPENDIX C

DECIMAL EQUIVALENT CHART

1/64......	.01562	17/64......	.26562	33/64......	.51562	49/64......	.76562
1/32......	.03125	9/32......	.28125	17/32......	.53125	25/32......	.78125
3/64......	.04687	19/64......	.29687	35/64......	.54687	51/64......	.79687
1/16......	.0625	5/16......	.3125	9/16......	.5625	13/16......	.8125
5/64......	.07812	21/64......	.32812	37/64......	.57812	53/64......	.82812
3/32......	.09375	11/32......	.34375	19/32......	.59375	27/32......	.84375
7/64......	.10937	23/64......	.35937	39/64......	.60937	55/64......	.85937
1/8.......	.125	3/8.......	.375	5/8.......	.625	7/8.......	.875
9/64......	.14062	25/64......	.39062	41/64......	.64062	57/64......	.89062
5/32......	.15625	13/32......	.40625	21/32......	.65625	29/32......	.90625
11/64......	.17187	27/64......	.42187	43/64......	.67187	59/64......	.92187
3/16......	.1875	7/16......	.4375	11/16......	.6875	15/16......	.9375
13/64......	.20312	29/64......	.45312	45/64......	.70312	61/64......	.95312
7/32......	.21875	15/32......	.46875	23/32......	.71875	31/32......	.96875
15/64......	.23437	31/64......	.48437	47/64......	.73437	63/64......	.98437
1/4.......	.250	1/2.......	.500	3/4.......	.750	1..........	1.000

APPENDIX D

PROBLEMS IN SHOP ARITHMETIC

The following examples of simple arithmetic operations are provided as a brief review of some basic rules.

Fractions: Two general rules on fractions should be recalled.

1. When fractions with different denominators are involved, convert them to fractions having the least common denominator.
2. Reduce all answers to the lowest terms.

Example 1: ADD 1 1/8, 2 1/4, 3 3/16

Solution: 1 1/8 + 2 1/4 + 3 3/16
= 1 2/16 + 2 4/16 + 3 3/16
= 6 9/16
= 6 3/4

Example 2: SUBTRACT 5 3/4 from 9 5/16

Solution: 9 5/16 − 5 3/4
= 9 5/16 − 5 12/16

Since 12/16 is greater than 5/16, it cannot be subtracted from it. To carry out the solution, "borrow" 16/16 from 9 and add it to 5/16.

Thus: 8 21/16 − 5 12/16
= 3 9/16

Example 3: MULTIPLY 3 1/3 by 4 1/8

There are two special rules in multiplying fractions that should be recalled:

1. Convert all mixed fractions to improper fractions before multiplying.

Thus: 3 1/3 x 4 1/8
= 10/3 x 33/8

2. Simplify fractions by reducing them to the lowest terms before multiplying. In the above example, the denominator 3 and the numerator 33 can be divided by the common factor 3. The numerator 10 and the denominator 8 can be divided by the common factor 2. This process is known as cancellation.

Thus: 10/3 x 33/8
= 5/1 x 11/4
= 55/4
= 13 3/4

Example 4: DIVIDE 4 2/3 by 1 1/6

The following rules concerning the division of fractions should be kept in mind:

1. Convert mixed fractions to improper frac-

tions before completing the operation.

Thus: 4 2/3 ÷ 1 1/6
= 14/3 ÷ 7/6

2. Invert the fraction which is to be divided into the other before completing the operation. After inverting this fraction called the divisor, change the sign of the operation from divide to multiply, and complete the operation by multiplying.

Thus: 14/3 ÷ 7/6
= 14/3 x 6/7
= 4/1
= 4

Decimals. A decimal is a way of expressing a fraction which has a denominator which is some multiple of ten. Thus 3/10 is expressed as a decimal by .3, 3/100 by .03, 3/1000 by .003, etc. The "denominator" of a decimal may be determined by counting the number of digits to the right of the decimal point. A decimal with one digit to the right of the decimal point is expressed in tenths, with two digits in hundredths, three digits in thousandths, etc.

Adding and Subtracting Decimals. The addition and subtraction of decimals is identical to the addition and subtraction of whole numbers with the exception that the decimal points in adding or subtracting decimals should be lined up vertically with each other. The numbers are then added or subtracted as they normally would be. The answer, however, is only correctly expressed by placing the decimal points in the same relative position it had when the numbers were lined up for the operation.

Example 1: ADD 3.245, 11.230, 191.4

Solution:
```
   3.245
  11.230
 191.4
 -------
 205.875
```

Example 2: SUBTRACT 385.890 from 418.375

Solution:
```
 418.375
 385.890
 -------
  32.485
```

Multiplying Decimals. The multiplication of decimals is identical with the multiplication of whole numbers except for the placement of the decimal point. In multiplying decimals, the correct answer is expressed only when the decimal is correctly placed. This is done by counting the number of decimal places in both the numbers being multiplied, and placing the decimal point in the answer as many places from the right as there are decimal places in the numbers being multiplied.

Example: MULTIPLY 3.14 by 5.2

Solution:
```
   3.14  (2 decimal places)
    5.2  (1 decimal place)
   ----
    628
  1570
 ------
 16.328  (3 decimal places)
```

Since there is a total of three decimal places in the numbers which were multiplied, there must also be three decimal places in the answer.

Dividing Decimals. Decimals are divided in exactly the manner that whole numbers are except for the placing of the decimal point in dividing decimals. The decimal point is correctly placed in the answer by first counting the number of decimal places in the divisor (the number being divided into the other). After you have determined the number of decimal places in the divisor, move the decimal point of the dividend (the number being divided by the divisor) as many places to the right as there are decimal places in the divisor. See the example below. The third step is to move the new decimal point of the dividend directly above to where the quotient (the answer) is to be written.

Example: DIVIDE 6.384 by 2.8

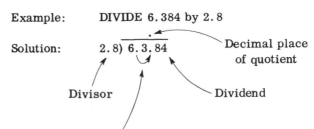

Solution:

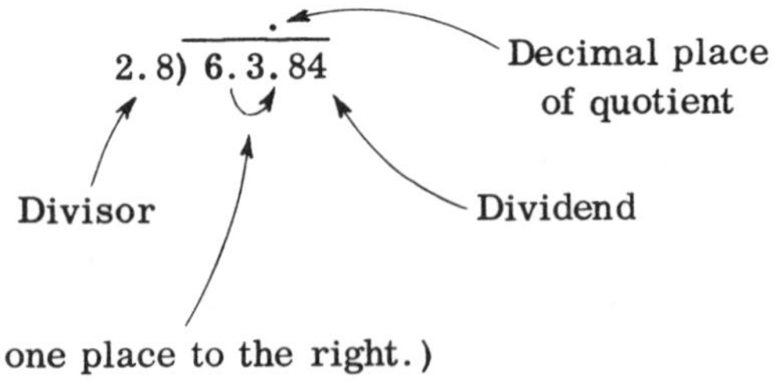

Now divide as you normally would with whole numbers. The decimal place of the answer is correctly placed without any further operation.

Thus:
```
      2.28
 2.8)6.3.84
     5 6
       7 8
       5 6
       2 24
       2 24
```

TRADE COMPETENCY TEST
Problems in Shop Arithemetic, 1

Student's Name____________________ Instructor's Name____________________

Questions	Answers
1. Add 543 + 751 + 843 + 921	1. __________
2. Add 8541 + 3632 + 4568 + 7893	2. __________
3. Add 7.510 + 8.437 + 9.652 + 7.856	3. __________
4. Add 8.500 + 7.687 + 8.312 + 4.625	4. __________
5. Subtract 3125 from 7854	5. __________
6. Subtract 5936 from 7867	6. __________
7. Subtract 58.45 from 78.26	7. __________
8. Subtract 3.125 from 643	8. __________
9. Subtract 1.7854 from 8.54	9. __________
10. Subtract 12.8 from 854.3	10. __________
11. Subtract 68.352 from 358.4	11. __________
12. Multiply 752 by 26	12. __________
13. Multiply 854 by 32	13. __________
14. Multiply 85.1 by 31	14. __________
15. Multiply 96.8 by 1.23	15. __________
16. Multiply 90.8 by 40.6	16. __________
17. Multiply 4.723 by 1.006	17. __________
18. Divide 7854 by 6	18. __________
19. Divide 7485 by 3.1	19. __________
20. Divide 8527 by 783	20. __________
21. Divide 7.875 by .125	21. __________
22. Divide 68.84375 by .015625	22. __________

TRADE COMPETENCY TEST
Problems in Shop Arithemetic, 2

Student's Name______________________________Instructor's Name__________________________

Note: In the following problems change all fractions to their decimal equivalents *before solving* the problems. Answers should be in decimals.

Questions	Answers
1. Add $\frac{1}{8} + \frac{1}{4} + \frac{1}{2} + \frac{3}{4}$	1. ____________________
2. Add $\frac{3}{8} + \frac{3}{4} + \frac{5}{8} + \frac{1}{32}$	2. ____________________
3. Add $1\frac{1}{2} + 3\frac{5}{8} + 2\frac{1}{64} + 5\frac{1}{16}$	3. ____________________
4. Add $\frac{1}{2} + 18\frac{3}{8} + 4\frac{3}{4} + 7\frac{1}{16}$	4. ____________________
5. Subtract $\frac{5}{16}$ from $\frac{1}{2}$	5. ____________________
6. Subtract $\frac{3}{16}$ from $\frac{7}{8}$	6. ____________________
7. Subtract $\frac{7}{8}$ from $2\frac{5}{16}$	7. ____________________
8. Subtract $1\frac{3}{8}$ from $74\frac{1}{16}$	8. ____________________
9. Subtract $2\frac{7}{8}$ from $10\frac{31}{32}$	9. ____________________
10. Subtract $\frac{21}{32}$ from $7\frac{1}{64}$	10. ____________________
11. Multiply $1\frac{3}{4}$ by $\frac{1}{8}$	11. ____________________
12. Multiply $\frac{3}{8}$ by $\frac{3}{4}$	12. ____________________
13. Multiply $\frac{7}{16}$ by $\frac{5}{32}$	13. ____________________
14. Multiply $45\frac{1}{4}$ by $2\frac{1}{8}$	14. ____________________
15. Multiply $1\frac{1}{32}$ by $2\frac{3}{8}$	15. ____________________
16. Divide .875 by .125	16. ____________________
17. Divide $18\frac{1}{4}$ by $\frac{1}{8}$	17. ____________________
18. Divide $34\frac{1}{16}$ by $2\frac{1}{8}$	18. ____________________
19. Divide .78125 by .15625	19. ____________________
20. Divide $25\frac{3}{8}$ by $1\frac{1}{8}$	20. ____________________
21. Divide $42\frac{1}{4}$ by .75	21. ____________________

APPENDIX E

GLOSSARY OF COMMON MACHINE TRADE TERMS

Abrasive: Crushed sharp crystals of grinding material used in the form of grinding wheels, polishing cloth, and lapping powder.

Align: To set mechanisms or machines in proper relation to each other.

Angle: The geometrical figure formed by two intersecting straight lines; also, the space between the intersecting straight lines.

Anneal: To heat a metal piece to its critical temperature and then allow it to cool slowly, thus reducing brittleness and increasing ductility.

Arc weld: To weld by the electric arc process.

Assembly: The finished separate pieces put together according to the assembly drawing.

Boring: A machine operation consisting of enlarging a hole to specified dimension with a boring tool.

Boss: A projection, usually cylindrical, on a casting or forging.

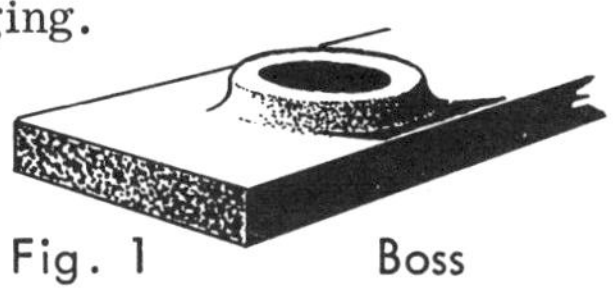

Fig. 1 Boss

Braze: To solder with hard solder made of brass (an alloy of copper and zinc).

Broach: To finish the inside of a hole to a shape usually other than round with a long cutting tool which has a series of teeth that gradually increase in size.

Buff: To polish with a buffing wheel; a general term applying to all soft wheels used in polishing, such as felt and cloth.

Burnish: To polish to a smooth finish by working a smooth tool against moving or revolving surfaces.

Bushing: A metal sleeve or bearing upon which a pulley or wheel turns; a liner for a hole to make it smaller.

Cam: A mechanical device used to change rotary motion to some other pattern of motion, such as reciprocating or intermittent.

Cam lobe: A rounded projection on a cam.

Carburize: To heat a low-carbon steel to approximately 2000° F for several hours in a box packed with carbonizing material, then allowing to cool slowly in preparation for heat treatment.

Caseharden: To harden the outer surface of carburized steel by heating to critical temperature and then quenching in oil or water.

Chamfer: The flat surface formed by cutting off a sharp corner, usually having an angle of 45°.

Chase: To cut threads in a lathe, as distinguished from cutting threads with a die.

Chatter: Noise produced by the rapid vibration of a machine tool against the work being machined.

Chill: To surface-harden cast iron by the sudden cooling of a piece of metal or a mold.

Chip: To cut or clean metal with a cold chisel.

Coin: To stamp or form a metal piece in one operation, usually with a surface design.

Collar: A ring around a shaft, used to inhibit motion at the end of the shaft.

Core: (1) That part of a mold which shapes the interior of a hollow casting. (2) The unaffected interior of a case hardened piece.

Counterbore: To enlarge an end of a hole cylindrically with a counterbore. (Fig. 2A)

Countersink: To enlarge an end of a hole conically with a countersink. (Fig. 2B)

Fig. 2 A

Fig. 2 B

Die: (1) One of a pair of hardened metal blocks or plates used to cut or form a required shape. (2) A tool for cutting external threads.

Die-casting: A method of producing castings to finished size by forcing molten metal into a suitable mold; an object or part formed by die-casting.

Diestock: A device used to hold threading dies, having two handles for turning the die on work, as when threading bolts by hand.

Dowel: A pin or other form of metal projection, anchored in one piece and fitting into a hole in an abutting piece to prevent motion or slipping, or to keep the pieces in their correct relative position.

Draft: The angle given a molding pattern to permit it to be withdrawn from the mold without disturbing the mold form.

Draw: (1) To stretch or shape metal by hammering. (2) To temper steel by gradual or intermittent quenching.

Drawplate: A hardened steel plate having a hole, or a series of conical holes, through which wire is drawn.

Drill: (1) To cut a cylindrical hole with a drill,

usually a twist drill. (2) A pointed cutting tool rotated under pressure.

Drop forging: A shaped object formed between dies by the use of a drop hammer; the process of forging with a drop hammer.

Eccentric: Deviating from the center or from the line of a circle.

Elongated hole: A hole which is not completely circular by reason of the inserting of a flat area. Dimensioned by giving the radii of the two halves plus the dimension of the flat area. (Fig. 3)

Fig. 3

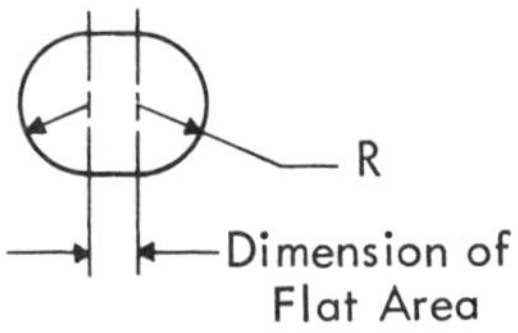

Face: To machine a flat surface on a metal piece by means of a machine tool.

Feather: A flat sliding key, partly sunk in a shaft and partly in a hub, to guide the motion of the hub lengthwise.

File: A tool made of tool steel having sharp cutting points or teeth across its surface, used for abrading or smoothing other substances.

Fillet: A concave surface or filling at the intersection of two surfaces to provide increased strength.

Fin: A thin projecting edge on a casting or other metal surface.

Finish: A fine cut made on a wide face with a square-ended machine tool; also to produce a smooth surface by filing, grinding, etc.

Fit: The type and closeness of contact between two surfaces.

Forge: To shape hot metal by hammering or pressing a piece of metal which has been formed while hot.

Fuller: A type of hammer used for forming grooves and spreading hot iron under hammer blows; a forging tool.

Galvanize: To coat a metal surface with zinc as a protection against corrosion.

Graduate: To mark in regular degrees of measurement, as in a scale or dial.

Grinding: A machine operation which consists of finishing a metal surface by means of abrasives.

Gusset: An angular piece of iron fastened in the angle of a metal frame to give strength or stiffness.

Hub: The central part of a wheel such as the part into which the spokes are inserted.

Jig: A fixture for locating, holding the work, and guiding the cutting tool in operations such as drilling, reaming, counterboring, and countersinking.

Kerf: The slit or groove produced in a piece of stock by a saw or cutting tool.

Key: A bar, pin, or wedge used to secure an object to a shaft to prevent relative movement between the two pieces.

Key seat: The groove which holds the key in position, as in a shaft. To form a seat for a key by cutting.

Knurl: To roughen or indent a turned surface, such as a knob or handle, to produce a better hand grip.

Lap: A soft metal surface impregnated with fine abrasive for use in polishing operations. To finish by polishing with a lap.

Lock nut: A nut constructed so that it automatically locks itself when screwed tightly; also, a nut screwed tightly on another to prevent slackening.

Lug: A projecting piece through which something is attached, bears on, or passes.

Malleable-iron casting: A casting toughened by annealing.

Mill: To shape or dress metal by means of a rotary cutter on a milling machine.

Normalizing: Heat treatment of steel to improve grain and structure, performed by heating it above its critical temperature and then cooling in air.

Pack-harden: To carburize and then to case harden.

Peen. To stretch or shape metal by hammering with the peen of a hammer; that is, with the end opposite the face of the hammerhead.

Pickle: A bath of diluted sulphuric or nitric acid, used to clean castings and aid in their seasoning.

Pinion: The smaller of two mating gears.

Planer: A machine used to machine flat surfaces of large size to a smooth finish.

Planish: To toughen and polish a metal surface by light hammering.

Profile: To machine an outline by moving a small rotating cutter usually controlled by a master cam or die.

Punch: A tool usually made of steel, having a shaped end for marking, making holes, forming indentations, etc. Also the male member of a die set.

Ream: A machine operation consisting of enlarging a hole slightly with a rotating fluted tool to provide greater accuracy.

Relief: A slight variance in the dimension of a machine part to allow for clearance.

Rivet: A metal pin with a head, used for fastening two pieces together by shaping the end opposite the head after the two pieces have been joined by the rivet.

Sandblast: To clean metal by means of sand blown by compressed air through a hose.

Shape: To machine with a shaper.

Shaper: A machine tool used for machining flat surfaces. The work is fed to the tool which moves with a backward and forward motion and cuts on the forward stroke.

Shear: To cut material by placing the material between two blades and exerting force on the blades in opposite directions.

Shim: A thin piece of metal placed between two parts to adjust the fit between the parts.

Slippage: Loss of motion in the transmission of power.

Spin: To shape sheet metal into a desired shape as it revolves on a roller or lathe.

Spline: A projection on a shaft or hub which fits into a slot or groove of another part.

Spot face: To machine a round spot on a rough surface, usually around a hole to give a good seat to a nut or bolt.

Spot weld: To join two pieces by welding separate spots rather than a continuous weld.

Steel casting: Material used in making machine parts. It is ordinary cast iron to which furnace scrap steel has been added in casting.

Surface roughness: Roughness refers to the finely spaced irregularities produced in fabricating a part. The height of the irregularities is rated in microinches.

Swage: To change the shape of a piece of metal by hammering, rolling, or otherwise forcing without cutting.

Swage block: A heavy block or iron having various shaped holes and grooved sides; used for shaping objects of large size.

Sweat: To unite two pieces of metal by heating after the joining surfaces have been coated with solder; also the condensation appearing on metal at certain temperatures when heated.

Tack weld: To fasten two metal pieces together by welding small spots at intervals to hold the position while the welding is being done.

Tap: A machine tool for forming internal screw threads; to cut internal threads with the use of such a tool.

Temper: To heat-treat hardened steel to make it harder by reheating it to below its critical temperature and allowing it to cool.

Template (or templet): A thin plate used as a guide for the form and shape of the piece to be made; also a gage for checking the form and shape of a piece.

Thumbscrew: A screw with a knurled or flat-sided head which may be turned by the thumb and the finger.

Tolerance: Permissible variation in the size of a machined part, usually specified as a maximum, or upper limit, and a minimum, or lower limit.

Torque: Twisting force set up in or by a revolving shaft.

Tumbler: A barrel-shaped rotating container in which parts are tumbled to clean them of scale and sand; also called a rumbler.

Turn: To shape a part by applying the cutting tool to the work while the work revolves on a shaft or spindle.

Upset: To shorten and thicken a metal bar or shaft by hammering on one of its ends.

Washer: A ring of leather, metal, lead, or other material used to secure tightness of joints, fill space, etc.

Waviness Height: Waviness refers to surface irregularities spaced too far apart to constitute roughness. Both height and width of waviness irregularities are rated in inches.

Wedge: A piece of material tapering to a thin edge; also to use such a piece in an operation.

Weld: To join pieces of metal by heating the surfaces to be joined until the molten metal fuses.

APPENDIX F

LIST OF COMMON ABBREVIATIONS

Adjust....ADJ
Allowance....ALLOW.
Alloy....ALY
Alteration....ALT
Alternating Current....AC
Aluminum....AL
American Standard....AMER STD
American Wire Gage....AWG
Approximate....APPROX
Arc Weld....ARC/W
Area....A
Assemble....ASSEM
Assembly....ASSY
Auxiliary....AUX
Babbit....BAB
Balance....BAL
Ball Bearing....BB
Base Line....BL
Base Plate....BP
Between Centers....BC
Bill of Materials....B/M
Blueprint....BP
Brass....BRS
Brazing....BRZG
Break....BRK
Bronze....BRZ
Bureau of Standards....BU STD
Bushing....BUSH.
Carburize....CARB
Cast Iron....CI
Cast Steel....CS
Center....CTR
Center Line....CL
Center to Center....C to C
Chamfer....CHAM
Cold Drawn Steel....CDS
Counterbore....CBORE
Countersink....CSK
Coupling....CPLG
Detail....DET
Diameter....DIA
Dimension....DIM.
Dowel....DWL
Drawing....DWG
Drill....DR
Drive Fit....DF
Drop Forge....DF
Eccentric....ECC
Finish All Over....FAO
Fixture....FIX.
Flat Head....FH
Forged Steel....FST
Forging....FORG
Foundry....FDRY
Gage....GA
Grind....GRD
Groove....GRV
Harden....HDN
Head....HD
Heat Treat....HT TR
High-Speed Steel....HSS
Inside Diameter....ID
Left Hand Thread....LH THD
Locate....LOC
Lubricate....LUB
Machine Steel....MS
Malleable Iron....MI
Material....MATL
Maximum....MAX
Micrometer....MIC
Minimum....MIN
National Coarse....UNC or NC
National Extra Fine....UNEF or NEF
National Fine....UNF or NF
On Center....OC
Open Hearth....OH
Outside Diameter....OD
Overall....OA
Pattern....PATT
Piece....PC
Pitch Diameter....PD
Plate....PL
Punch....PCH
Radius....R
Ream....RM
Required....REQD
Right Hand Thread....RH THD
Rivet....RIV
Screw....SCR
Section....SECT
Set Screw....SS
Shaft....SFT
Spotface....SF
Stainless Steel....SST
Stock....STK
Teeth Per Inch....TPI
Threads Per Inch....TPI
Tolerance....TOL
Tool Steel....TS
Wrought Iron....WI

ANSWERS TO SELF STUDY SUGGESTIONS

SELF STUDY SUGGESTION NO. 2

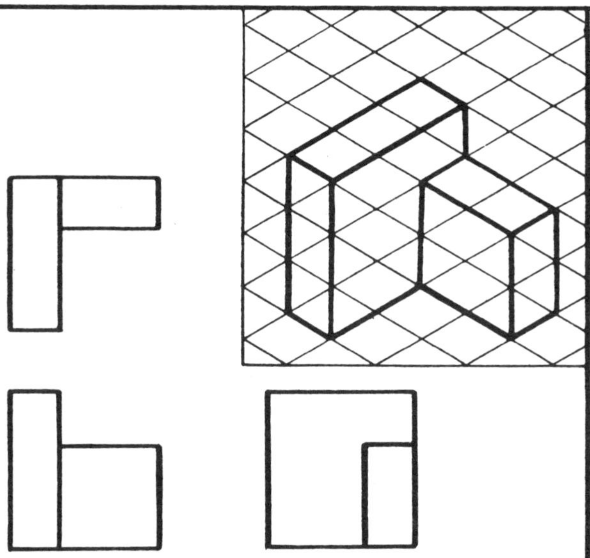

Fig. 19

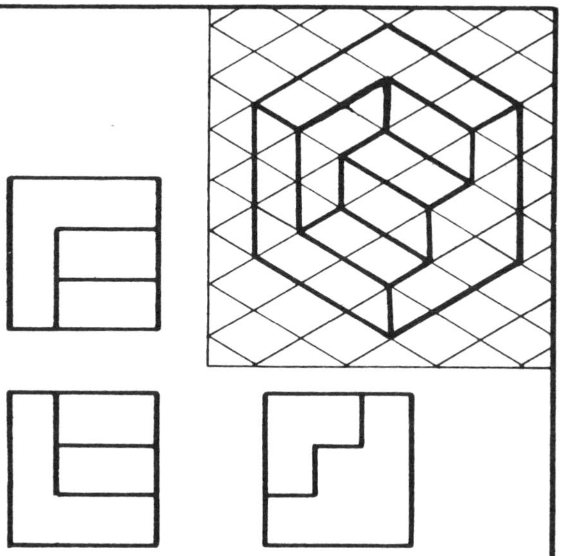

Fig. 20

SELF STUDY SUGGESTION NO. 3

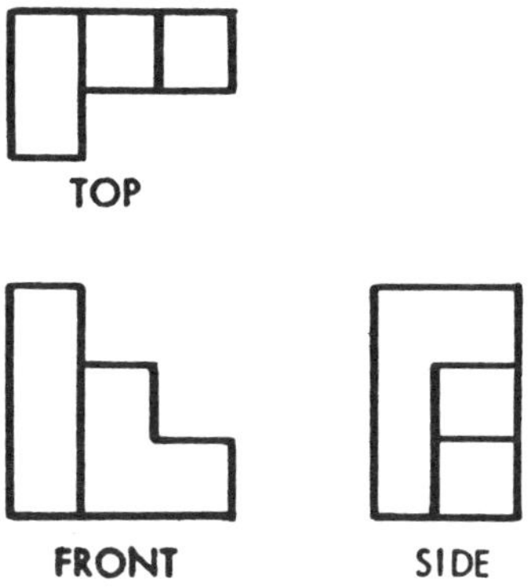

Fig. 16

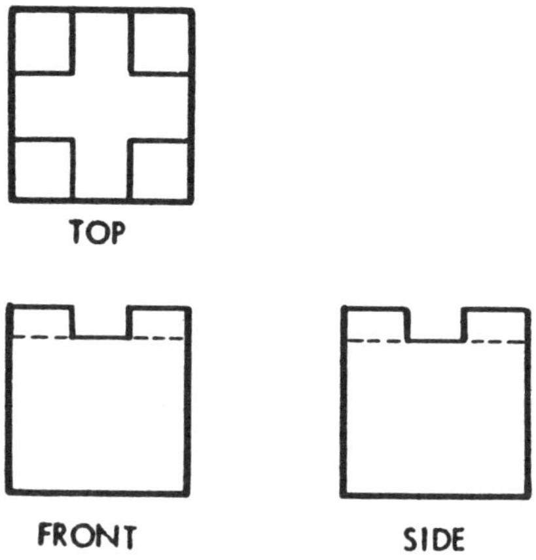

Fig. 17

EXERCISE 1

SELF STUDY SUGGESTION NO. 3 (CONT'D)

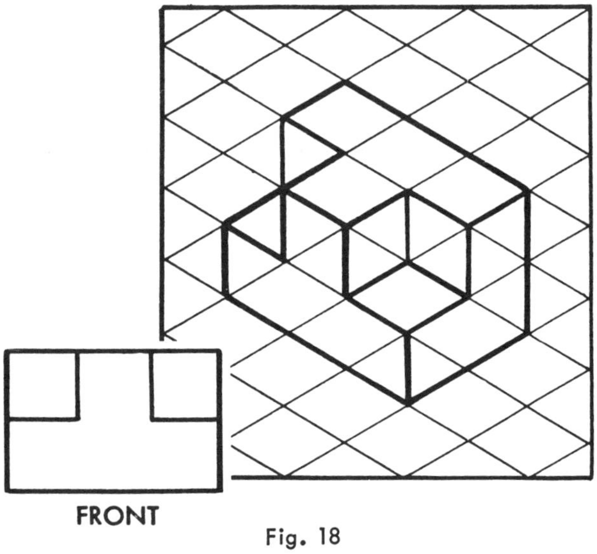

Fig. 18

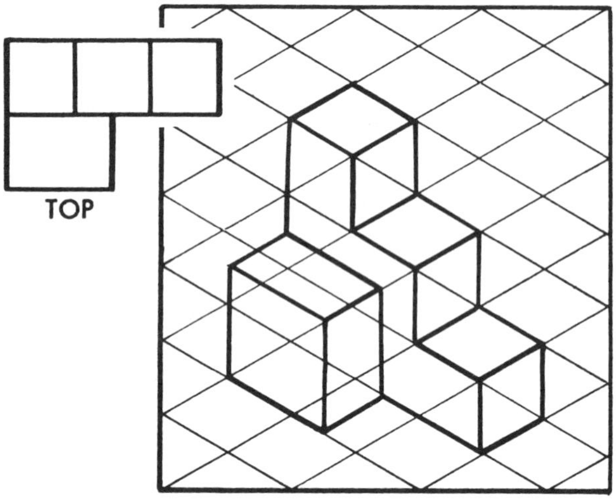

Fig. 19

EXERCISE 2

SELF STUDY SUGGESTION NO. 4

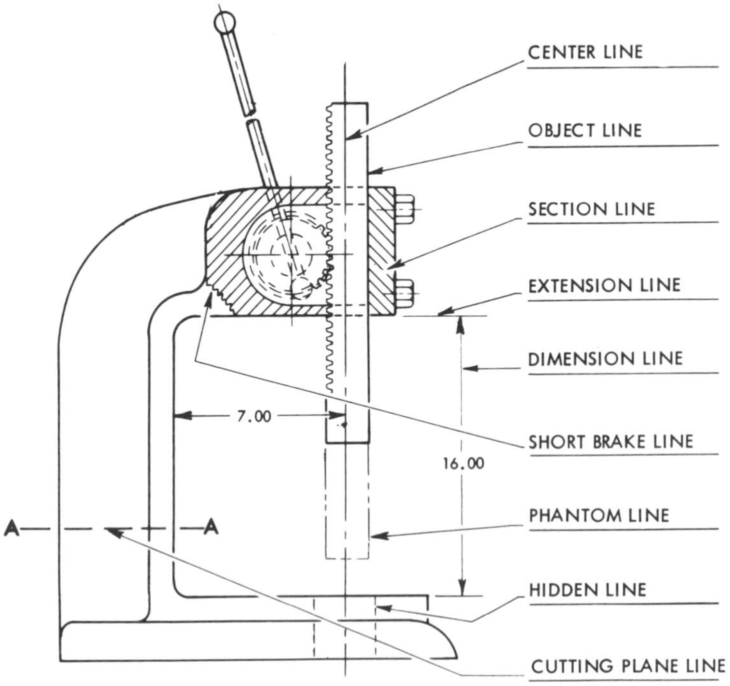

EXERCISE 1

3.06 = 3 1/16	2.50 = 2 1/2	.69 = 11/16
.28125 = 9/32	3.750 = 3 3/4	.13 = 1/8
.31 = 5/16	.781 = 25/32	1.875 = 7/8

EXERCISE 2

SELF STUDY SUGGESTION NO. 5

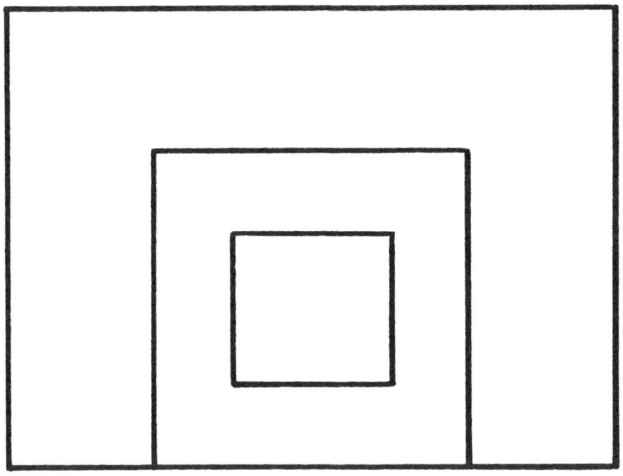

FRONT

Fig. 11

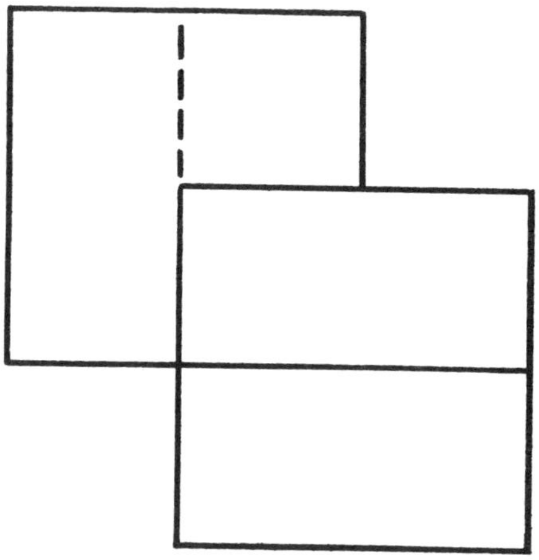

TOP

Fig. 12

SELF STUDY SUGGESTION NO. 6

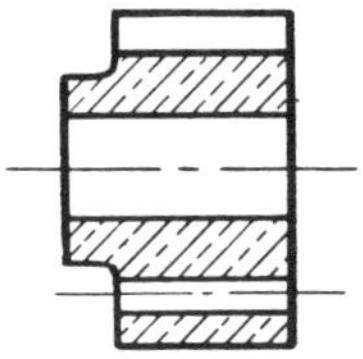

SIDE

Fig. 9

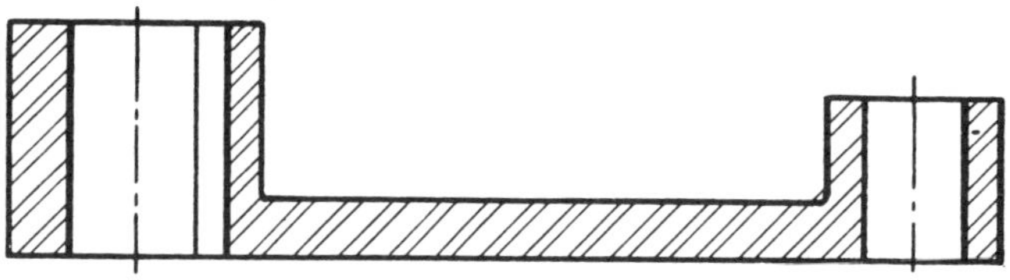

FRONT

Fig. 10

SELF STUDY SUGGESTION NO. 7

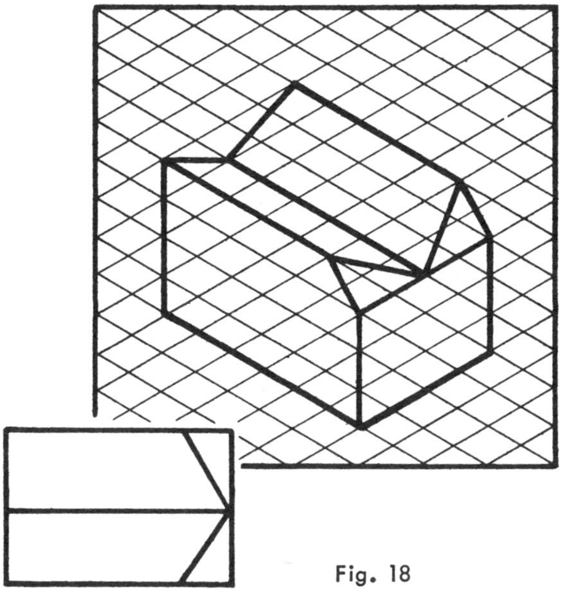

TOP

Fig. 18

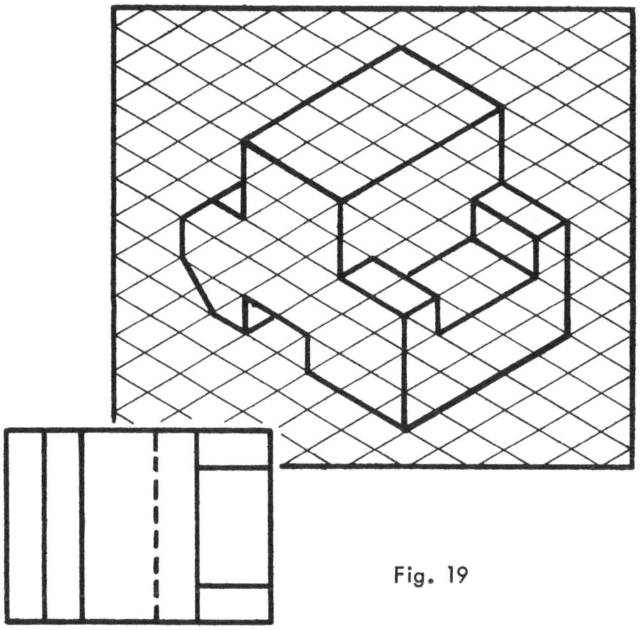

TOP

Fig. 19

THIS PAGE FOR STUDENT NOTES

GRAPH PAPER FOR DRAWINGS

STUDENT'S NAME ______________________ INSTRUCTOR'S NAME ______________________

GRAPH PAPER FOR DRAWINGS

STUDENT'S NAME________________________ INSTRUCTOR'S NAME____________________

GRAPH PAPER FOR DRAWINGS

STUDENT'S NAME ______________________ INSTRUCTOR'S NAME ______________________

GRAPH PAPER FOR DRAWINGS

STUDENT'S NAME_________________________ INSTRUCTOR'S NAME___________________

GRAPH PAPER FOR DRAWINGS

STUDENT'S NAME_________________________ INSTRUCTOR'S NAME_____________________

GRAPH PAPER FOR DRAWINGS

STUDENT'S NAME__________________________ INSTRUCTOR'S NAME____________________

GRAPH PAPER FOR DRAWINGS

STUDENT'S NAME______________________ INSTRUCTOR'S NAME____________________

GRAPH PAPER FOR DRAWINGS

STUDENT'S NAME__________________________ INSTRUCTOR'S NAME______________________

GRAPH PAPER FOR DRAWINGS

STUDENT'S NAME_________________________ INSTRUCTOR'S NAME___________________

GRAPH PAPER FOR DRAWINGS

STUDENT'S NAME__________________________ INSTRUCTOR'S NAME_____________________